S. Luthi
Geological Well Logs

Springer

*Berlin
Heidelberg
New York
Barcelona
Hongkong
London
Mailand
Paris
Singapur
Tokio*

Stefan M. Luthi

Geological Well Logs

Their Use in Reservoir Modeling

With 234 Figures and 28 Tables

HALLIBURTON ENERGY SERVICES
TECHNICAL INFORMATION CENTER
P.O. BOX 60070
HOUSTON, TEXAS 77205

Springer

Prof. Dr. Stefan M. Luthi
Delft University of Technology
Mijnbouwstraat 120
2628 RX Delft
The Netherlands

ISBN 3-540-67840-9 Springer-Verlag Berlin Heidelberg New York

Cataloging-in-Publication data applied for
Die Deutsche Bibliothek – CIP-Einheitsaufnahme

Luthi, Stefan M.:
Geological well logs : use in reservoir modeling / Stefan M. Luthi. -
Berlin ; Heidelberg ; New York ; Barcelona ; Hongkong ; London ;
Mailand ; Paris ; Singapur ; Tokio : Springer, 2000
  ISBN 3-540-67840-9

This work is subject to copyright. All rights are reserved, whether the whole or part of the material 15 concerned, specifically the rights of translation, reprinting, re-use of illustrations, recitation, broadcasting, reproduction on microfilms or in any other way, and storage in data banks. Duplication of this publication or parts thereof is permitted only under the provisions of the German Copyright Law of September 9, 1965, in its current version, and permission for use must always be obtained from Springer-Verlag. Violations are liable for prosecution under the German Copyright Law.

Springer-Verlag Berlin Heidelberg New York
a member of BertelsmannSpringer Science Business Media GmbH

http://www.springer.de

© Springer-Verlag Berlin Heidelberg 2001
Printed in Germany

The use of general descriptive names, registered names, trademarks, etc. in this publication does not imply, even in the absence of a specific statement, that such names are exempt from the relevant protective laws and regulations and therefore free for general use.

Typesetting: Mitterweger & Partner GmbH, D-68723 Plankstadt
Printed on acid-free paper   SPIN: 10635263   32/3130 as 5 4 3 2 1 0

Foreword

When I joined Schlumberger in 1982 I was surprised to find very few geologists in the company, and the few there were worked more as log analysts than geologists. The reason for this became soon clear to me: Except for the dipmeter there was no tool, and no other service, that was considered "geological". Schlumberger geologists were supposed to work with dipmeters, and, if they had a taste for it, the natural gamma-ray spectroscopy logs.

It turned out that my timing was fortunate. At Schlumberger's research center, in Ridgefield, Connecticut, a prototype electrical imaging tool had been designed, and after having spent three years in the Middle East I was transferred there. The first field test results were just coming in, and the images were startling. We could see geological details that nobody had ever seen from a log: cross-beds, unconformities, pebbles, fractures, folds, faults. No cores were needed to confirm the reality of these data; they were too real to be artifacts.

The guesswork had been taken out of dipmeters and had been replaced by downhole imaging. I immediately realized that geologists now had a very good reason to work for Schlumberger: We were needed to put these measurements to good use, to help building geological models of the subsurface. These measurements were not only comparable to cores, they were sometimes superior to them because they were oriented and continuous. When put onto an interactive workstation, the user could determine the dip and azimuth of relevant bedding surfaces, experiment with image processing algorithms to zone the well, or analyze the fractures that were – prominently visible on the images.

"Imaging" became so popular in logging that many other measurements were given the attribute: wellbore seismic, sonic waveforms, and even array induction logging. But more geological measurements also were developed: nuclear spectroscopy logs, ultrasonic images, paleomagnetic logging, nuclear magnetic resonance logging, but mechanical sidewall coring. They all probed properties of the rocks that had little to do with the classic "petrophysical" well logging approach. Instead, they carried a strong signal related to the rock. Well logging became an important source of geological information that let the geologists do facies mapping, paleocurrent analysis, chronostratigraphy, well correlation, and much more, with a great degree of confidence.

The need for this book arose from the impression that no – textbook adequately recounts the value of these measurements to the geologists, or the geoscientist in general. Returning to Europe after more than ten years in the

V

Americas, I found a publisher who supported my cause. Meanwhile, however, emphasis had shifted from geological well logs to geological software. Contractors competed to develop and market programs that did well correlation, zonation, mapping, property modeling, cross-sections, etc. This was a welcome development. The new measurements needed appropriate, sophisticated software to become useful. The modern geologist not only should be computer-literate, he also should disrespect old boundaries and interact closely with petrophysicists, geophysicists, reservoir engineers, petroleum and drilling engineers. In this exercise, mutual understanding, synergy and integration grow and bring out the best in all.

The book, therefore, had to transcend preexisting boundaries all the while focussing on the major geological principles. It addresses itself to the practicing petroleum geologist, to students, and to researchers who need an introduction to the topic. It is not a compendium, but tries to generate enthusiasm within geologists for new well log technology. Therefore, the title states "logs", not"tools", and "reservoir modeling", not "applications". The book is divided into three parts, first a review of traditional concepts, then a description of the geological logs, and lastly a selection of case studies. There may have been many other ways of organizing the book, but this seemed appropriate to demonstrate the value of these new (and sometimes not so new) measurements.

This book has been written while I was on various assignments in Paris, Luanda, and now Delft. In the hope that the readers find it useful, I look forward to many more new and exciting developments that let us understand the subsurface even better.

Delft, May 2000   Stefan M. Luthi

Acknowledgements

First I want to thank Schlumberger for having allowed me the time and resources to work on this book. Although warned by friends, I never imagined how time-consuming such an endeavor could be, and I appreciate therefore so much more the patience I was given.

I am particularly indebted to the following people for reviewing chapters of this book: S. Whittaker, P. Theys, A. Sibbit, R. Nurmi, J. Gartner, T. Bornemann, A. Hayman, P. Cheung, R. Kleinberg, D. Ellis, J. Thibal, A. Malinverno, I. Le Nir, G. Ruiz and F. Shray.

The following people have gratefully provided material for the book or helped in other ways: S. Hansen, M. Grace, C. Suter, C. Boyeldieu, A. Etchecopar, B. Newberry, J. van Doorn, T. Bornemann, L. Sonneland, S. Menger, M. Prammer, J.-C. Trouiller, R. Young, H. Géhin, J. Harper, D. Bergt, H. Anxionnaz, R. Rosthal, N. Mazzawi, C. Straley, R. Chandler, W. Borland, K. Saito, U. Pons, M. Herron, S. Herron, K. Castelijns, P. Lloyd, L. Thompson, R. Piggin, T.S. Ramakrishaan, G. Gubelin, L.S.D. Onuigbo, T. Hagiwara, A. Gupta, J. Sinclair, K. Schwab, A. Konovalov, H. Barrow, P. Sen, I. Bryant, A. Dumont, M.-Y. Chen, F. Montoya, D. Allen, J. Singer, Y. Aubin, H. Edmundson. Figure credits are given at the end of this book.

Nizier Sicard has been the graphic designer for most of the book. He spent endless nights behind his massive array of computers clouded in cigarette smoke, invariably emerging in the morning unshaved but proudly waving the finished figures. I appreciate his effort more than anything else. Merci beaucoup, Nizier!

And finally, I want to thank my family for having accepted me working on this book on so many weekends, rather than spending them in a more congenial way.

Table of Contents

1	Introduction	3
1.1	Overview	3
	Scope of Book	3
	Well Logging Companies	8
	Geological Service	9
	Professional Societies	9
	Periodicals	10
1.2	A Brief History of Logging	12
1.3	The Petrophysical Approach	21
1.3.1	Archie's Law	21
1.3.2	Mobility	24
1.3.3	Clays and Saturation Determination	26
1.3.4	Simultaneous Solutions for Lithologies and Saturations	29
1.3.5	Permeability	31
2	Geological Measurements	35
2.1	Dipmeter	37
2.1.1	Tool Principles	37
2.1.2	Dip Calculation	41
2.1.3	Graphical Representation	46
2.1.4	Dipmeter Interpretation	49
	Human Expertise	51
	Dipmeter Interpretation with Expert Systems	60
	Dipmeter Interpretation Using Geometric Constructions	61
	Statistical Methods on Dip and Curve Characters	65
2.1.5	Summary	71
2.2	Electrical Borehole Imaging	74
2.2.1	Tool Principles	74
	Microelectrical Imaging	76
	Azimuthal Resistivity Imaging	78
	LWD Resistivity Imaging	80
	Summary	82
2.2.2	Image Processing	83

2.2.3	Graphic Representations	89
2.2.4	Dip Calculations	91
2.2.5	Image Interpretation	94
	Bedding Analysis	96
	Structural Analysis	104
	Fracture Analysis	108
	Heterogeneities	116
2.2.6	Summary	118
2.3	**Acoustic Borehole Imaging**	**124**
2.3.1	Tool Principles	124
2.3.2	Tool Response	129
2.3.3	Processing and Graphic Representations	132
2.3.4	Dip Calculations	136
2.3.5	Image Interpretation	137
	Fractures	137
	Hole Shape Analysis	139
	Bedding Analysis	143
2.3.6	Summary	144
2.4	**Density Borehole Imaging**	**147**
2.4.1	Measurement Principles	147
2.4.2	Applications	151
2.5	**Optical Borehole Imaging**	**154**
2.5.1	Measurement Principles	154
2.5.2	Applications	155
2.6	**Nuclear Magnetic Resonance Logging**	**159**
2.6.1	Introduction	159
2.6.2	Measurement Principle	161
2.6.3	Applications	171
	Bound and Free Fluids	171
	Heavy Hydrocarbons	173
	Pore Size Distribution	174
	Permeability	175
2.6.4	Summary	180
2.7	**Nuclear Spectroscopy Logging**	**183**
2.7.1	History of Lithology Computations	183
2.7.2	Origin and Nature of Natural Radioactivity	188
2.7.3	Total Gamma Ray Logging	191
2.7.4	Natural Gamma Ray Spectroscopy Logging	194
2.7.5	Applications of Natural Gamma Ray Measurements	197
	Volume of Clay	197
	Correlation and Pattern Analysis	198
	Mineral Identification	200

2.7.6	Induced Gamma Ray Spectroscopy	201
2.7.7	Applications of Spectroscopy Measurements	206
	Mineral Concentrations	206
	Permeability	210
	Other Applications	212
2.7.8	Summary	213
2.8	**Paleomagnetic Logging**	**216**
2.8.1	Basics of Paleomagnetism	216
2.8.2	Measurement Principle	222
2.8.3	Processing	224
2.8.4	Interpretation	228
2.8.5	Summary	233
2.9	**Core Sampling**	**236**
2.9.1	Introduction	236
2.9.2	Percussion Coring	238
2.9.3	Rotary Coring	240
2.9.4	A Brief Comparison	244
3	**Applications and Case Studies**	**245**
3.1	**Structural Modeling**	**247**
3.1.1	General Comments	247
3.1.2	Combining Seismic and Dip Data	248
3.1.3	Modeling Cross-Beds	250
	Deterministic Approach	251
	Statistical Approach	252
3.1.4	Remarks	258
3.2	**Bedding and Reservoir Zonation**	**259**
3.2.1	Introduction	259
	Unsupervised Classifications	260
	Supervised Classifications	266
	Reservoir Zonation Using Geological Logs	267
3.2.2	"Thin Beds"	268
	Bed Thickness and Tool Resolution	268
	Thresholding for Sand Count	271
	A Turbidite Case Study	271
	Sharpening the Tool Response	284
3.2.3	Bedding from Image Texture	290
3.2.4	Conclusions	294
3.3	**Fractured Reservoirs**	**297**
3.3.1	Introduction	297
3.3.2	A Fractured Carbonate Case Study	298

	Fracture Analysis 300
	Results 305
	Discussion 308
3.3.3	A Fractured Basement Case Study 311
	Introduction 311
	Data Analysis and Results 312
	Discussion 314
3.4	**Well Correlation** 317
3.4.1	Introduction 317
3.4.2	Lithostratigraphic Correlation Using Spectroscopy Logs 318
3.4.3	Chronostratigraphic vs. Lithostratigraphic Correlation 321
3.4.4	Correlation with Neural Networks 325
3.4.5	Correlation with Dynamic Programming 330
3.4.6	Correlation from Facies Zonation 333
3.4.7	Sequence Analysis 337
3.4.8	Summary 340
3.5	**Geological Drilling** 342
3.5.1	Directional Drilling 342
	Directional Drilling Techniques 345
3.5.2	Geosteering 350
	Technique 350
	Borehole Images: The Future of Geosteering? 353
	Geostopping 355
3.5.3	Summary 357

Conclusions 358

Figure Credits 364

Abbreviation Index 366

Subject Index 367

Part 1: Introduction

1 Introduction

1.1 Overview

Scope of Book

Logging, or more precisely wireline logging, is a technique developed mostly in the oil industry that measures properties of the rocks surrounding the borehole. This is done with a sonde containing one or several sensors that is pulled uphole on a cable with a winch (Figure 1.1.1). A major purpose of logging is to identify

Figure 1.1.1. Schematic sketch of wireline logging operation (left) and logging-while-drilling (LWD) operation (right).

and evaluate hydrocarbon-bearing strata. Classically, three types of measurements are distinguished: Electrical, acoustic, and nuclear. In the 1980s, a significant number of textbooks have been written on the subject, notably by Dewan (1983), Hallenburg (1984), Serra (1984, 1986), Desbrandes (1985), Jorden & Campbell (1985; 1986), Labo (1986), Tittman (1986), and Ellis (1987). Additionally, the oil service companies have published documents on log interpretation principles and logging chart books (e.g. Schlumberger, 1989a, 1989b, and 1992).

Some of these books are compendium-like references, while others are textbooks for the graduate student or the professional. Most of them address primarily the petrophysicist, focussing on the classical problem of determining rock porosities and fluid saturations, while the solid rock properties themselves are often only of secondary interest. Textbooks specifically addressing the geologists have been written by Asquith (1982), Doveton (1986) and more recently by Doveton (1994a, 1994b) and Rider (1996). Several conferences held by the Geological Society in London reviewed new developments in geological logging, core analysis, and petrophysical methods. The proceedings from these conferences were compiled in books that came out in fairly regular intervals (Hurst et al., 1990; Hurst et al., 1992; Lovell & Harvey, 1997; Harvey & Lovell, 1998; Lovell et al., 1999).

These publications responded to a need from the geologic community, but were also the result of new technological advances that brought more useful measurements to the earth scientists. A noticeable acceleration in logging developments occurred in the last fifteen years in the wake of the oil crisis of the 1980s. It was spurred by several external factors:

- **Electronics and Computing**. The modern high-data-rate tools – imaging tools, array sensor tools etc. – could not have been developed without the electronic revolution. Very Large Scale Integration (VLSI) chips are deployed on the downhole sensors and are used for data multiplexing, permitting high data acquisition and transmission rates. Downhole microprocessors control the acquisition and process the data in the tools' electronic cartridges. Downhole memories store data in the tool for efficient packaging prior to transmitting them uphole. Computers in the surface acquisition unit permit real-time processing, quality control and display on color monitors, therefore facilitating rapid and educated decisions on the wellsite (Theys, 1999). Data transmissions via satellite bring the data to the user's office practically in real time, where he can evaluate them and make rapid decisions.

- **Drilling Technology**. Directional and particularly horizontal wells have challenged the logging industry in many ways. The most significant development is a full line of Logging-While-Drilling (LWD) tools, which are deployed in the bottom-hole assembly and which acquire and transmit data in real time, i.e. while the well is being drilled (Figure 1.1.1). The data is recorded at an early stage when fluid invasion and borehole damage effects are often minimal, which facilitates the identification and evaluation of hydrocarbon-bearing zones. LWD tools are also used for geosteering, a procedure where well drilling is monitored and adjusted, usually by comparing the downhole data with a computed response based on a geological model (Prilliman et al., 1995). LWD is also used to decide on coring or casing points ("geostopping") by way of a resistivity sensor positioned directly at the drill bit.

- **New Targets.** Exploration activities have continued to shift in the ever-ongoing quest for new plays. In the last ten to twenty years, the most significant move is towards deeper water targets, a development linked to advances in drilling technology, but also to new geological insights (Anderson, 1998). Discoveries in deep water offshore Brazil, in the Gulf of Mexico and offshore West Africa have been of impressive sizes and continue to fuel a race towards deeper water plays. These reservoirs are often geologically young, poorly consolidated, highly porous, and thinly bedded. This challenged the logging industry to develop robust sensors of sufficiently high vertical resolution that work even in the poor borehole conditions often encountered in these environments. Modern logging tools often make a measurement every 1.2 inches (3 cm), rather than the conventional 6 inches (15 cm). The highest vertical resolution is achieved by an electrical imaging measurement and is about one centimeter. Other new or non-conventional targets challenging the standard logging procedures are fractured basement rocks (for geothermal energy, but also for oil), coal beds (for methane), environmental logging (to detect pollutants), ultradeep research wells, and mineral mining wells. Oftentimes, the logging tools and the interpretation methods used in these projects can be adapted from oil-field technology, but in some instances entirely new techniques have to be developed.

The original logging methods have broadened their scope significantly and today address a much wider range of users with more adequate tools. Figure 1.1.2 lists the currently used techniques in surface geophysics, logging, sampling and flow measurements in the left column, and in a simplified and schematic way the basic reservoir properties in the second column. The latter are mostly derived quantities, although they are in many cases strongly related to the original measurement, as is the case, for example, with the fluid pressure measurement. The third column contains the "elements" of reservoir characterization, or those overall aspects of a reservoir that allow us to understand its composition and behavior: the architecture and internal layering, the reserves and the flow units. Finally, in the last column, there is the only tangible of ultimate relevance in the oil industry: production, or the return on investment. The gray shaded area covers those measurements that are the topic of this book. It can be seen that their major contributions are in the field of reservoir architecture and reservoir zonation. Figure 1.1.2 is valid for reservoirs whose permeabilities are governed by the rock matrix. In other, less conventional reservoirs a different set of properties is required.

Measurement	Property		RC Element	Tangible
Seismic Gravimetry	Structure	Geophysics		
Borehole Imaging	Bedding		Reservoir Architecture	
Paleomagnetic Log	Age	Geology		
Coring Nuclear Spectroscopy	Mineralogy			
Downhole Video	Fabric		Reservoir Zonation	
NMR Logs	Pore Geometries	Petrophysics		Production ($)
Acoustic Logs Nuclear Logs	Porosity		Reserves	
Electrical Logs	Saturation			
Formation Testing Well Testing	Pressure	Reservoir Engineering		
	Permeability		Flow Units	
Fluid Sampling	Fluid Properties			
Flow Measurements	Flow Rates			

Figure 1.1.2. From measurement to production. The gray area covers the measurements discussed in this book. The relationship between measurements and reservoir quantities is approximate and therefore no connecting lines are drawn. The same holds true for the relationship between reservoir quantities and elements of reservoir characterization. The fields of geophysics, geology, petrophysics, and reservoir engineering are indicated approximately and obviously overlap to some degree.

Fractured reservoirs, for example, require the knowledge of stresses and rock strengths.

Many of the measurements in Figure 1.1.2 are new or have been considerably improved in the last few years due to the external factors mentioned above. They give the geoscientists and engineers better tools to perform their tasks, resulting in a better, more complete assessment of reservoirs. Geological measurements in particular have undergone a significant development. These tools are rarely standard items on a logging program, and some of them have been built in very small numbers only, often referred to as "specialty tools" or "niche tools". The major purpose of this book is to demonstrate the use of these tools in reservoir modeling, but also in other fields where these techniques are found to be useful[1] (Goldberg, 1997).

---

[1] I apologize to readers from the mining, environmental, and academic professions for the strong oil-industry flavor of this book and hope they still will get enough useful information.

Figure 1.1.3. Sketch of an offshore field development with vertical, deviated and horizontal wells (Gullfaks field, North Sea). The complex geological architecture of this field requires a drilling strategy that allows all oil-bearing compartments to be optimally drained.

New drilling technology, together with new logging and detailed 3-D seismic surveys, allow the petroleum professional to undertake field developments with almost surgical precision. He can target oil in bypassed zones, drain narrow fluvial point bars along the optimum direction, or intersect fractures at the best angle to maximize production. Real-time measurements during drilling, fast data processing and the possibility to make on-site decisions are important elements in this process. Drilling and evaluating wells in a way that honors geological reality can be contrasted with the previously common "geometric drilling", and undoubtedly represents a major advance in reservoir development. Figure 1.1.3 shows a field where a multitude of deviated and horizontal wells was carefully designed and drilled into the various compartments of the reservoir with the goal of optimizing reservoir drainage.

This book gives in its first part a look at the history of logging and the traditional uses of well logs for petrophysical evaluation. The reader can thus appreciate where the techniques come from, how they evolved over time, and in what aspects the newer techniques differ from the older ones.

The second part introduces the measurement principles and applications of the major geological logging techniques. No deeper physics or mathematics background is needed here, and equations are merely used to satisfy the curious.

What qualifies as a geological measurement is, to a great degree, arbitrary. Here the focus is placed on those methods that greatly improve the geological characterization of reservoirs, or, put more accurately, those that are a primarily sensitive to the rock, and only secondarily to the fluids (the only exception being the nuclear magnetic resonance technique which has been included because of its unique sensitivity to small-scale textural features). Not included are near-wellbore imaging techniques, including sonic imaging with long transmitter-receiver spacings, seismic-while drilling, vertical seismic profiling, borehole radar, and cross-well tomography. This group of measurements is left out because it is a field in full development that merits its own textbook. Applications mentioned in this second part of the book are for single tools only: how to reconstruct cross-bedding from borehole imaging, how to calculate mineralogy from nuclear spectroscopy, how to estimate grain size from nuclear magnetic resonance log, etc.

The third part of the book is dedicated to demonstrating how new logs, and new interpretation techniques, can help the petroleum geologist (and the geoscientist in general) obtain better reservoir models. The topics are grouped into the major tasks of the petroleum geologists: Structural analysis, reservoir zonation, fracture analysis, well correlation, and well placement using directional drilling. Emphasis is put on how to integrate multiple measurements in the reservoir modeling process: What combination of tools is best used to obtain a lithostratigraphic zonation? How can different logs improve the characterization of fractured reservoirs? How can seismic interpretation be improved with high-resolution images obtained in boreholes? An effort has been made to include recent field examples so that the practicing geologist can evaluate how much he stands to gain from these new technologies.

Well Logging Companies

Well logging in either wireline or logging-while-drilling mode is provided by the three major oil service companies (Schlumberger, Halliburton, Baker Hughes), some mid-sized and smaller companies as well as national entities. Complete lists of tools, their specifications and mnemonics can be found in their service catalogs and on their websites.

Schlumberger Oilfield Services
(Wireline, Anadrill)
http://www.slb.com

Halliburton Energy Services
(Halliburton Logging and Perforating, NUMAR, Sperry-Sun)
http://www.halliburton.com

Baker Hughes
(Baker Atlas, Baker Hughes INTEQ)
http://www.bakerhughes.com

Geoservices
http://www.geoservices.com

Reeves Wireline
(formerly BPB Wireline Services)
http://www.bpbwireline.com

Geological Services

Interpretation of well logs and other data pertinent to geological reservoir modeling is offered by a large number of companies. Here are some of the larger contractors.

Schlumberger Oilfield Services
(GeoQuest)
http://www.slb.com
Halliburton Energy Services
(Landmark)
http://www.halliburton.com
Baker Hughes
(Z&S, Baker Atlas)
http://www.bakerhughes.com

Professional Societies

There is no professional society that dedicates its efforts solely to geological well logging. Several organizations deal with the topic through special sessions at their annual conferences, topical seminars, and papers in their professional journals. Most of these societies have regional chapters in numerous parts of the world. Their websites generally contain information on publications, conferences and other events, industry news, and useful oil industry links.

Society of Professional Well Log Analysts SPWLA
8866 Gulf Freeway, Suite 320
Houston, TX 77017, USA
Phone: 1-713-947-8727 Fax: 1-713-947-7181
Email: spwla@spwla.org
Website: http://www.spwla.org

Society of Petroleum Engineers SPE
P.O. Box 833836
222 Palisades Creek Dr.
Richardson, TX 75080, USA
Phone: 1-972-952-9393 Fax: 1-972-952-9435
Email: service@spe.org
Website: http://www.spe.org

American Association of Petroleum Geologists AAPG
P.O. Box 979
1444 S. Boulder Avenue
Tulsa, Oklahoma 74101-0979 USA
Phone: 1-918-584-2555 Fax: 1-918-560-2665
Email: info@aapg.org
Website: http://www.aapg.org

The Geological Society
Burlington House, Piccadilly
London W1V 0JU, UK
Phone: 44 (0)171-434-9944 Fax: 44 (0)171-439-8975
Email: enquiries@geolsoc.org.uk
Website: http://www.geolsoc.org.uk?tpb=-5dd›

Periodicals

No professional journal is entirely dedicated to geological well logging, but the following contain more or less regularly papers of interest in this field.

The Log Analyst
from January 2000 onwards: *Petrophysics*
(A bimonthly journal edited by the SPWLA and dedicated entirely to well logging)

American Association of Petroleum Geologists Bulletin
(A monthly journal by the AAPG with a wide range of papers in all areas of petroleum geology)

Journal of Petroleum Technology
(A monthly publication by the SPE featuring recent advances in petroleum engineering)

SPE Reservoir Evaluation & Engineering
(A bimonthly combining the former SPE Reservoir Engineering and SPE Formation Evaluation, with a focus on multidisciplinary reservoir characterization)

Oilfield Review
(A Schlumberger publication featuring articles on new advances in oil field technology, published quarterly by Elsevier)

References

Anderson RN (1998) Oil production in the 21st century. Scientific American 278, 3: 68–73.
Asquith G (1982) Basic well log analysis for geologists. Am Assoc Petrol Geol, Tulsa.
Desbrandes R (1985) Encyclopedia of well logging. Gulf, Houston.
Dewan JT (1983) Essentials of modern open-hole log interpretation. PennWell, Tulsa.
Doveton J (1986) Log analysis of subsurface geology. John Wiley Sons, New York.
Doveton J (1994a) Geologic log analysis using computer methods. Am Assoc Petrol Geol, Computer Applications in Geology No 2, Tulsa.
Doveton J (1994b) Geologic log interpretation. SEPM Soc Sed Geol Short course 29, Tulsa.
Ellis DV (1987) Well logging for earth scientists. Elsevier, Amsterdam.
Goldberg D (1997) The role of downhole measurements in marine geology and geophysics. Rev Geophys 35, 3: 315–342.
Hallenburg JK (1984) Geophysical logging for mineral and engineering applications. PennWell, Tulsa.
Harvey PK, Lovell MA (1998) Core-Log Integration. Geol Soc Spec Publ 136.
Hurst A, Lovell MA, Morton AC (1990) Geological applications of wireline logs. Geol Soc Spec Publ 48.
Hurst A, Griffiths CM, Worthington PF (1992) Geological applications of wireline logs II. Geol Soc Spec Publ 65.
Jorden JR, Campbell FL (1985) Well logging I – Rock properties, borehole environment, mud, and temperature logging. Soc. Petrol. Eng. Monograph Series: 9.
Jorden JR, Campbell FL (1986) Well logging II – Electric and acoustic logging. Soc. Petrol. Eng. Monograph Series: 10.
Labo J (1986) A practical introduction to borehole geophysics. Soc Explor Geophys, Tulsa.
Lovell MA, Harvey PK (1997) Developments in petrophysics. Geol Soc Spec Publ 122.
Lovell MA, Williamson G, Harvey PKJ (1999) Borehole images: applications and case histories. Geol Soc Spec Publ 159.
Prilliman JD, Allen DF, Lehtonen LR (1995) Horizontal well placement and petrophysical evaluation using LWD. 70th Ann Conf Soc Petrol Eng: Paper 30549.
Rider M (1996) The geological interpretation of well logs. Gulf, Houston.
Schlumberger (1989a) Log interpretation principles/applications. Schlumberger Educational Services, Houston.
Schlumberger (1989b) Log interpretation charts. Schlumberger Educational Services, Houston.
Schlumberger (1992) Logging while drilling. Schlumberger Educational Services, Houston.
Serra O (1984) Fundamentals of well-log interpretation. 1. The acquisition of logging data. Elsevier, Amsterdam.
Serra O (1986) Fundamentals of well-log interpretation. 2. The interpretation of logging data. Elsevier, Amsterdam.
Theys P (1999) Log data acquisition and quality control. 2nd edition. Editions Technip.
Tittman J (1986) Geophysical well logging. Academic Press, Orlando.

1.2 History of Logging

Professor Forbes from the Edinburgh Observatory was perhaps the first person to make well log measurements, when, from 1837 to 1842, he lowered temperature sensors into three shafts up to 24 feet deep in order to record temperature variations with depth and time. His results were later analyzed by the physicist Lord Kelvin (Thomson, 1861), who was able to determine secular variations in temperature and heat flow, which he needed for his calculation of the earth's age. The separation between data acquisition and interpretation in this largely academic exercise seemed to presage a distinction still present in today's logging operations. Temperature measurements are still routinely done in modern well logging, but they are rarely used for anything else than correcting other sensors' responses for temperature effects[1].

The beginning of commercial well logging is entirely attributed to the initiative of two French brothers, Conrad Schlumberger (1878–1936), a physicist who graduated from the Ecole Polytechnique France, and Marcel Schlumberger (1884–1953), an engineer from the Ecole Centrale de Paris (Allaud & Martin, 1977). Encouraged and supported by their father Paul, a relatively wealthy businessman, they experimented with electrical measurements at the surface to locate iron and copper deposits. Initial successes in several countries led them to create the Société de Prospection Electrique (SPE). In 1921, Marcel Schlumberger tried for the first time to make resistivity measurements in a borehole in the coal basin of Bessèges (France) with the purpose of validating the surface electrical surveys. The strong resistivity variations they observed convinced the two brothers to pursue this further. In 1927, Conrad outlined the principle of "electrical coring" as well logging was initially called[2], in a technical note. Henri Doll was hired to develop the equipment and conduct the first operation in an oil well, which took place on September 5, 1927, in Pechelbronn in Alsace, where at the time a small but thriving oil industry existed (Figure 1.2.1).

The resulting log (Figure 1.2.2), hand-drawn with measuring points spaced one meter apart, represents a turning point in oil exploration. It showed the layered nature of the subsurface and allowed easily identifying the major geological formations such as the top of the Hydrobia Marl, an important regional marker. Subsequent logging in nearby wells made clear that the method could be used for correlation purposes with greater accuracy than drill cuttings and at lower costs than core drilling (Schlumberger et al., 1932). Furthermore, the technique showed potential for detecting hydrocarbon-bearing layers, but the pay zones in Pechelbronn were too thin to demonstrate this. The Schlumberger brothers were able to convince Royal Dutch Shell that the technique was reliable,

---

[1] Russian specialists, however, are very skilled at extracting production and geological information from interpreting temperature logs, particularly in fractured reservoirs.

[2] *Carottage électrique* in French.

Figure 1.2.1. The first logging operation in an oil well. Pechelbronn, well 2905, on September 5, 1927.

after a repeat log done by the client in person (Allaud & Martin, 1977), and they were awarded a logging contract in Venezuela. Following a successful campaign, the company expanded to India and Russia (Figure 1.2.3), and, after unsuccessful initial trials, established themselves in the United States (Figure 1.2.4). They were helped by the almost serendipitous discovery of the spontaneous potential (SP), a voltage observed on two of the resistivity electrodes when no external current was applied to the device. The two measurements combined were shown to locate permeable hydrocarbon-bearing layers, and well logging soon became an industry on its own, far beyond the supportive role to surface geophysics that it had been designed for. In 1934, the Schlumberger Well Services company was founded in Houston, Texas, and it immediately expanded at a rapid pace. Logging proved successful in other parts of the world, notably Mexico, Ecuador, Argentina, Germany, Austria, and Borneo. The number of crews grew rapidly as

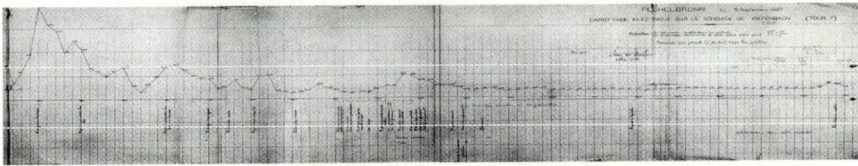

Figure 1.2.2. The first continuous well log, Pechelbronn, well 2905, on September 5, 1927. Resitivities increase from left to right. Depth interval is from about 170 to 270 meters. Geological descriptions are annotated on the left, with the Hydrobia marl marker approximately located in the middle.

Figure 1.2.3. Henri Doll, one of the first engineers hired by the Schlumberger brothers, shown here in 1932 in the Baku oilfield (Azerbeijan) calibrating a teleclinometer to measure the azimuth and angle of the borehole drift.

Figure 1.2.4. Logging operations in 1932 in California. Engineers Deschatre and Legrand (left) record two curves, the resistivity and the spontaneous potential, using potentiometers.

illustrated by table 1.2.1, with a particular acceleration after the spontaneous log was introduced in 1931.

Table 1.2.1. The number of Schlumberger logging crews in the early years and in the year 2000.

Year	No. of Crews
1928	1
1929	3
1930	7
1931	10
1932	14
1933	24
1934	45
1935	65
2000	1100[a]

[a] The estimated worldwide number of all logging units is 4100 (Russia 600, China 700, Rest of World 2800)

In the early 1930s the first tools for sidewall coring and perforating were developed. Well geometry surveys using first the teleclinometry, then the photoclinometry system, were introduced by Sperry Sun and Schlumberger, and were crucial in accurately locating the positions of deviated wells. In 1938 the Wells Survey Company performed the first gamma ray logging in Oklahoma and commercialized it one year later by promoting it for accurate perforating. In 1942, Schlumberger introduced the first dipmeter tool, initially as an "anisotropy tool", then based on a spontaneous potential (SP) principle with three arms. It was replaced ten years later by passively focussed micro-electrical devices. The growing number of tools required a parallel effort in cable technology and surface acquisition systems, which was mostly done by service companies. Research labs from oil companies, among others Mobil, Shell, and Chevron, significantly contributed to the development of many logging tools and in many cases built prototype tools themselves. In 1949, after intensive research, Doll was able to prove the feasibility of induction logging. This opened the market to the large number of wells drilled with freshwater or oil-based muds. In the early 1950, Schlumberger developed several new electrical measurements such as the microlog, and brought an early version of the neutron porosity log onto the market. Efforts by Mobil and Esso (then Magnolia and Humble) led to the introduction of the acoustic log in the mid-50s, initially without and then with borehole compensation. It provided the first accurate and relatively direct porosity measurement, a quantity needed for the proper assessment of hydrocarbon reservoirs. Already in 1942, Gus Archie from Shell had proposed empirical relationships between porosity, fluid resistivity, rock resistivity, and water saturation, which seemed to hold true in a large variety of rock types.

Now, with a porosity log, a shallow, and a deep resistivity measurement, it became possible to determine fluid saturations from logs, as had been envisioned by early proponents of quantitative log interpretation (Wyllie & Rose, 1950). The original use of logs – geological correlation and location of hydrocarbon-bearing zones – was thus slowly but steadily supplanted by "formation evaluation", or petrophysics. The 1950s also saw the first formation fluid sampling tool, the first shaped charges for perforating, and initial nuclear magnetic resonance logging tests. In the 1960s, density logging and sophisticated focussed resistivity devices were put in commercial use. Mobil's borehole televiewer, an ultrasonic borehole-scanning tool, was the first downhole imaging measurement that could be used in oil wells. With this, most of the basic logging techniques known today were in place.

Much of the logging development in the 1970 was related to the growing use of electronics. Tool modeling and design was increasingly done with the aid of computer simulations. Downhole sensors, transmission, and surface acquisition were increasingly controlled by microprocessors. The analog-digital conversion moved steadily towards the sensors, to a point where today's logging systems are essentially all digital in nature. Data processing became entirely a task performed by computers, and data interpretation was increasingly done on computers using specialized software. With drilling reaching greater depths, logging cables, data transmission rates, tool specifications and tool reliability all had to be adapted to the changing environments, and surface acquisition systems had to handle rapidly increasing data volumes (Figures 1.2.5 and 1.2.6).

Figure 1.2.5. Logging operations in 1997 in Tyler, Texas. A fully computerized MAXIS* truck is ready for logging while a satellite dish is installed for data transmission to the oil-company base (Photo N. Russel).
* Mark of Schlumberger

Figure 1.2.6. Interior view of a MAXIS Express truck, with the engineer supervising operations on a monitor (Photo N. Russel).

The 1980s saw the emergence of logging-while-drilling (LWD), mostly fuelled by the increased use of directional drilling after steerable downhole motors had been introduced by Smith International and Eastman-Christensen. This development had started off in the 1960s with downhole measurements of the well deviation, torque and weight on the bit and simple formation parameters. Initially it was done in stationary mode whenever drilling was stopped, and then in measurement-while-drilling (MWD) mode, whereby the data was sent to the surface by mud pulses, or, in the case of turbodrilling, by a wireline cable. Teleco, Sperry Sun, and Anadrill, a Schlumberger company, introduced single-resistivity downhole measurements in the early 1980s, where the data was stored in a downhole memory and later retrieved. The need to have immediate information from the drill bit made development proceed at a rapid pace, and in the mid-1980s Sperry-Sun introduced several tools that allowed formation evaluation from LWD measurements. In 1988, real-time LWD saw the light with Schlumberger's compensated dual-resistivity tool, which was combinable with a gamma ray, density, and neutron tool. Their measurements were transmitted with pulses in the borehole mud which propagated in the borehole from the downhole sensor to the surface. This paved the way to evaluate the formation penetrated by directional drilling in real-time and, therefore, to adjust the well trajectory immediately when needed. Other LWD measurements were developed at a rapid pace and today most wireline techniques are also available in LWD mode. Together with the MWD measurements, LWD has created a positive feedback loop on well design, since these measurements allowed for sophisticated wellbore positioning of sidetracks, multiple targets, horizontal or extended reach wells.

The 1980s also saw a strong acceleration of borehole imaging technology, a field of particular importance for geologists. It was spurred by micro-electrical imaging, a method of recording an image of the borehole wall with numerous small electrical sensors (Figure 1.2.7). The term "imaging" was soon applied to any measurement with two independent variables, such as depth/radius, depth/azimuth, or depth/time, and it expanded to other electrical as well as acoustic and nuclear measurements. This move away from the one-dimensional logs towards visualizing the subsurface helped geoscientists unravel the complexities of reservoirs better than ever. In the early 1990s, Schlumberger's RAB* tool performed the first LWD electrical imaging using electrodes on the rotating bottom hole

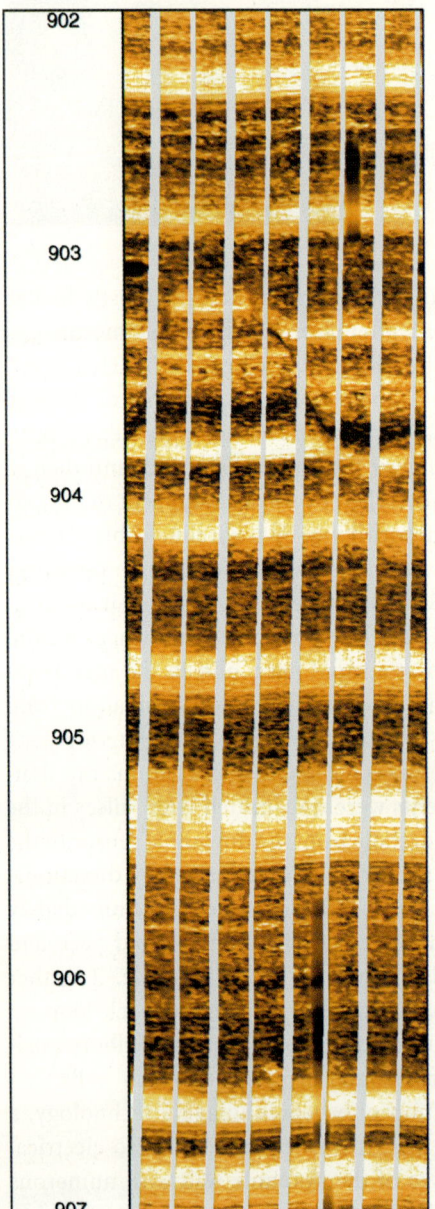

Figure 1.2.7. A modern imaging log from the Fullbore Formation Microimager (FMI*), a tool recording resistivity variations on the borehole wall. The image corresponds to an unwrapped cylinder. This sequence is a fine-grained calcareous sandstone, with lighter (more resistive) layers corresponding to limestone stringers. A small-scale fault is seen in the upper part. Lower Paleozoic Riley Formation, Texas. Depths area in feet.

* Mark of Schlumberger

assembly. Also in the 1990s, nuclear magnetic logging, a technique somewhat dormant since its inception by Chevron in the 1950s, became available as a robust wireline measurement, first from Numar, then from Schlummberger. By measuring the pore size distribution, it provides the volume of capillary-bound fluid – a quantity long sought after by petrophysicists – and an estimate of permeability.

Today there are numerous service companies active in well logging, led by Schlumberger, Baker-Atlas and Halliburton. Desbrandes (1985), Serra (1984), and the logging company catalogs contain compilations of the currently available well logging services. A list of all major aspects of well log applications is compiled by the Society of Professional Well Log Analysts (SPWLA) and published in their trade journal "The Log Analyst" (starting in 2000: "Petrophysics") in annual updates (Prensky (1997). More recently, most larger oil service companies have developed into general contractors, or integrated service companies, by providing a wide array of services from seismic surveys, drilling, mud chemicals, cementing, completions, wireline and LWD logging, production and other downstream facilities, data processing and interpretation. Such a concentration of services allows these companies to tender for entire reservoir development and management contracts, whereby the integration of services under one umbrella can provide a synergistic and cost-effective operation – an aspect of particular importance in today's complex field developments and economic environment. Well logging is an integral part in these as well as in conventional, operator-controlled projects. It is one of the principal means of collecting relevant downhole data upon which many field development decisions are based.

The history of well logging depicts how the technique has developed from its humble beginnings to a wide range of services and recently into a powerful combination of real-time measurements with directional drilling. While the first measurements were simple tools for geological correlation, they soon developed into petrophysical tools permitting reliable and repeatable formation evaluation. Today's tools allow drilling progress to be monitored from the surface, with the direction adjusted according to the geological situation encountered in the wellbore. Formation analysis can be done in real time with advanced analytical techniques. Rock layering can be visualized on surface monitors, and fluid as well as rock samples can be taken accurately at locations of interest. Logging in all its modern forms thus contributes greatly to improving the success rate of oil and gas wells, and to produce the reserves more efficiently and more effectively.

In conclusion, the history of well logging can be divided, perhaps somewhat arbitrarily, into four distinct phases:

- **The conceptual phase (1921–1927):** Well logs were intended to support and complement surface geophysical surveys.

- **The acceptance phase (1927–1949):** Well logging became increasingly useful for layer correlation, identification of hydrocarbon-bearing zones, well surveying and perforating.

- **The maturity phase (1949–1985):** The advent of several porosity and resistivity logs rings in the golden age of petrophysics. Well logging is primarily used for quantifying oil and water saturations. Additionally, numerous new non-petrophysical logging methods are developed.

- **The reinvention phase (since 1985):** Numerous technological breakthroughs such as logging-while-drilling, borehole imaging, and nuclear magnetic resonance redefine well logging as an important aid for efficient reservoir development and management.

References

Allaud LA, Martin M (1977) Schlumberger, history of a technique. John Wiley Sons, New York.
Archie GE (1942) Electrical resistivity log as an aid in determining some reservoir characteristics. Trans Am Inst Min Eng 146: 54–62.
Desbrandes R (1985) Encyclopedia of well logging. Gulf, Houston.
Doll HG (1949) Introduction to induction logging and application to logging of wells drilled with oil-base muds. Petr Trans Am Inst Min Eng: Tech Paper 2641.
Hallenburg JK (1984) Geophysical logging for mineral and engineering applications. Penn-Well, Tulsa.
Prensky SE (1997) Bibliography of well-log applications: 1996 Update. Log Analyst 38: 17–88.
Schlumberger C, Schlumberger M, Doll HG (1932) Electrical coring: A method of determining bottom-hole data by electrical measurements. Amer Inst Min Eng Techn. Paper 462.
Serra O (1984) Fundamentals of well-log interpretation. 1. The acquisition of logging data. Elsevier, Amsterdam.
Thomson W (1861) On the reduction of observations from underground temperature. Trans Roy Soc Edinburgh 22: 405–427.
Tittman J (1986) Geophysical well logging. Academic Press, Orlando.
Wyllie MRJ, Rose WD (1950) Some theoretical considerations related to the quantitative evaluation of the physical characteristics of reservoir rocks. J Petr Tech 189.

1.3 The Petrophysical Approach

1.3.1 Archie's Law

Oil and gas are contained in pores of reservoir rocks such as carbonates and sandstones. In order to assess a possible hydrocarbon reservoir, the porosity and hydrocarbon saturations need to be known, which together define the amount of hydrocarbon per unit volume of rock. Well logging has substantially contributed to the evaluation of these two quantities. The reserves of a particular geological structure are additionally controlled by the geometry of the reservoir, i.e. its thickness and lateral extent, as well as by the producible or movable portion of the hydrocarbon volume. To evaluate the producibility of the reservoir, however, dynamic properties such as the permeability of the reservoir rock, the viscosity of the fluid and the pressure need to be known. This entire field is known as petrophysics. The term "formation evaluation" is used synonymously, although it generally covers a much broader range of topics. Notice that the term "formation" for a petrophysicist is equivalent to "rock", a habit that at times may be somewhat confusing for a geologist. In the following the two terms are used interchangeably[1]. Dewan (1983), Ellis (1987), and Schlumberger (1989) contain good discussions of the essentials of petrophysics. The following is an overview of the main principles of relevance for this book.

Petrophysics, in its simplest form, is the calculation of porosity and fluid saturations as a function of depth in a well. Archie (1942), through careful experimentation, has given the basic relationship for this. He observed that in fully water-saturated rocks the rock resistivity R_o is proportional to the resistivity of the saturating brine R_w through a resistivity-independent formation factor F in the simple form

$$R_o = F \cdot R_w \quad (1.3.1)$$

F remains constant not just for one rock specimen measured at different water resistivities, as long as $R_w < 1$ Ωm (above which surface conductance becomes important), but is also the same for rocks of the same porosity and with a similar pore structure. Thus, F is determined by the porosity (expressed as a fraction of unity) and the pore structure. In general, it was found that

$$F = a \cdot \phi^{-m} \quad (1.3.2)$$

---

[1] The logging literature features other, rather nonchalant uses of geological terminology. "Sandstone", for example, often designates the mineral quartz, "shale" is used interchangeably with "clay", and more recently in the borehole imaging literature "texture" is used to designate "bedding".

 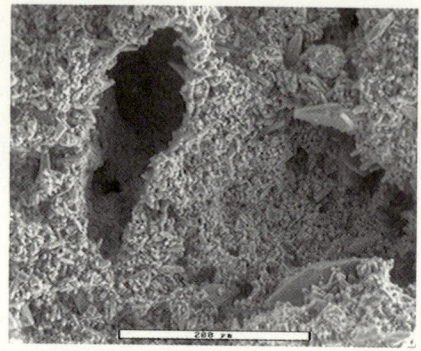

Figure 1.3.1. On the left a crystalline dolomite with $\phi = 47\,\%$ and $m = 1.95$. On the right a moldic bioclastic packstone with $\phi = 36\,\%$ and $m = 3.27$. This large variation in m illustrates the importance of rock texture on petrophysical evaluation. Permeability is 3160 mD for left sample, and 7.7 mD for right sample. Environmental scanning electron microscope images, scale bar on left is 100 μm, on right 200 μm.

such that

$$R_o = a \cdot \phi^{-m} \cdot R_w \quad (1.3.3)$$

where m is known as the cementation exponent and is found to be in the range between 1.5 (mostly for sandstones) and about 3 (for certain carbonates). The cementation exponent, in fact, reflects the tortuosity and connectivity of the pore system: For perfectly straight tubes it has theoretically been shown that $m = 1$. Values around 2 are typical for rocks with intergranular and intercrystalline porosity, while higher values are found in carbonates with vuggy, intergranular or moldic pores (Figure 1.3.1). The coefficient a is in general around 1 (in which case the equation is referred to as Archie's Law), with some sandstones showing values as low as 0.6. In the oil industry, the usefulness of equation 1.3.3 is limited because it applies only to water-saturated rocks. For example, it may be used to determine porosities if the water resistivity is known and good estimates of a and m can be made.

The practical application of resistivity measurements only came after Archie analyzed data by Leverett (1938) on partially water-saturated rocks. He plotted his relative resistivity R_t/R_o against the water saturation S_w on log-log paper and obtained a straight line, suggesting a law of the form

$$S_w^{-n} = R_t/R_o \quad (1.3.4)$$

Here, R_t is the "true" resistivity of the rock at water saturation S_w. The saturation exponent n was found to be generally around 2. Therefore, the resistivity of a partially saturated rock can be described by

$$R_t = a \cdot \phi^{-m} S_w^{-n} \cdot R_w \quad (1.3.5)$$

This equation is still widely used because it lets us determine S_w if the true formation resistivity known, if there is an independent measurement of the porosity, and if R_w, a, m and n are known. Clearly, the requirements are numerous, but equation 1.3.5 illustrates the relevance of resistivity and porosity logging. If we assume $m \approx 2$, $a \approx 1$, and $n \approx 2$ we obtain

$$R_t \approx (S_w \cdot \phi)^{-2} R_w \qquad (1.3.6)$$

Comparing this with the fully water-saturated Archie equation we see that the hydrocarbon-bearing portion of the pore space is treated as part of the rock matrix. Since m and n are usually not exactly equal, the equivalence is of course not perfect, presumably because the distribution of oil in the pores alters the tortuosity for the remaining conducting water. Solving for the water saturation, we get

$$S_w = \frac{1}{\phi} \cdot \sqrt{\frac{R_w}{R_t}} \qquad (1.3.7)$$

which is a simple and widely used formula to estimate the hydrocarbon saturation $S_o = (1 - S_w)$. The full solution for the water saturation using Archie's equation, however, is

$$S_w^n = \frac{a}{\phi^m} \cdot \frac{R_w}{R_t} \qquad (1.3.8)$$

an equation which requires determination of m, a and n, usually through laboratory measurements on cores or local knowledge. Additionally, the porosity and the "true" or uninvaded resistivity have to be known. These two parameters are usually measured or derived from well logs.

The porosity can be quite accurately determined from nuclear or acoustic logs. When compared with porosities measured on cores, there are generally only minor differences that are attributed to decompaction of the cores, the different types of measurement, or an incomplete correction of the lithological effects on the logs. Very often, a combination of neutron, density, and acoustic logs is used to calculate the formation porosity, since this allows correcting for lithological effects. Nuclear magnetic resonance logs are a potentially better technique and are gaining increased acceptance in the industry (chapter 2.6). The uninvaded resistivity R_t is more difficult to obtain because all resistivity measurements are influenced by the resistivity of nearby layers and by the resistivity of the invaded zone in the immediate vicinity of the borehole wall. Generally, R_t is calculated through a combination of shallow, medium and deep resistivities from induction or laterolog tools. The choice of resistivity tool used in a logging operation depends on the rock and the mud resistivities. Logging-While-Drilling measurements have the significant advantage that they record the rock resistivity shortly after the drill bit has passed, when the invasion of mud filtrate into the rock is still small compared to when the wireline logging is done.

While equation 1.3.8 is useful in estimating oil in place, it has two significant shortcomings

- It does not tell us how much of the oil is producible
- It is not valid for rocks with surface conductance or electrolytes other than the formation fluid (for example, clays)

There is a vast amount of literature on these two subjects, and we will only take a cursory glance at the most important aspects.

1.3.2 Mobility

The invasion process hampers the process of obtaining the true, uninvaded formation resistivity, but it can be used to our advantage. In fact, it constitutes an experiment which separates hydrocarbons that can be flushed by the invading mud filtrate from those that are immovable or residual (Figure 1.3.2). This admittedly simplified picture allows us to write an expression for the resistivity of the invaded zone R_{xo} analogous to equation 1.3.1 in the form of

$$R_{xo} = F \cdot R_{mf} \qquad (1.3.9)$$

where R_{mf} is the resistivity of the invading fluid (the mud filtrate). Since s_w is the water saturation before the invasion, and s_{xo} that after invasion, $s_{xo} - s_w$ is the movable, or producible oil saturation.

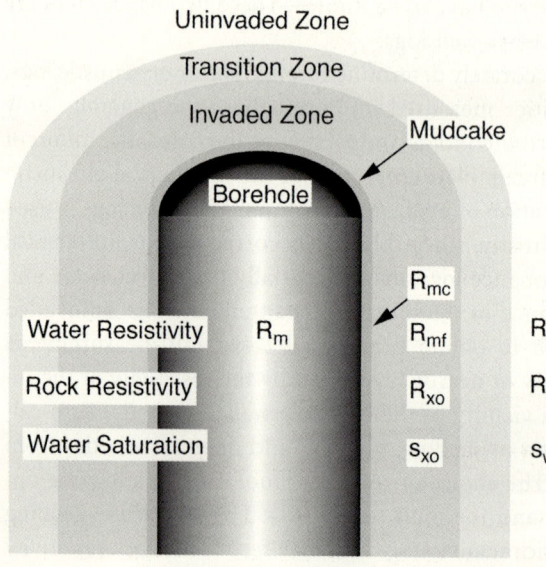

Figure 1.3.2. Sketch of the borehole and its surrounding, showing the notation used in the text.

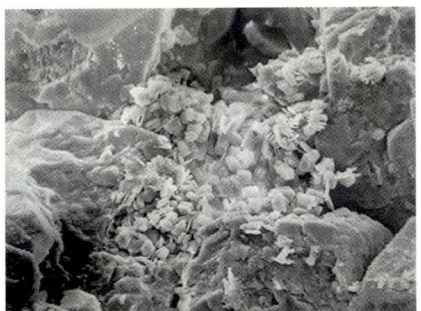 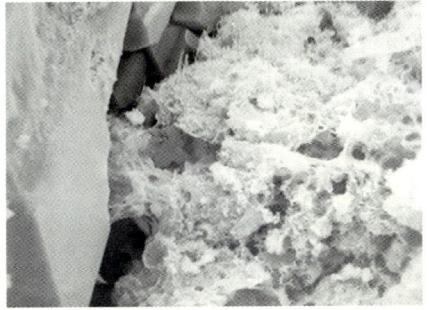

Figure 1.3.3. Scanning electron microscope (SEM) images of two reservoir sandstones from the Ceuta field, Venezuela, showing kaolinite (left) and mixed-layer illite/smectite (right) on the grain surfaces and in the pore throats.

In a manner analogous to equation 1.3.7 we can write

$$s_{xo} \approx \frac{1}{\phi} \cdot \sqrt{\frac{R_{mf}}{R_{xo}}} \qquad (1.3.10)$$

and for the movable oil relative to a unit volume of rock we obtain

$$\phi(s_{xo} - s_w) \approx \sqrt{\frac{R_{mf}}{R_{xo}}} - \sqrt{\frac{R_w}{R_t}} \qquad (1.3.11)$$

where the same assumptions as for equation 1.3.7 are made. Equation 1.3.11 contains three quantities measurable by well logs (R_t, R_{xo} and ϕ), and two that are best measured from fluid samples (R_w and R_{mf}). With it, an approximation of the movable or producible oil saturation can be made.

Clearly, there are numerous simplifications in the approach outlined here. One is the assumption that the salinity of the invaded zone water is the same as that of filtrate. Although this is unlikely, it is reasonable as long as $R_w \approx R_{mf}$. With mud filtrate resistivities much higher than formation water resistivities, however, this assumption is no longer valid, resulting in R_{xo} being too low and, therefore, the oil mobility to be overestimated. Another problem with equation 1.3.11 is that it is not applicable to argillaceous rocks or "shaly formations". Figure 1.3.3 shows two scanning electron microscope (SEM) images of reservoir rocks containing clays that not only reduce the permeability by plugging the pore throats, but also affect the electrical conductivity of the rock.

1.3.3 Clays and Saturation Determination

That clays influence the rock resistivities has been observed a long time ago, but to this date researchers strive to find simple physical models for this effect. The presence of clays adds in essence an extra term (or several terms) to the resistivity equations. This is particularly obvious at higher water resistivities, where the rock resistivity becomes increasingly lower than would be predicted by equation 1.3.1, as has been amply demonstrated by the extensive data collection of Worthington (1985). In the world of shaly formation evaluation, however, equations are usually formulated in terms of conductivities, a tradition that we will follow here as well.

Patnode & Wyllie (1950) proposed one of the first conductivity models for clay-bearing rocks. It is a variant of equation 1.3.1 and has the form of

$$C_o = \frac{C_w}{F} + C_s \qquad (1.3.12)$$

The constant conductivity term C_s is independent of the saturating fluid and accounts for the surface conductance attributed to clays, acting in parallel with the fluid conductivity. Winsauer & McCardell (1953), by contrast, proposed

$$C_o = \frac{1}{F}(C_w + C_z) \qquad (1.3.13)$$

where the "double-layer conductivity" C_z is a function of C_w and goes essentially to zero for $C_w \to 0$, correcting one of the shortcomings of equation 1.3.12. Both models are fairly abstract in the sense that the added conductivity term has no concrete physical significance. Waxman & Smits (1968) tried to amend this using an extensive data set from Shell's laboratory. They proposed

$$C_o = \frac{1}{F}(BQ_v + C_w) \qquad (1.3.14)$$

which is similar to equation 1.3.13, with B the counterion equivalent conductance (proportional to the ion mobility) and Q_v the counterion charge concentration (in milliequivalents per milliliter). At high brine concentrations, they found that $B = 3.83$ S/m per meq/ml. Q_v is high for clays such as smectites, and low for kaolinites. For partial saturation, Waxman & Smits (1968) adapted their formulation as follows:

$$C_t = \frac{S_w^n}{F}\left(\frac{BQ_v}{S_w} + C_w\right) \qquad (1.3.15)$$

This implies that the conduction due to clay is a weaker function of water saturation than the one due to the pore fluid. The rock conductivity according to equation 1.3.15 is, therefore, controlled by the porosity, the brine conductivity, the water saturation, and Q_v, which is related to the cation exchange capacity CEC (in meq/100g) by

$$CEC = 100 \cdot Q_v \phi [p_g(1-\phi)] \qquad (1.3.16)$$

where p_g (in g/cc) is the grain density. Generally, the CEC is estimated or obtained from laboratory analyses of the clay types in the rock. Table 1.3.1 lists typical values for the common clay minerals in sedimentary rocks. As can be seen, the clays of the montmorillonite group have the highest values, followed by illite, kaolinite and chlorite. CEC is a fundamental property of clay minerals and is due to a negative surcharge of the clays themselves, caused in general through substitution of Al^{3+} by other, lower-valence cations. This positive charge on the clay surface creates two very thin aqueous layers: One immediately on the clay surface that is essentially ion-free (the "Stern Layer"), overlain by one with a surplus of counterions, which are generally Na^+ (the "Gouy Layer"). These two layers create a separate conductive path, often referred to as "double layer", in which enhanced ion transport increases the conductivity. Newer data indicate that geochemical logging using nuclear spectroscopy logs might be able to provide a fairly good estimate of CEC and thus of Q_v (Herron, 1986).

Table 1.3.1. The Cation Exchange Capacity CEC as a function of clay type (from Dewan, 1983; Weaver, 1989; van Olphen & Fripiat, 1979).

Clay type	CEC (meq/100 g)
Chlorite	0–10
Kaolinite	2–7
Illite	10–40 (pure: 15)
Montmorillonite	70–120

The Waxman-Smits model, although widely used by log analysts, has the shortcoming of not considering the geometry of the clay distribution and of considering the same formation factor for brine and clay counterion conduction. Clavier et al. (1977) tried to address some of these problems by proposing the "dual-water" model in the form

$$C_t = \frac{S_w^n}{F}\left[C_w + \frac{v_Q Q_v}{S_w}(C_{cw} - C_w)\right] \qquad (1.3.17)$$

where v_q is the near-water volume per charge site and C_{cw} the clay-water conductivity. The authors thus envision an entirely different fluid to be present in the double layer around the clays. By analogy with equation 1.3.16 we see that $B = v_q C_{cw}$, where C_{cw} was determined to be 6.8 S/m at ambient room temperature. The model is very similar to the Waxman-Smits model. In both models, Q_v and, therefore, the clay type are the parameter of greatest importance.

A final model to be mentioned in this context is based on work by Sen et al. (1988), who express the rock conductivity in the fully water-saturated case as

$$C_o = \frac{1}{F}\left[C_w + \frac{AQ_v}{\left(1 + \dfrac{CQ_v}{C_w}\right)}\right] + EQ_v \qquad (1.3.18)$$

The parameters E and C are geometrical factors, and A is equivalent to B in equations 1.3.16 and 1.3.17. As in the previous case, for clean sands with $Q_v = 0$, equation 1.3.18 reduces to Archie's law (equation 1.3.3). This model has some attractive features, particularly because it combines the surface-conductance model of equation 1.3.12 with the double-layer model of Waxman-Smits. Additionally, the slope of the relationship C_w versus C_o changes from high to low brine concentration in accordance with observations on laboratory data. Schwartz & Sen (1988) expanded equation 1.3.18 to partial saturations. The major drawback of this model is that it requires one more parameter than both the dual-water model and the Waxman-Smits model. However, as illustrated by figure 1.3.4, it can provide better fits to laboratory data than either the Archie or the Waxman-Smits equation.

This progression of clay-conductivity models illustrates how complicated the models become if one wants to describe all aspects adequately. However, in many cases the simpler equations may be quite appropriate. Common to all models is the requirement to determine accurately the types and volumes of clays present in the rock.

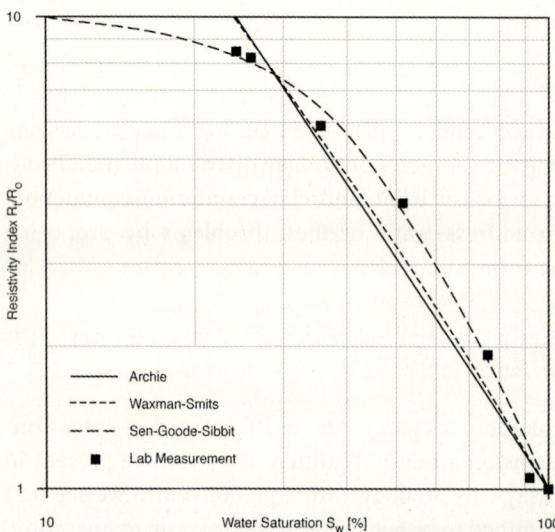

Figure 1.3.4. Rock resistivities as a function of partial saturation of a micaceous sandstone from an oil field in the Sergipe-Alagoas Basin, Brazil. Porosity of sample is 14.7 %, and water resistivity R_w is 0.17 Ohmm. The trend of the laboratory data is best described by a model such as Sen et al. (1988) which has a double-layer term as well as a surface conductance term (after Negrão & Luthi, 1993).

1.3.4 Simultaneous Solutions for Lithologies and Saturations

In petrophysical formation evaluation, there is hardly ever a direct correspondance between the measurements – the logs – and the desired properties such as fluid volumes, producibility, etc. Logs are sensitive to several fluids and rock parameters, and the best approach, therefore, is to simultaneously solve a set of equations that relate the measurements to the desired properties as tool response equations. Ellis (1987) mentions the following example of a rock containing two minerals, calcite and dolomite. Assume that we measure the neutron porosity ϕ_N and the bulk density ρ_b of a rock containing volumetric proportions V_c and V_d of calcite and dolomite respectively. We can write three equations in the form

$$\rho_b = \phi\rho_f + (1-\phi)(V_c\rho_c + V_d\rho_d) \tag{1.3.19a}$$

$$\phi_N = \phi H_f + (1-\phi)(V_c H_c + V_d H_d) \tag{1.3.19b}$$

$$1 = \phi + V_c + V_d \tag{1.3.19c}$$

Aside from the unity constraint in eq. 1.3.19c, these are the response equations for the neutron and density tools, whereby ρ designates the density and H the hydrogen index of the components calcite *(c)*, dolomite *(d)*, and fluid *(f)*. Equations like these are often represented as crossplots. Particularly common is the neutron-density crossplot, but numerous other types are also in use. The M/N plot (Burke et al., 1969) or the MID plot (Clavier & Rust, 1976) use the sonic travel time in addition to the density and neutron and aim at identifying the mineral volumes. The system above is determined but leaves no room for variations in the fluid composition. Doveton (1986) puts these equations into matrix form, which is possible as long as all response equations are of the same form. If the water saturation S_w is included as an unknown, an additional response equation has to be added, for example the Archie equation 1.3.5. The uninvaded "true" resistivity R_t is generally determined from the shallow, medium and deep resistivity measurement, from which the invaded resistivity, uninvaded resistivity, and the diameter of invasion can be calculated. This is yet another set of simultaneous equations, however with much more complicated response equations than the one above. It is often represented as crossplots, known as tornado charts, which are available on most chart books.

There is a complex interrelationship between the measured quantities in figure 1.1.3 and the desired petrophysical properties, often controlled by the geological characteristics of the rock. Doveton (1994) gives a fairly comprehensive review on the subject. Early approaches relied on sequentially solving a set of linear equations such as the ones discussed above, often in a determined system wherein the number of knowns is equal to the number of unknowns. More recent approaches involve the simultaneous solution for the relevant desired quantities using all available logs. The GLOBAL* program developed by Meyer & Sibbit (1980) uses a Bayesian approach to estimate the most probable solution. It start, from a set of tool response equations relating the log measurements to the desired properties such as porosity, fluid saturations and mineral volumes, examples of which are given in equations 1.3.19. The system is preferably over-

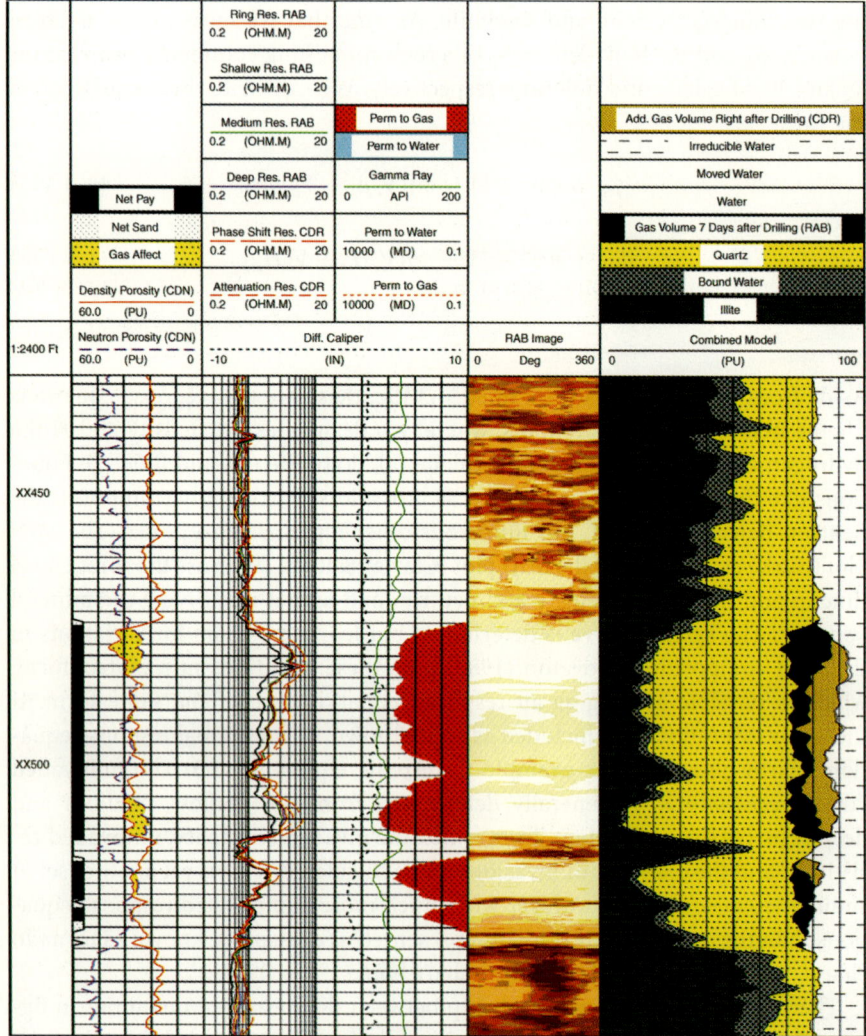

Figure 1.3.5. Formation evaluation using LWD logs and the ELAN* program in a gas field of the Gulf of Mexico. The three tracks on the left are the LWD logs, followed by electrical borehole images from the RAB* tool and ELAN results on the right.

* Marks of Schlumberger

determined, i.e. there are more measurements than unknowns. Global then estimates the most probable solution by minimizing a cost function, the so-called "incoherence function". It takes into account uncertainties in the inputs, tool response equations, and geological constraints such as the minerals expected within a given geological interval. A particularly interesting aspect of this approach is the reconstruction of the logs from the results, allowing the user to perform a quality check and to spot potential zones where the mineral model may need to be changed. In a more recent implementation, Schlumberger's ELAN program performs a similar task as GLOBAL, but includes a linear as well as a non-linear solver and is adapted to newer logs. An example of ELAN obtained from LWD logs is shown in Figure 1.3.5, together with the input logs and a LWD electrical borehole image highlighting the layered nature of the reservoir.

Formation evaluation requires significant expertise and skills because rock systems are complex and not fully described by logging measurements, either for lack of appropriate measurements, or for simplification in the tool responses. However, recent developments in nuclear spectroscopy, borehole imaging and nuclear magnetic resonance have allowed log analysts to make important progress, much of which will be the subject of later chapters in this book. Among these, chapter 2.7.1 contains a more detailed review of the lithology computations from logs.

1.3.5 Permeability

Permeability is the ease at which fluids flow through a material, in this case the reservoir rocks. It is principally controlled by the pore throats, i.e. the size of the connections between pores, but also by the tortuousity and connectivity of the pore networks. In clastic rocks with good sorting, the permeability should roughly scale with the square of the grain size, since permeability has the dimensions of a surface. Figure 1.3.6 shows the permeability/porosity relationships for a large laboratory data set (Chilingar, 1964), wherein the strong dependence of permeability on grain size becomes evident in a form roughly as expected. For example, particles in coarse-grained sandstones are about twice as large as in medium-grained sandstones, and the permeabilities seem to be about four times larger. For carbonates, Lucia (1983) demonstrated the dependence on texture as classified by Dunham's (1962; see also figure 1.3.1). The grain-supported samples show a much higher permeability for a given porosity because their interparticle pore throats are considerably larger than the ones in the micrite matrix.

Permeability is probably the single most important quantity in a hydrocarbon reservoir, and there is a large amount of literature on its dependence on rock texture, fluid pressures and saturations, and other factors. It is measured or es-

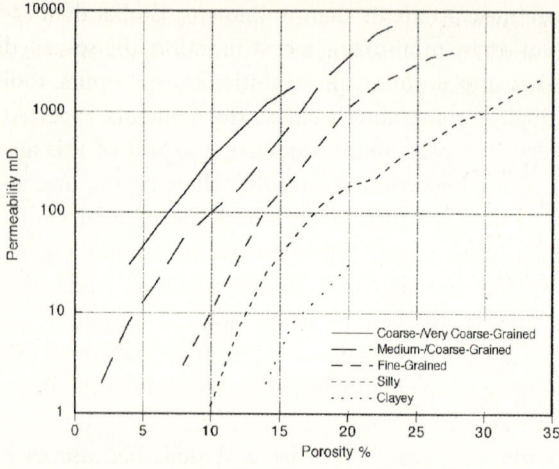

Figure 1.3.6. The permeability/porosity relationship from core measurements on various types of clastic rocks, illustrating the strong dependence on grain size (Chilingar, 1964).

timated from laboratory measurements on cores, from well tests such as drill stem tests, or from logging measurements. Among the latter, there are several approaches:

- Determination from a formation test during which there is a pressure drawdown and a pressure buildup phase (Stewart & Wittmann, 1979). This is a relatively direct estimation of permeability, since it involves donwhole fluid flow. Modern formation testers such as the Modular Formation Dynamic Tester (MDT*) are designed to determine both the vertical as well as the horizontal permeability.

- Estimation of permeability from logging measurements assumed to be more or less directly related to permeability. These are static measures since no fluid flow takes place. An example is the irreducible water saturation S_{wi}, first proposed by Tixier (1949), and for which a large number of variants have since been developed. They are generally of the form

$$k = c\phi^x S_{wi}^{-y} \qquad (1.3.20)$$

where x ranges from 2 to 3, and y is around 2. These relationships pose the important problem of estimating S_{wi}.

- Estimation of permeability using the transverse relaxation time T_2 from nuclear magnetic resonance, a method discussed in chapter 2.6. It is related to the previous method since it involves estimating the pore size and from this the bound water in the rock, assumed to be contained in the micropores.

---

* Mark of Schlumberger

- Estimation of permeability from the Stoneley wave attenuation (Burns & Cheng, 1986). This method is still under development and has the advantage that it is a dynamic measurement that probes the behaviour of the rock under a pressure pulse.

- Calculation of permeability from several logs, using in general empirically determined transform functions. Examples are the geochemical logs derived from nuclear spectroscopy (Herron 1987, see also chapter 2.7), or just a suite of standard open-hole logs such as the neutron porosity, the density and gamma ray. In these cases, multiple regressions or more recently neural networks are used to establish the relationship with permeabilities measured on core. An example of the latter can be seen in the middle track of figure 1.3.5.

None of these methods are direct measurements of permeability, a fact well known to logging companies who consistently hear about the need for such a measurement. Although the dynamic techniques mentioned above (formation testers and Stoneley wave attenuation) are perhaps closest to a direct measurement, they are often influenced by formation damage in the vicinity of the wellbore, and they often reflect the mobility of the mobile phase in the rock rather than the intrinsic permeability. Nonetheless, the recent advances in wireline formation testing and nuclear magnetic resonance measurement are important improvements in the quest for formation permeability.

References

Archie GE (1942) The electrical resistivity log as an aid in determining some reservoir characteristics. Petr Trans AIME, 146: 54–62.
Burke JA, Campbell RL, Schmidt AW (1969) The lithoporosity cross plot. The Log Analyst,
Burns DR, Cheng CH (1986) Determination of in-situ permeability from tube wave veolocity and attenuation. Trans. 27th Symp. Soc. Prof. Well Log Analysts.
Clavier C, Coates G, Dumanois J (1977) Theoretical and experimental bases for the "Dual-Water" model for interpretation in shaly sands. 52nd Ann Conf Soc Petr Eng: Paper 6859; also (1984) J Soc Petr Eng 24: 153–168.
Clavier C, Rust DH (1976) MID-plot: A new lithology technique. The Log Analyst, 17, 6.
Chilingar GV (1964) Relationship between porosity, permeability and grain-size distribution. In: van Straaten LMJU (ed) Deltaic and shallow marine deposits. Elsevier, Amsterdam.
Dewan JT (1983) Essentials of modern open hole log interpretation. PennWell, Tulsa.
Doveton JH (1986) Log analysis of subsurface geology: Concepts and computer methods. Wiley & Sons, New York.
Doveton JH (1994) Geologic log analysis using computer methods. Applications in Geology No. 2, Am Assoc Petrol Geol, Tulsa.
Dunham RJ (1962) Classification of carbonate rocks according to depositional texture. In: Ham WE (ed) Classification of carbonate rocks. Mem. 1, Am Ass Petrol Geol.
Grim RE (1968) Clay mineralogy. McGraw-Hill, New York.
Ellis DV (1987) Well-logging for earth scientists. Elsevier, Amsterdam.
Herron MM (1986) Mineralogy from geochemical well logging. Clays and Clay Minerals 34: 204–213.
Herron MM (1987) Future applications of elemental concentrations from geophysical logging. Nucl Geophysics 1: 197–211.

Leverett MC (1938) Flow of oil-water mixtures through unconsolidated sands. Petr. Trans AIME 132.

Lucia JF (1983) Petrophysical parameters estimated from visual description of carbonate rocks: A field classification of carbonate pore space. J Petrol Technol 35: 626–637.

Mayer C Sibbit AM (1980) GLOBAL, A new approach to computer processed log interpretation. 55th Ann Conf Soc Petr Eng: Paper 9341.

Negrão DS, Luthi SM (1993) Determining the hydrocarbon saturation in reservoir rocks with fresh formation water, Pilar field, Alagoas-Sergipe Basin. Proc 3rd Int Congr Braz Geophys Soc, 336–339.

Olphen van H, Fripiat JJ (1979) Data handbook for clay materials and other non-metallic minerals. Pergamon Press, Oxford.

Patnode HW, Wyllie MRJ (1950) The presence of solids in reservoir rocks as factor in electric log interpretation. Petr Trans AIME 189: 47–52.

Schlumberger (1989) Log interpretation, principles and applications. Schlumberger Educational Services, Houston.

Schwartz LM, Sen PN (1988) Electrical conduction in partially saturated shaly formations. 63rd Ann Conf Soc Petr Eng: Paper 18131.

Sen PN, Goode PA, Sibbit AM 1988) Electrical conduction in clay bearing sandstones at low and high salinities. J Appl Physics 63: 4832–4840.

Stewart G, Wittmann M (1979) Interpretation of the pressure response of the repeat formation tester. 54th Ann Conf Soc Petr Eng: Paper 8362.

Tixier MP (1949) Evaluation of permeability from electric-log resistivity gradients. Oil and Gas Jour (June 16, 1949)

Waxman MH, Smits LJM (1968) Electrical conduction in oil-bearing shaly sands. J Soc Petr Eng 8: 107–122.

Weaver CE (1989) Clays, muds and shales. Developments in Sedimentology 44, Elsevier, Amsterdam.

Winsauer WO, McCardell WM (1953) Ionic double-layer conductivity in reservoir rocks. Petr Trans AIME 198: 129–134.

Worthington PF (1985) The evolution of shaly-sand concepts in reservoir evaluation. The Log Analyst, Jan–Feb.

Part 2: Geological Measurements

Part 2: Geological Measurements

2.1 Dipmeter

2.1.1 Tool Principle

The main purpose of the dipmeter measurement is to obtain structural dips of the layers traversed by the borehole. This is achieved by recording variations in the rocks' electrical properties along tracks of the borehole wall, and by correlating these curves. Dips are measured with respect to the borehole ("apparent dips") or with respect to geographic north and the earth's vertical at a given location ("true dips").

The first dipmeter was the "anisotropy tool", a stationary electromagnetic measurement with four electrodes on the tool mandrel (Schlumberger & Doll, 1933; Schlumberger et al., 1935; Schlumberger & Doll, 1937). It was followed by a dipmeter based on spontaneous potential (SP) curves, which were measured on three pads applied to the borehole wall (Doll, 1943). A photoclinometer measured the orientation of one of the pads with respect to magnetic north as well as the azimuth and deviation of the well. In order to do this the tool had to be stopped at convenient intervals. Because the SP curves lacked character in certain environments, this system was abandoned in 1945 for an electrical design based on a lateral resistivity in which the potential between two closely spaced electrodes on three arms each was recorded. In 1952, resistivity methods gained the upper hand: First a microlog (de Chambrier, 1953), then a microlaterolog tech-

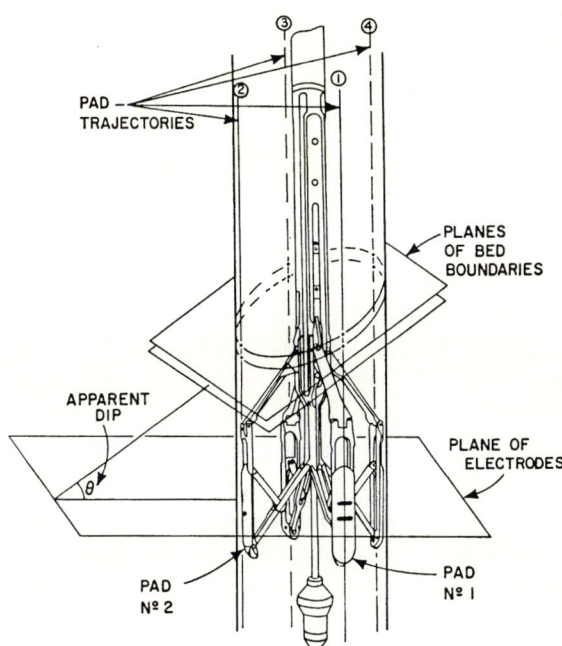

Figure 2.1.1. Sketch of a four-arm dipmeter tool (HDT) with the trajectories the pads take when the tool is pulled uphole (from Hepp & Dumestre, 1975).

nique (Bricaud & Poupon, 1959) was used to record three resistivity curves of high vertical resolution in Schlumberger's continuous dipmeter (CDM*). A breakthrough was the introduction of the "poteclinometer" which allowed a continuous measurement of the deviation and azimuth of the well, and additionally of the orientation of a reference pad ("pad 1") with respect to magnetic north. This setup allowed for the first time to record continuous dipmeter logs. In 1967, the high-resolution dipmeter (HDT*) was introduced (Allaud & Ringot, 1969). It featured four arms at 90° to one another, each with an elongate electrode (Figure 2.1.1). The arms were powered and could be opened and closed from the surface. This design had the advantage of being well suited to elliptical holes, which are common in deviated wells. Additionally, the tool provided a much-needed redundancy in dip computation, because in the previous design malfunctioning or poor recording of one pad rendered the entire log useless. Samples were recorded every 0.2 inches (0.5 cm). A second electrode on one of the pads allowed correcting for irregular downhole speeds of the tool through cross-correlation with its twin electrode, thereby assuring proper depth positioning of the recording.

The measurement principle of the HDT is still in use in today's dipmeters and electrical imaging tools. It consists of emitting a low-frequency alternating current from the lower, conductive part of the sonde, where the arms are placed (Figure 2.1.2). The current takes a path through the formation to a return electrode placed in the upper part of the sonde, separated from the lower part by an insulated middle section. This arrangement is called passive focusing, because the entire lower part of the sonde is at the same potential, including all pads and much of the borehole mud surrounding them. This, in turn, causes the borehole surface around the dipmeter arms to be at the same potential. Because electrical current is at a right angle to equipotential lines, it therefore flows perpendicular to the borehole surface into the rock. The actual measurement of dipmeters is the current flowing out of each electrode as a function of depth. It is sometimes loosely referred to as resistivity curve, since its inverse often correlates well with shallow laterolog resistivities. The engineer can adjust voltage and total current (called EMEX) to suit the logging conditions. Gain and offset of the current measurement are adjusted so as to avoid saturation of the circuit and thus to maintain good activity of the curves.

In the subsequent generation, the stratigraphic high-resolution dipmeter tool (SHDT*), two electrodes are placed on each of the four pads, providing even higher redundancy of dip measurement and allowing dip calculations of small-scale bedding features which do not cross the entire borehole (Figure 2.1.2; Chauvel et al., 1984). For the first time, the inclinometry unit contains no moving parts. A fluxgate magnetometer and a solid-state accelerometer measure tool orientation and tool movement with great precision, allowing

---

* Marks of Schlumberger

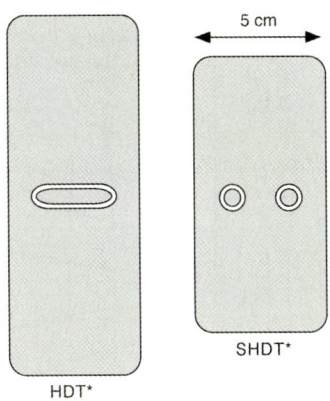

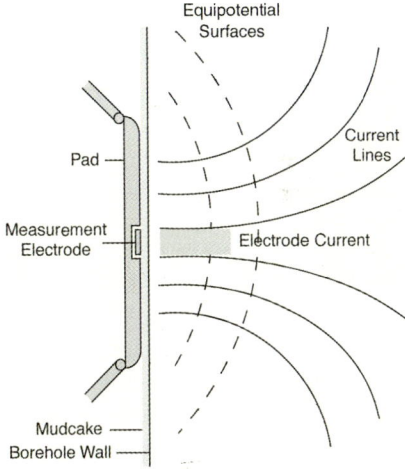

Figure 2.1.2. Top: Front view of two dipmeter pads (HDT and SHDT) with approximate scale. Conductive parts in grey, insulators in white. The measurement electrodes are in the center of the pads. Bottom: Side view of pad/electrode configuration for dipmeter and electrical imaging tools.

an accurate determination of the tool position. Sampling is increased to 0.1 inch, or 2.5 mm and vertical resolution is around 1 cm. Halliburton's SED** and Western-Atlas' HDIP*** diplogs are six-arm dipmeters which operate on the same physical principle as the Schlumberger tools (Morrison & Thibodaux, 1984). They have independently actuated arms and swivel joints at the pads to better adjust to non-circular holes. Since they record six dipmeter curves, they also provide ample redundancy in dip calculations.

Soon after these latest-generation dipmeter tools the first electrical imaging tools arrived on the market and increasingly replaced them (chapter 2.2). However, there are still many dipmeters in use today, and they are unlikely to be phased out entirely in the near future. Imaging tools can also serve as dipmeter

---

** Mark of Halliburton
*** Mark of Western Atlas

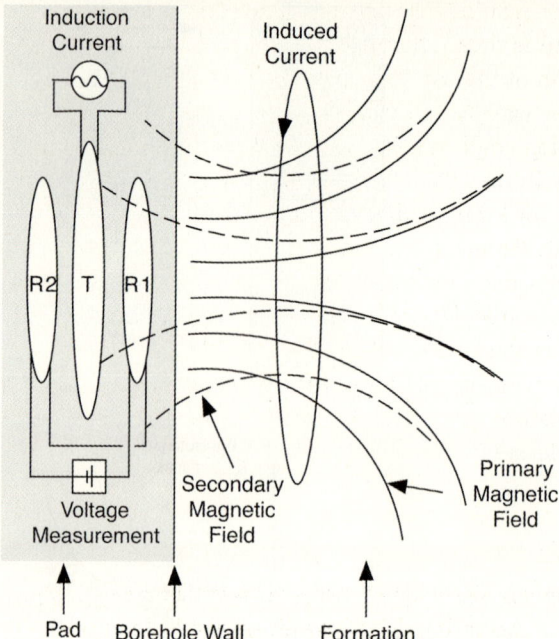

Figure 2.1.3. Principle of the oil-based mud dipmeter (OBDT). The tool measures small induction currents in front of the pads.

tools. Dips can be picked interactively from the images using a graphic workstation (Luthi & Banavar, 1988; Cheung & Heliot, 1990) or automatically using tracing algorithms on the images (Antoine & Delhomme, 1990; Faivre & Catala, 1995; Ye et al., 1997). Borehole images can also be downgraded to dipmeters by extracting a few image columns and feeding them as dipmeter curves into the automatic dip calculation programs.

In *oil-based mud*, which is widely used in certain areas to provide better drilling performance and more stable holes, conventional dipmeters cannot be used because the borehole mud is not conductive. Protruding metal scratchers are sometimes mounted onto the dipmeter electrodes in order to establish a direct conductive path with the rock, but the method is relatively unreliable. The specifically designed oil-based mud dipmeter tool (OBDT*) is based on the induction principle and as such does not rely on conductive borehole mud. It contains in each of its four pads a coplanar transmitter coil and two coupled receiver coils (Dumont et al., 1987; Kleinberg et al., 1987). The electric current of the transmitter coil produces a magnetic field that induces eddy currents in the conductive formation (Figure 2.1.3). These currents in turn create a secondary magnetic field that induces currents in the two receiver coils. The OBDT measures the difference in voltage between the two receiver coils; it is found to be proportional to the formation conductivity. The measurement is very shallow, with over 90 % of the

---

* Mark of Schlumberger

signal coming from within one inch (2.5 cm) of the pad. Its vertical resolution lies in the range of 2 to 3 cm and is thus inferior to the SHDT.

The OBDT is an adaptation of the conventional dipmeter technique to non-conductive mud similar to the way the induction log was developed to measure resistivities where the laterolog could not be used. However, the inclinometry and transmission technology in the OBDT is the same as for the SHDT. There is no dipmeter tool available for logging-while-drilling. However, electrical and nuclear imaging allow dip calculations from images (chapters 2.2 and 2.4). This can be done in the downhole microprocessor using specially designed algorithms. The results are then sent uphole with the mud pulse transmission system to provide dips in real time at the surface (Rosthal et al., 1995).

Figure 2.1.4 shows an example of dipmeter acquisition data and table 2.1.1 lists the "fast" and "slow" data obtained from dipmeter tools. Dipmeters are generally run in combination with at least one other tool, typically the gamma ray.

2.1.2 Dip Calculation

Common to practically all dip calculations is the assumption that the bedding is locally planar. The bedding plane is defined in space by the depth at which it intersects the borehole, its *dip*, and its *azimuth*. To define these two terms, it is best to imagine a line following the steepest descent on the bedding plane. The dip is the angle between this line and the horizontal, measured in a vertical plane, and the azimuth is the angle between this line and geographic north, measured in a horizontal plane (Figure 2.1.5). Dip values range from 0° to 90°, or from horizontal to vertical, and azimuth values range from 0° to 360°.

Early dip calculations were done by hand-correlating the three or four dipmeter curves. In a first step, a relative or *"apparent dip"* is obtained with the borehole axis as a reference frame[1]. It is then used to obtain the *"true dip"* with the earth's geographic coordinates as a reference frame by compensating for the borehole deviation and azimuth. Today all dip calculations are done with computers, but hand-calculating a dip from the raw curves can be still be very useful for controlling the quality of computer-processed dips. Serra (1984, p. 276) describes a method to do this.

There are currently two common ways of obtaining computer-processed dips:

- By cross-correlating the dipmeter curves
- By pattern recognition programs

Cross-correlation of dipmeter curves is done by comparing a segment of one dipmeter curve g to a segment of the same length z of another curve f, shifted

---

[1] This definition differs from the one commonly used in structural geology (see later).

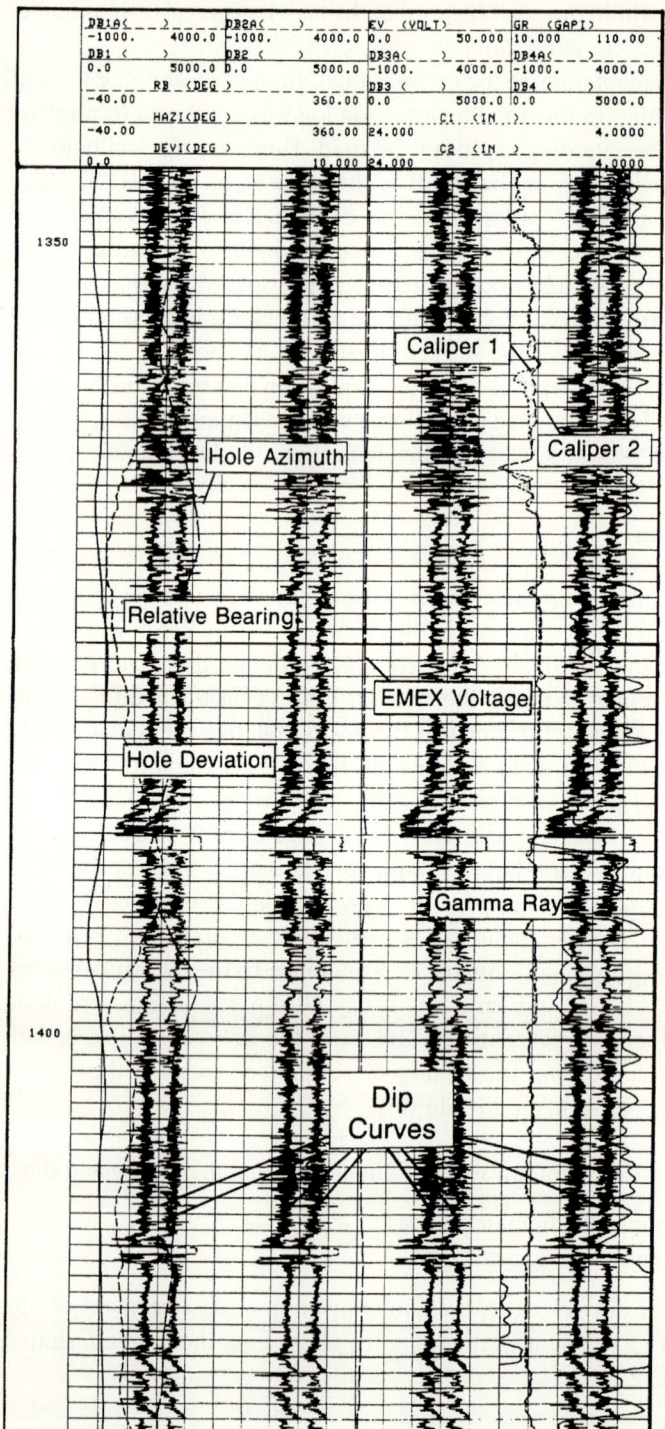

Figure 2.1.4. Field print of the data acquired by a SHDT dipmeter survey.

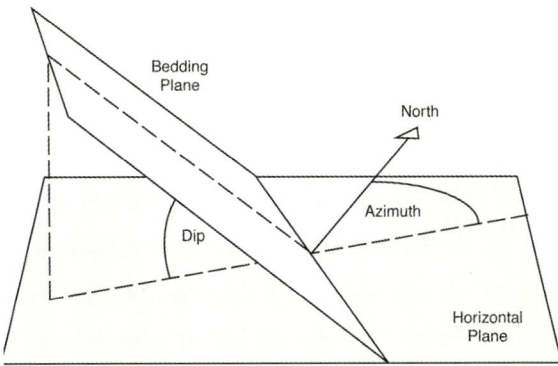

Figure 2.1.5. Definition of dip and azimuth of a bedding plane.

in depth by displacement h. The correlation C between the two curves is the similarity in shape and is computed in general through an equation of the form

$$C(h) = \frac{1}{n}\sum_z f(z)g(z+h) \quad (2.1)$$

where n is the number of samples over the correlation interval. This gives a correlogram, in which the correlation coefficient C is plotted against the displacement h of the two curves. The value of h where C is largest is taken as the most likely correlation of the two curves (figure 2.1.6). Theoretically, only two cross-

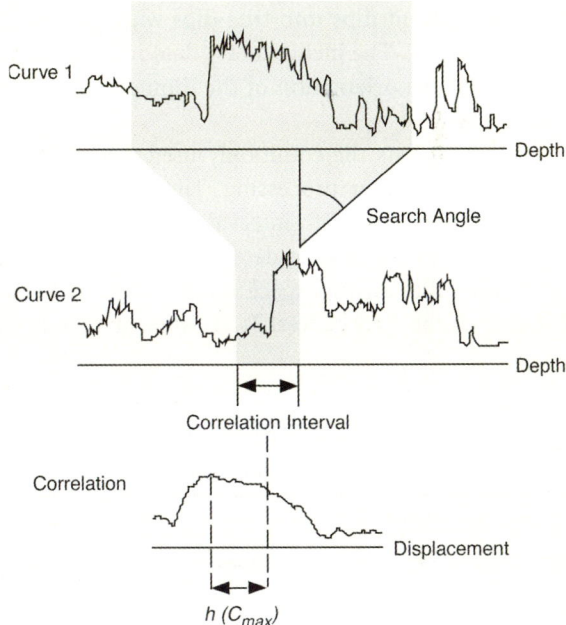

Figure 2.1.6. Cross-correlation of two curves and the resulting correlogram computed from eq. 2.1.1.

Table 2.1.1. Data commonly obtained from dipmeter tools. Borehole azimuth, deviation, tool orientation, and tool speed are calculated from the magnetometer and accelerometer data.

"Slow" Channels	"Fast" Channels
Borehole Deviation	Dipmeter Curves (4, 6 or 8)
Borehole Azimuth	1 or 2 Speed-Correction Curve(s)
Tool Orientation	
Tool Speed	
Calipers	
EMEX Voltage	
EMEX Current	
Cable Tension	

correlations are needed to define a plane and thus to obtain a dip and azimuth. However, in all modern dipmeter tools many more cross-correlations are possible, and procedures to determine the most likely dip at a given depth have to be employed (e.g. Hepp and Dumestre, 1975). The quality of fit of the plane to all calculated displacements determines the quality of the dip, displayed by various symbols on the dipmeter results. The "slow channels", i.e. those data that are sampled at a slower rate than the "fast" dipmeter curves (table 2.1.1), are used in the dip calculation in two ways. First, the acceleration data are used to estimate the downhole tool speed and to position each sample of the dipmeter curves at the most likely depth. Secondly, the inclinometry data are used to convert the apparent dips into true dips with the Earth geographic coordinates as reference frame. The inclinometry data include the hole deviation and azimuth, and the relative orientation of the dipmeter sonde with respect to north or the borehole axis.

Table 2.1.2 lists the commonly used values of input parameters into cross-correlation dipmeter processing. The *correlation interval* is the length z. The *step distance* is the depth interval at which the computational procedure is repeated. The *search angle* a determines the maximum value h by which the two curves are shifted with respect to each other, by taking into account the physical distance of the two curves. It is often at first kept relatively low, but if the

Table 2.1.2. The commonly used dipmeter processing parameters.

	English Units	Metric Units
Correlation Interval	4 feet	1 meter
Step Distance	2 feet	0.5 meter
Search Angle	35°× 2	35°× 2

correlation quality is insufficient it is doubled. These parameters have to be adapted to local conditions and to the goal of the interpreter. In a structurally complex area where overthrusting is expected, for example, the search angle might have to be increased. If fine sedimentary detail is required, the correlation interval might have to be reduced, thereby increasing the number of dip calculations, but in return the search angle might be kept low.

Cross-correlation methods are pure mathematical procedures aiming to produce dips at regular intervals regardless of the curve character. They have proven useful, particularly for large-scale geological structures, and easy to implement on computers. Most of today's dipmeter interpretation methodology is based on results from cross-correlation dip calculations.

Pattern recognition methods are often more suitable for detailed sedimentary interpretations than cross-correlation. When bedding planes are abundant, the dipmeters curves show much activity and, therefore, more dips can be ob-

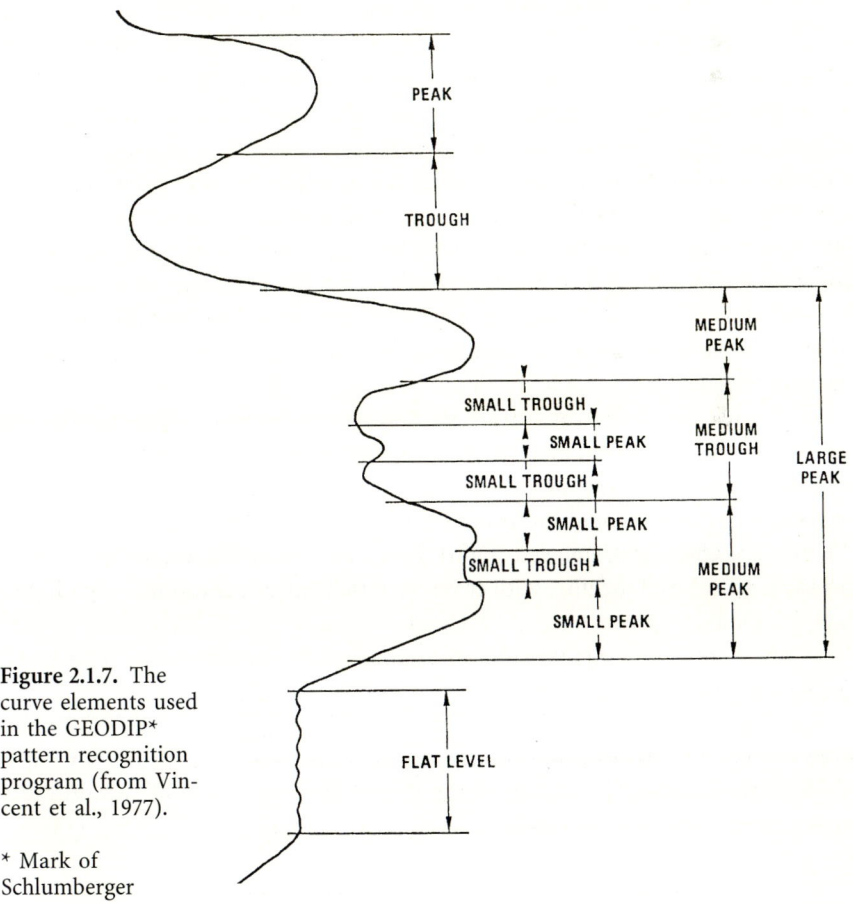

Figure 2.1.7. The curve elements used in the GEODIP* pattern recognition program (from Vincent et al., 1977).

* Mark of Schlumberger

tained. By contrast, cross-correlation methods only calculate one dip per step distance regardless of curve activity. The pattern recognition program GEODIP* by Vincent et al. (1979) is designed to calculate one dip for each bed boundary. An involved procedure is followed, whereby the dipmeter curves are first divided into elements of features, such as small, medium or large peaks or troughs, plateaus or spikes (figure 2.1.7). These elements can be hierarchically superposed; for example, a small trough can be contained within a large peak. The program then correlates tops and bottoms of these elements in a hierarchical manner with prescribed rules and, if successful, it computes a dip. If poor planarity is observed but the elements correlate well, four dips for the four possible planes computations (in the case of the HDT) are output. This program honors several geological correlation principles, notably that larger elements (thicker beds) are correlated first and that the density of dips is a function of the layering. There is also a no-crossing rule stipulating that correlations cannot cross each other even if they are of good quality. Pattern recognition results of dipmeters are often displayed and interpreted together with cross-correlation results.

For the oil-based mud dipmeter tool OBDT, pattern recognition programs are only rarely applied because of the inherent lower resolution of the measurement.

2.1.3 Graphical Representations

Dipmeter results consist of one dependent variable (depth) and two independent variables (dip and azimuth), which complicates the graphical representation. A commonly used plot is shown in Figure 2.1.8. The vertical axis is the depth, as is common for most logs. The horizontal axis shows the dip values on a scale that, for traditional reasons, is logarithmic to 60° and then linear to 90°. Dips are usually displayed as circles or triangles, and a barb attached to them indicates the azimuth, with geographic north pointing up and south pointing down. This presentation is referred to as *"arrow plot"* – from the times when the barb had an arrow – or *"tadpole plot"* ever since it lost the arrow. The plot commonly also contains caliper, borehole deviation and azimuth as well as gamma ray data. In the case of pattern recognition program, the dipmeter curves as well as the correlation lines are also displayed (figure 2.1.9). Occasionally, pattern recognition results are displayed together with cross-correlation dips, for example in Schlumberger's DualDip presentation. This may be useful for quality control and in cases where subtle sedimentary dip variations need to be extracted within a larger geological structure.

There are numerous other ways of plotting dipmeter results. The most common of these are *stereographic* and *equal-area plots* (Schmidt projection), *azimuth frequency plots*, projection of the dip results onto a line of section (*stick plots*), or projection of the dip results onto the borehole cylinder. Among these,

---

* Mark of Schlumberger

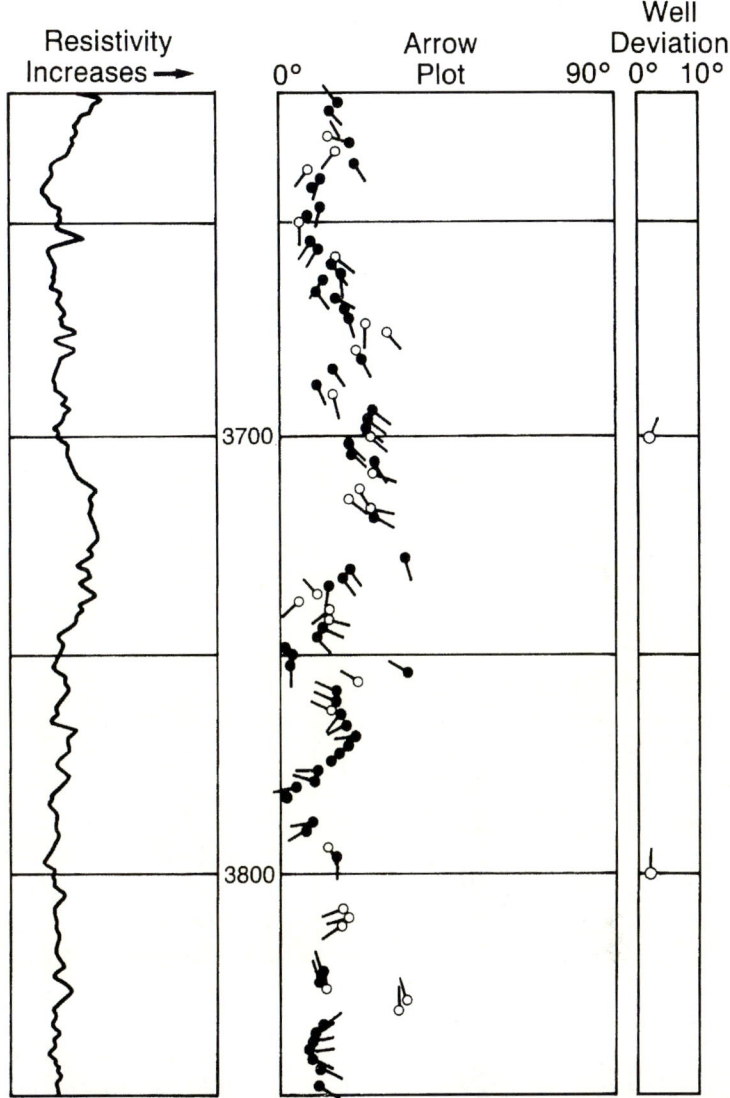

Figure 2.1.8. The tadpole or arrow plot presentation of dipmeter results from the CLUSTER* program. Black circles are high quality, empty circles lower quality dips. Borehole deviation is measured with respect to the vertical and also presented as tadpoles.

stereoplots are the most frequently used graphic aid and have been used since the early days of dipmeter interpretation (de Witte, 1956; Schlumberger, 1981) as they are particularly useful in the study of folds. Figure 2.1.10 shows an example of an equal-area Schmidt plot of a fold in which the two limbs, the fold plane and the axis can be determined.

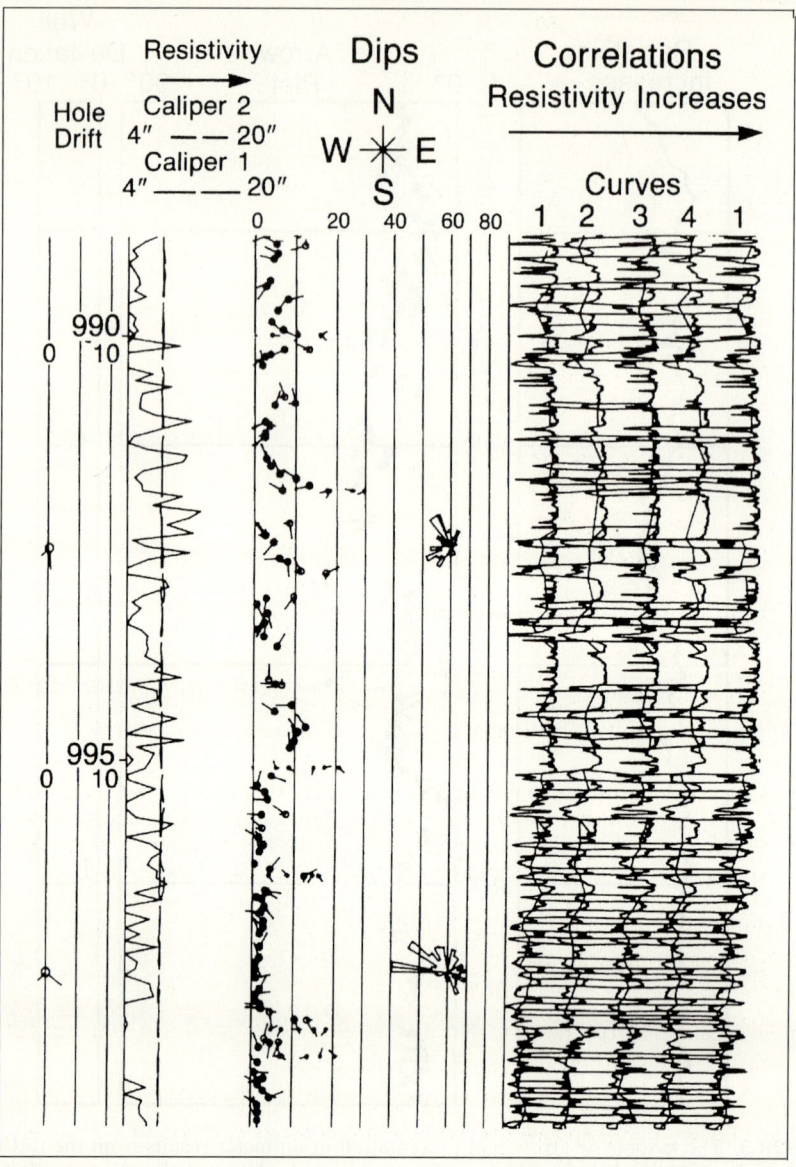

Figure 2.1.9. Dipmeter results of the GEODIP pattern recognition program, showing in addition to the dips the dipmeter curves and the correlations, indicated by lines. Display is generally on an expanded vertical scale, here at 1/40.

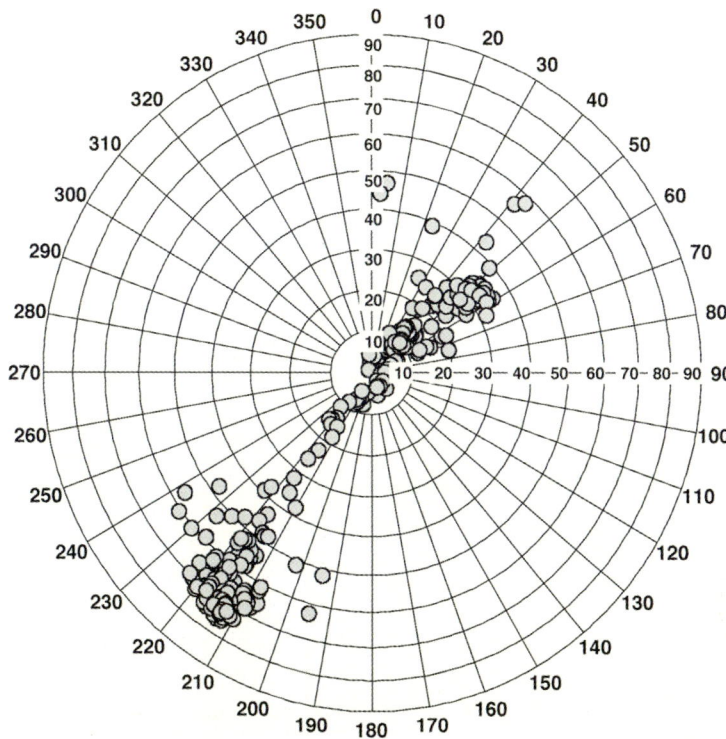

Figure 2.1.10. Schmidt equal-area plot in polar projection. Dips plotted are from a folded sequence, with the two clusters representing the two limbs of the fold and their midway point defines the tilt of the fold axis, estimated to be 20°NE. The great circle on which the dips fall is essentially vertical, indicating little to no axial plunge.

2.1.4 Dipmeter Interpretation

Today's dipmeter interpreters work mostly on workstations, where they can perform the processing routines mentioned above, and numerous graphic presentations are available. The workstations often also contain more sophisticated interpretation aids, such a special projections or geometric construction programs, which assist the interpreter in his task.

Early dipmeter interpreters had to limit themselves to extracting the overall structural picture from the few dips available to them. Figure 2.1.11 shows an old dipmeter survey in the original presentation display. The well was drilled in the proximity of a salt dome in the Gulf Coast of the United States, and the dipmeter survey was performed in order to determine the distance to the salt dome and details of the structure around it. Dips were hand-calculated at intervals of about 500 feet, or 150 meters, and show clear patterns despite the long gaps. Figure 2.1.12 is a proposed interpretation according to which a fault was crossed

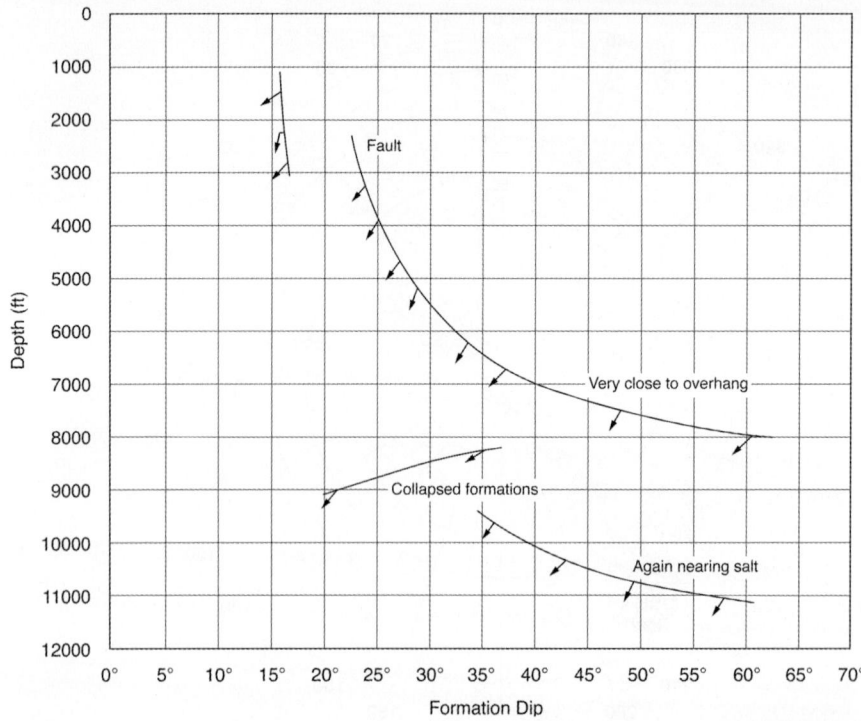

Figure 2.1.11. An early dipmeter survey illustrates the fundamental task of the dipmeter interpreter: To infer the near-wellbore geological structure (redrawn from unpublished internal documents, Schlumberger, c. 1940).

just below 3000 feet, and the salt overhang was narrowly missed at about 8000 feet. The example illustrates the main goal of dipmeter interpretation: to infer the geological structure in the vicinity of the wellbore from a sequence of dips (figure 2.1.13). This inversion process is non-unique, i.e. there are more than one possible solutions. Sometimes geological reasoning can rule in favor of one of the solutions, but even for an experienced geologist this is not always an easy task. Additionally, the presentation of two variables in one plot is visually demanding and requires extensive training. For these reasons, computer-aided methods have been developed and are now indispensable tools for dipmeter interpreters.

We can distinguish the following dipmeter interpretation approaches:

- Interpretation from the processed data using human expertise alone
- Interpretation aided by computational expert systems
- Interpretation assisted by computational geometric constructions
- Interpretation using statistical methods on dip and curve characters

In the following, an overview of these approaches is given.

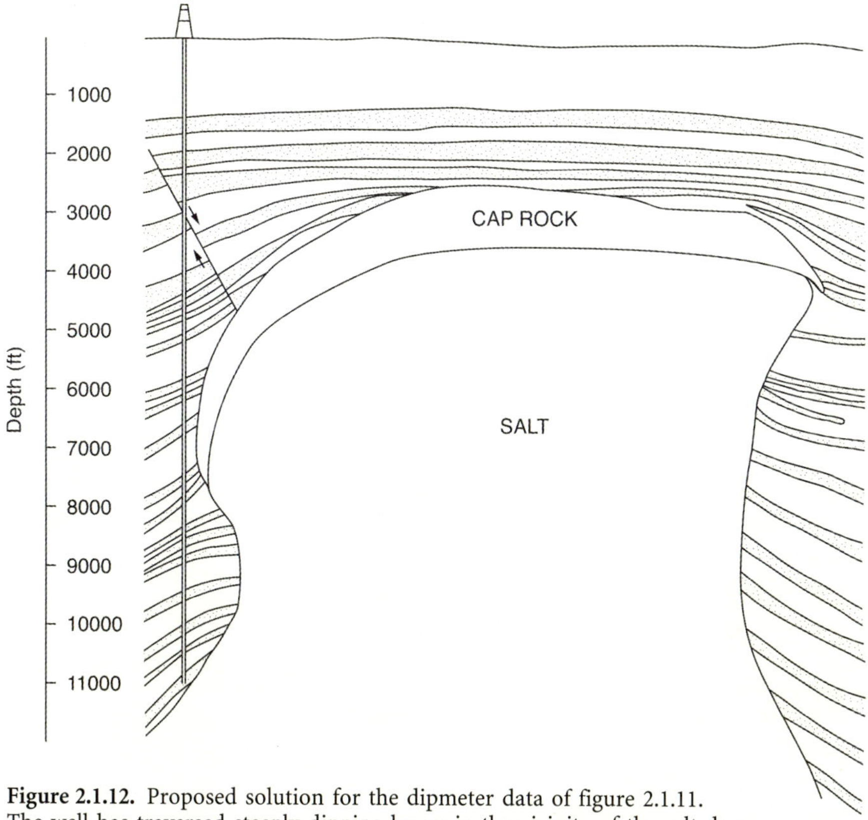

Figure 2.1.12. Proposed solution for the dipmeter data of figure 2.1.11. The well has traversed steeply dipping layers in the vicinity of the salt dome, missing the salt just barely (redrawn from unpublished internal documents, Schlumberger, c. 1940).

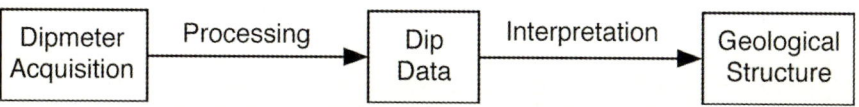

Figure 2.1.13. Simple flow chart for dipmeter analysis.

Human Expertise

The basic dipmeter interpretation principles have been developed mostly after robust, computer-processed dipmeter data became available. A division was established between *"structural"* and *"stratigraphic"* interpretation, whereby "structural" refers to dips attributed to tectonic deformations, and "stratigraphic" to those related to sedimentary processes.

Structural interpretation is a primary use of dipmeters, and many of the standard methods from classical structural geology (Ramsay & Huber, 1987; Billings, 1972) can directly be applied. For example, the fold axis of a similar fold can be determined as the intersection of the two planes corresponding to the fold limbs on a stereonet. In a parallel fold, the pole of the great circle described by the dips defines the fold axis. The books by Schlumberger (1981), Serra (1986) and Rider (1996), and the papers by Rønningsland (1990), Devilliers & Werner (1990) and Adams et al. (1992) contain a good number of structural dipmeter examples from various fields as well as schematic and synthetic examples based on outcrop analogues.

Structural dipmeter interpretation is usually done at a large scale, where the display scale is strongly compressed. The dips are often processed with long correlation intervals (table 2.1.2) or filtered so that scattering caused by sedimentary structures and other factors is reduced. Figure 2.1.14 sketches classical structural situations and their corresponding dip patterns. The structures are two-dimensional, i.e. they do not change for different cross-sections cut parallel to the ones

Table 2.1.3. Structures and their common expression on dipmeter results. Measurable elements refer to features derived from the dip sequence.

Structure	Dipmeter Expression	Measurable Elements
Monocline	Constant dip/azimuth with depth	
Kink fold	Monoclines above and below fold plane	Fold axis and fold plane
Similar fold (figure 2.1.14a)	Constant dip/azimuth on each limb; if tilted, rapid transition from one dip to the other	Dip/azimuth of one or both limbs; fold axis and fold plane
Concentric fold (figure 2.1.14b)	Increasing dip with depth unless well is on crest; azimuth may change monotonically with depth	Fold axis and fold plane
Rollover or growth fault (figure 2.1.14c)	Slowly increasing dips with depth to ~30°, then sudden return to low dips	Strike of structure, depth and in special cases dip of fault plane
Normal fault (figure 2.1.14d)	In incompetent rocks a "drag" pattern occurs as an increase in dip towards the fault, then a decrease	Where drag occurs, depth, strike and minimum dip of fault plane
Reverse fault	A more or less pronounced dip change towards the fault both from above and below	Strike of fault; in some cases depth and dip of fault
Thrust fault, ramped anticline (figure 2.1.14e)	For some well locations an increase in dip towards the thrust; below thrust plane usually low dips	Strike of structure; depth and in some cases dip of thrust plane
Unconformity (figure 2.1.14f)	In case of angular unconformity often low dip above and high dips below unconformity	Depth of unconformity

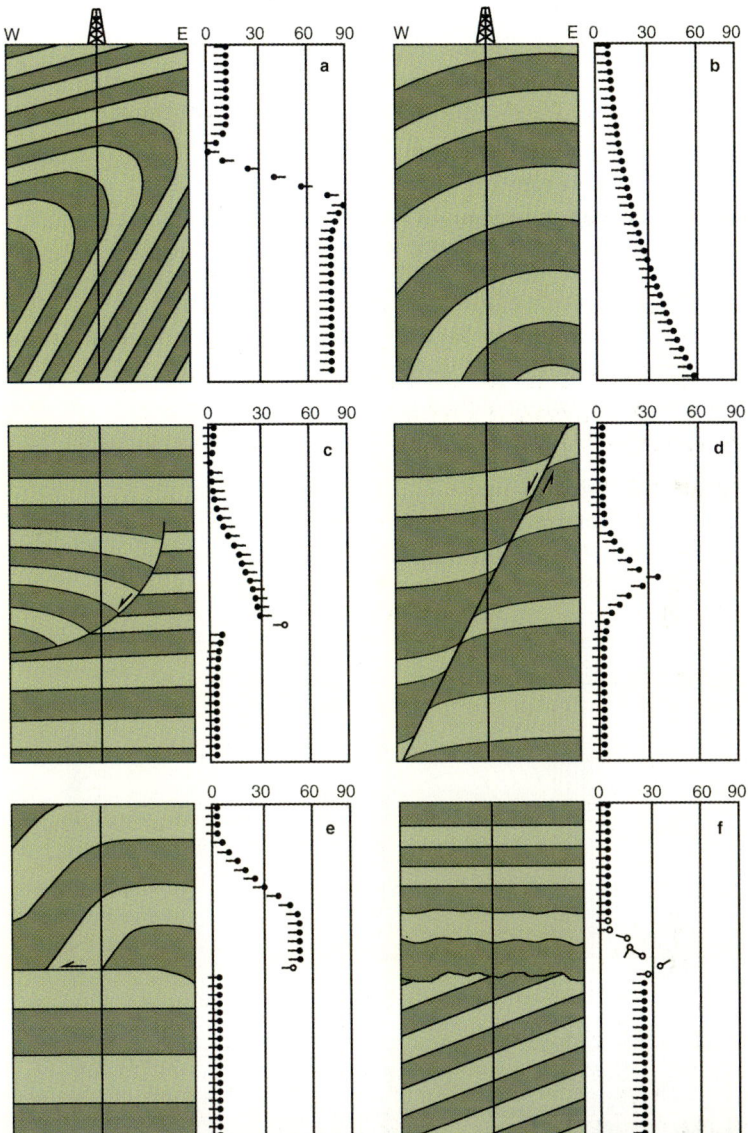

Figure 2.1.14. Schematic sketches of six geologic structures and their corresponding dip patterns. No scale is implied; all sections are east-west. a: Similar fold in which layer thickness parallel to the fold axis is preserved. Fold plane is tilted. b: Concentric fold, in which the layer thickness orthogonal to bedding is constant. c: Rollover or growth fault, whose dip and fault displacement typically increase with depth. d: Normal fault with drag of the layers in the vicinity of the fault. Drag in the lower (upthrown) block is often less pronounced. Sometimes a substantial fault breccia is present. e: Ramped anticline (partial view) with a subhorizontal thrust plane. Wells at other locations may show significantly different dip patterns. f: Angular unconformity. Sedimentation immediately above the unconformity is often coarse with poor layering, resulting in scattered dips. Downlaps may be similar but with much gentler dips.

shown. All sections are east-west and at a right angle to the strike direction of the structures. There are numerous variants to these situations, depending on the amount of tectonic deformation, sometimes showing dip patterns that differ significantly from the ones shown in figure 2.1.14. The main characteristics for the common structures are listed in table 2.1.3, together with those structural features ("measurable elements") that can be quantified. For the latter, recourse to the stereonet is often needed. In view of the ambiguity inherent to structural dipmeter interpretation, it is useful to keep the following basic rules in mind:

- Dipmeter interpretation is model-dependent. For example, a kink fold and an angular unconformity can have the same expression on dipmeters and only the geological context can differentiate the two;
- Dipmeter interpretation should preferably be done on several wells simultaneously and/or in conjunction with seismic or other supporting data.

This can be illustrated by figure 2.1.15, a dipmeter example from central Italy. Although it does not exactly correspond to one of the situations of figure 2.1.14, one can still identify two intervals with constant, steep dips, separated by an

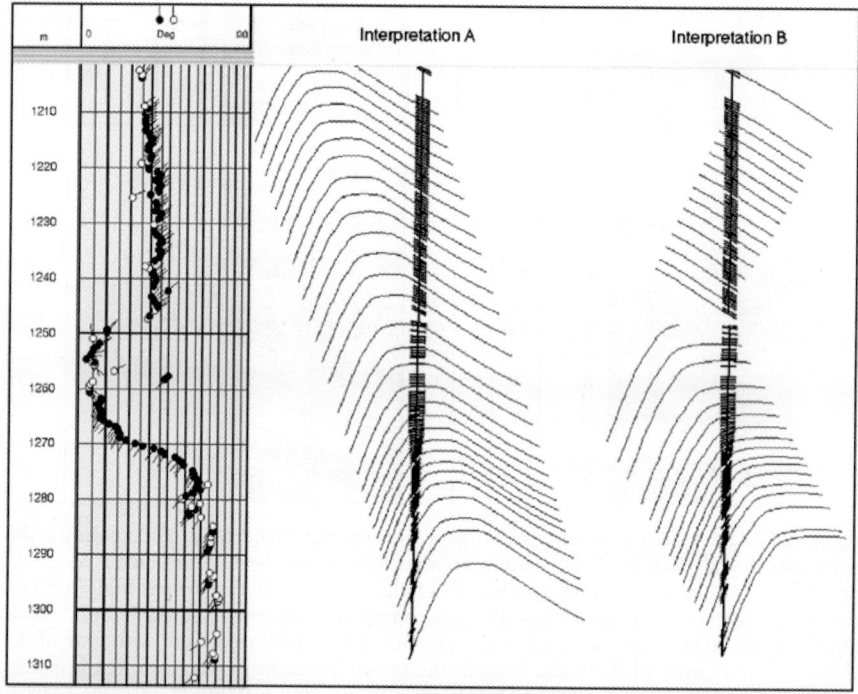

Figure 2.1.15. Dip results from a well on the Adriatic Coast in central Italy in Miocene and Pliocene clastics. The similar fold interpretation A, although temptingly simple, has to be ruled out in favor of interpretation B that includes a known angular unconformity at 1248m, overlain by young sediments draping the overthrust fold. See text for more details.

interval where the dip changes rapidly but continually. This is a telltale sign of a similar fold, characterized by two limbs of more or less constant dip and a narrow fold zone (Interpretation A). However, the sediments above 1248 m are *known to be much younger* than the underlying sequence, and they gradually flatten higher up in the well. The structure, therefore is a similar fold with a tilted fold plan in the lower part, unconformably overlain by younger sediments at 1248 m that drape over the underlying structure (Interpretation B). The example, therefore, contains elements of the two structures shown in figures 2.1.15a and 2.1.15f. The two interpretations shown in this figure have been computer-generated using the construction method by Etchecopar & Bonnetain (1992) described later in this chapter.

Structural dipmeter interpretation requires considerable expertise and experience, but fortunately it has received a large boost with recent developments of computer-aided geometric and visualization techniques.

Sedimentary interpretation has its roots in the Gulf Coast and mid-continent region of the USA, where Gilreath (1964) and Campbell (1968) have developed the basic methodologies (although they called it "stratigraphic interpretation"). The cornerstones are dip patterns, or sequences where the dip changes with depth in a typical manner. Four patterns are distinguished[2] (figure 2.1.16):

- An upward decrease in dip *(red pattern)*
- An upward increase in dip *(blue pattern)*
- A constant dip *(green pattern)*
- Scattered dips *(yellow pattern)*

Common to the first three patterns is that the azimuth changes little or not at all. These patterns are caused by, and reflect, the two- and three-dimensional geometry of the bedding, a concept germane to all dipmeter interpretation. They can be observed at scales ranging from a few tenths of a meter to several hundred meters. The green pattern is used to define the structural dip, while the red and blue patterns are indicative of the sedimentary structures sketched in figure

---

[2] Named after the pencil colors Al Gilreath marked them with.

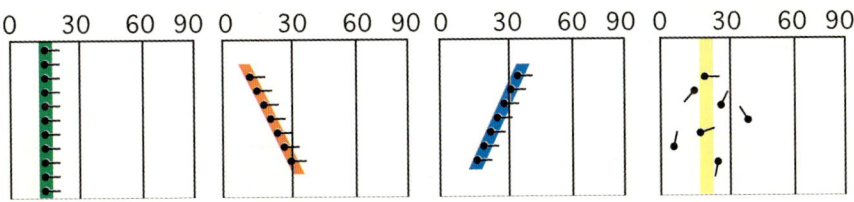

Figure 2.1.16. The standard dip patterns: Green, red, blue and – less commonly used – yellow.

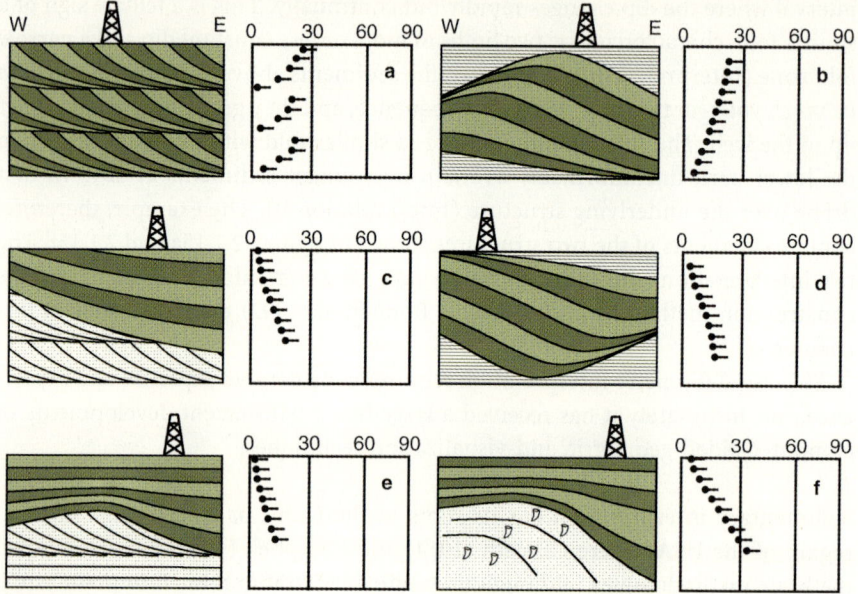

Figure 2.1.17. Schematic sketches of sedimentary structures and related dip patterns. a: Cross-bedded sets showing typical blue patterns, with lower dips in bottom sets and foreset dips rarely exceeding 30°. b: A prograding sand bar exhibiting a large-scale blue pattern. c: A trough fill showing a red pattern. d: A channel fill showing a large-scale red pattern caused by the passage from lateral accretion to vertical aggradation and shale infill. e: Shale draping over a sand bar leading to a red pattern. f: Sediment draping over a reef leading to a large-scale red pattern. Dip patterns may change as a function of well location, indicated on the sketches by the rig.

2.1.17, and listed in table 2.1.4. Because the patterns may vary considerably with the exact well location, an approximate well position is indicated in figures 2.1.17. One can imagine how the dip patterns change as the well is moved to the left or right in the examples. The yellow pattern does not contain any geometric information and is only rarely used.

Sedimentary dipmeter interpretation does not have to rely on the pattern approach. Methods to obtain and interpret detailed sedimentary dips are documented in Vincent et al. (1979), Payre & Serra (1979), Nurmi (1984), Serra (1986), Höcker et al. (1990), Cameron (1992), and Hocker et al. (1990). They all use small correlation intervals, and many favor simultaneous inspection of the dipmeter curves themselves, since they carry useful information on bedding boundaries and bedding types.

For a successful sedimentary dipmeter interpretation, the underlying structural dip has to be carefully compensated for by subtracting it from the entire interval to be analyzed. After having done so, all remaining dips can be attributed to sedimentary processes. The process of structural dip removal consists of ro-

Table 2.1.4. Dip patterns and examples of associated sedimentary structures at the scale of bedding and of entire sedimentary bodies.

Dip Pattern	"Small": Bedding Types	"Large": Sedimentary Bodies
Upward Decrease (Red Pattern)	Small-scale trough fill Hummocky cross-stratification	Channel fill Drapes over bar, slumps, carbonate build-up, etc.
Upward Increase (Blue Pattern)	Cross-bedding (tangential foresets)	Prograding barrier bar Channel accretion Prograding reef
Constant dip/azimuth (Green Pattern)	Parallel bedding Thin bedding with structures below tool resolution (e.g. ripples)	Vertical aggradation in low-/medium-energy environments
Scattered dips/azimuths (Yellow Pattern)	Conglomerates Distorted bedding Massive bedding	Mass flow deposits Carbonate build-ups

tating every measurement by the amount of the structural dip on small circle arcs such that the structural dip becomes a horizontal plane (Schlumberger, 1981). Defining structural dip can be a delicate matter, since it can change with depth and thus the amount of dip subtraction varies. Additionally, few sediments are laid down on a perfectly horizontal plane. In fact, many reservoir rocks are high-energy deposits where it may be difficult to find a reference bed that had been deposited horizontally. Geological reasoning is important here. Simple filtering of the dipmeter data is rarely good enough, but identification of such deposits as floodplain muds, lagoonal or pelagic fine-grained sediments may help identify the structural tilt of a sequence. A user-controlled subtraction of the structural tilt is therefore often preferable, and on modern workstations the user can do this in an interactive and accurate manner.

With a vertical sampling rate of 2.5 mm and an electrode size of 5 mm, the modern dipmeters have a resolution (i.e. an ability to separate two different events) of one centimeter. Therefore, the smaller sedimentary structures such as ripples cannot be resolved, since the ripple laminae are too closely spaced and no meaningful correlation between curves is possible. The same is true for parallel laminations in fine-grained clastics, which may therefore be undistinguishable from massive bedding. At the scale of larger sedimentary structures such as cross bedding, however, the cross-bed laminae are often spaced widely enough to allow correlations between dipmeter curves. Some of the most spectacular examples are from eolian cross-bedded sandstones, notably from the Rotliegendes in the Southern North Sea (figure 2.1.18). Cross-bed azimuths are indicative of local flow directions and therefore help define overall sediment transport directions, depositional mechanisms, and reservoir geometries. Several examples of this are discussed in chapter 3.1.

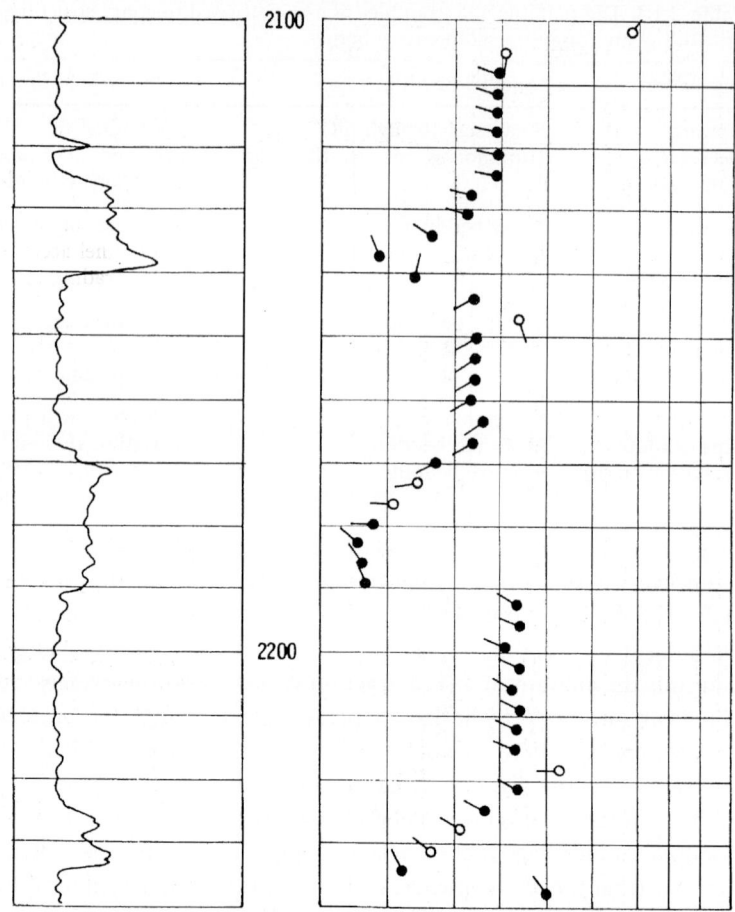

Figure 2.1.18. Dipmeter results obtained from the cross-correlation program CLUSTER over an eolian sandstone sequence. Several cross-bedded sets can be distinguished, characterized by blue patterns in the bottom set, and constant dips of around 30° in the foresets. Foreset azimuth is generally towards west.

At the larger scale of entire sedimentary bodies, internal patterns due to progradation of a sand bar or a carbonate build-up, lateral accretion of a fluvial sand bar, aggradation within a channel etc. can lead to characteristic patterns (Gilreath, 1964; Phillips, 1987). But not only internal bedding can be analyzed from dipmeter; the external shape of a sedimentary body can also affect dip patterns. For example, a sedimentary body may have an uneven upper surface. Subsequent sedimentation can level the topography gradually, leading to a characteristic draping that shows up as a red pattern (figure 2.1.17e). This is particularly important in the case of barrier bars and carbonate reefs, which are common reservoir targets and where the azimuth of these drape patterns can be indicative

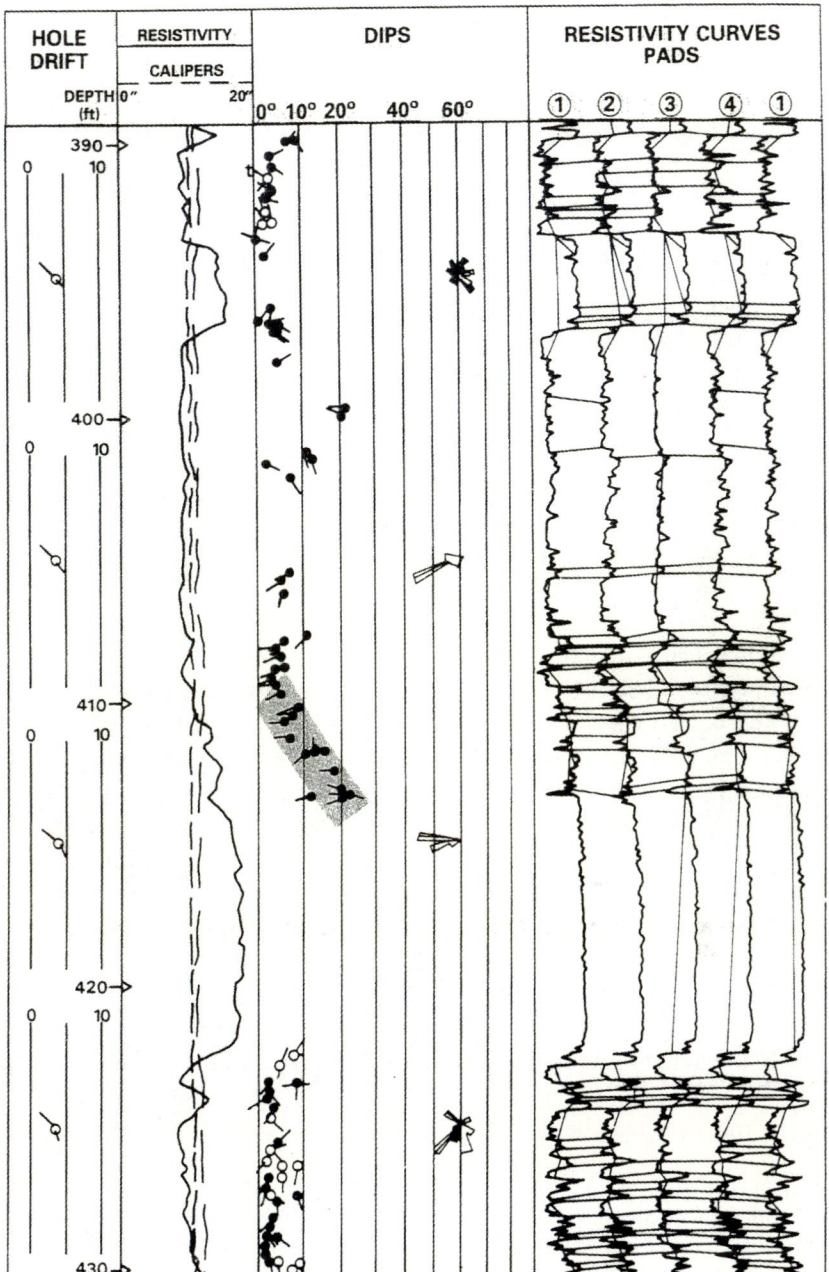

Figure 2.1.19. Red draping pattern over a sandstone layer (422–413m) is interpreted as caused by a sand bar (cf. figure 2.1.17d) in this fluvio-deltaic setting. Well is located to west of the bar as indicated by the eastward dips. Above and below the sandstone a low structural dip is evident. Dipmeter results from the pattern-recognition correlation program GEODIP.

of the maximum thickness of the sedimentary body (figure 2.1.19). This is because the azimuth of the draping pattern points away from the highest topography and the dip magnitude will be highest somewhere on the flank, while on top of the sedimentary body and far away from it the draping pattern will be less pronounced or absent. Cox (1968) and Bigelow (1973) demonstrated this effect on Paleozoic reefs in North America and developed rules of how these patterns can be used to locate the center of the reef (cf. figure 2.1.17f).

At even larger scales such as seismic stratigraphic units, the dips involved are generally only in the order of a few degrees, except perhaps in some carbonate systems. This can be deduced from published examples (Mitchum & Vail, 1977; Brown & Fisher, 1977). On dipmeters, therefore, only a very careful analysis of the large-scale dip patterns may give clues on potential baselaps or toplaps, but so far no examples have been published.

Dipmeter Interpretation with Expert Systems

Dipmeter interpretation is empirical and requires a mixture of geological expertise, pattern recognition, and three-dimensional visualization skills. The scarcity of experts who combine all these skills led to one of the first commercial implementations of an expert system on a computer. The rules of an experienced dipmeter interpreter (Al Gilreath) were coded into a database in such a manner that the system, called Dipmeter Advisor[*], could be used by less experienced interpreters as an assistant (Davis et al., 1981). The system imitates how the expert proceeded in his interpretation: It first identifies green patterns, equivalent to structural dips with no folding or faulting, and then large red and blue patterns. The structural analysis then locates possible faults and folds and quantifies them as outlined in table 2.1.3. Following this, a sedimentary analysis is performed in which the dipmeter patterns attributed to sedimentary processes are analyzed in conjunction with other well logs. User and machine closely interact during this process, with the user constraining the choices by defining, for example, the tectonic setting of the well, or the sedimentary environments.

Extensive field testing showed that the system can be used in situations covered by the rule base, but that it has difficulties outside its intended range and that the rule base may be flawed. According to Doveton (1994), it was

"...noted that the original expert sometimes appeared to abandon his own rules when faced with a markedly different example, and reasoned from geological and geometric models.... Dipmeter analysis stretches the envelope of tractable problems when experts disagree among themselves; some facts are highly interpretative assertions and correct behavior by the system is difficult to evaluate."

---

[*] Mark of Schlumberger

While not in use today, experience with this expert system has shown that a better knowledge base of the relationship between dips and common sedimentary structures is needed. Such data is not available in the current literature. This expert system – and others not discussed here – has, however, impacted further workstation development of dipmeter interpretation software significantly. Most notably this happened through the great amount of interactivity provided on most systems, and the wide variety of graphic aids.

Dipmeter Interpretation Using Geometric Constructions

From the earliest days, geologists have sought to present dipmeter data in ways they are more familiar with, and to integrate them with other data. A common use of dipmeters is for constructing *contour maps*. For this, a suitable dip, preferably from a green pattern in the vicinity of the horizon of interest, has first to be determined. From this dip with magnitude a the contour spacing CS for a map at scale S is obtained as

$$CS = \frac{CI}{\tan a} \cdot S \qquad (2.2)$$

where CI is the contour interval. Thus, for example, the contour spacing for a dip of $10°$ on a map at scale $1:10,000$ with contour intervals of 10 m is $CS = 10\text{m}/10,000 \cdot \tan 10° = 5.7\text{mm}$. The contours have to be drawn perpendicular to the dip azimuth. In this way, dipmeters can refine maps that are based on seismic lines and other well data.

One of the most important uses of dipmeters, as illustrated in figures 2.1.11 and 2.1.12, is their use in constructing *geological cross-sections*. If incorporated into well-to-well correlations along a line of section, the apparent dip[3] can be calculated and be used as help in the layer correlation process. The "stick plots", i.e. dip projections onto a number of vertical planes in different azimuths, have been proposed by Schlumberger (1981) as a help in constructing cross-sections. Aside from helping in the layer correlation process, this method also allows identifying faults, and significant layer thickness changes. An example for this will be discussed in chapter 3.1.

An alternative use of structural dipmeter interpretation was proposed by Bengtson (1981), who suggests to break the dipmeter presentation up into five different plots in order to facilitate the geometric construction around the borehole (figure 2.1.20). In his statistical curvature analysis technique (SCAT) he first determines the principal direction of dip variation on a plot of dip versus azimuth, which he then designates as the transverse direction.

---

[3] Here apparent dip refers to the common usage in structural geology: The angle that a structural surface makes with the horizontal, measured in a vertical cross-section plane.

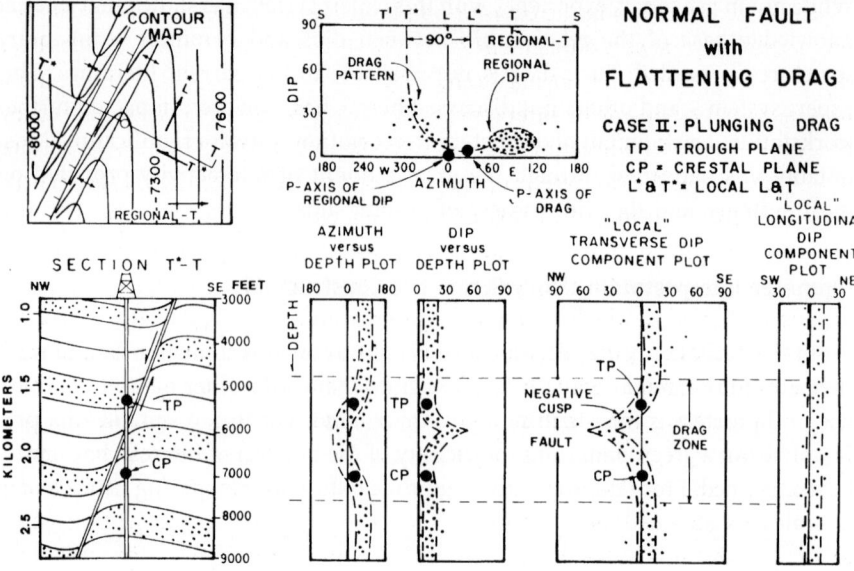

Figure 2.1.20. SCAT plots of a normal, antithetic fault. TP = trough plane; CP = crestal plane. The transverse dip component is used to construct the transverse cross section on the left (from Bengtson, 1981).

The longitudinal direction is at a right angle to it and should show only minimal dip magnitudes; it is in essence a strike direction, while the transverse direction is the plunge direction. Bengtson (1981) then plots the azimuth and dip magnitude versus depth, and the transverse and longitudinal dip component versus depth. The patterns exhibited on the plots are characteristic for each of the seven major tectonic structures he proposed. In the final step, the transverse and longitudinal dip component versus depth are used to construct transverse and longitudinal cross-sections, which can then be integrated into contour maps of relevant horizons.

Etchecopar & Bonnetain (1992) use a computer-based method to construct cross-sections from dipmeters. They apply the similar fold model, in which each layer geometry is identical and obtained through translation along a plane. The parallel fold model, where layer thickness is preserved and which is widely used in balanced cross-sections, is found less convenient when dealing with well data. The authors found it difficult to apply in small-scale folds, fault drags, and rollover structures. Etchecopar & Bonnetain (1992) first filter the dips in order to reduce the scattering and then subdivide the well into sequences for which a particular structure is valid. The translation plane is then determined as the axial plane for a fold, the fault plane for a fault, and the detachment plane for a rollover structure. Folds and fault drag zones are assumed to be cylindrical structures, with their bedding dips aligned on a great circle. The pole of this great circle is the structural axis and lies in the translation plane. For folds, the bisector of the two limbs provides the second element to define the translation plane. For faults, a dip of 65° is assumed for normal faults, and 25° for reverse faults in a first iteration and can afterwards be interactively modified. By defining the cross-section plane, the translation direction is determined from its intersection with the translation plane, and the dips are propagated from the wellbore away in the translation direction determined in this manner. Figure 2.1.21 illustrates that in this procedure the translation only reaches as far as there are dips that can be laterally linked up to the structure. Etchecopar & Bonnetain (1992) also propose a concentric fold model, which is applied where a similar fold gives unsatisfactory results. Additionally, they include a rollover model, for which a new parametrization based on field data was developed. For the latter, top and bottom of the zone affected by growth faulting have to be determined, and are used to define the extremes of a circular arc along which the listric fault movement took place. A field example for this is shown in figure 2.1.22. It is from the Transsylvanian Basin in Romania, where growth faulting is thought to be prevalent based on analogy with the nearby Pannonian basin. Another example of this technique can be found in chapter 3.1.2.

Geometric constructions for *sedimentary features* have received less attention, mostly because there is no history of quantifying the geometry of sedimentary

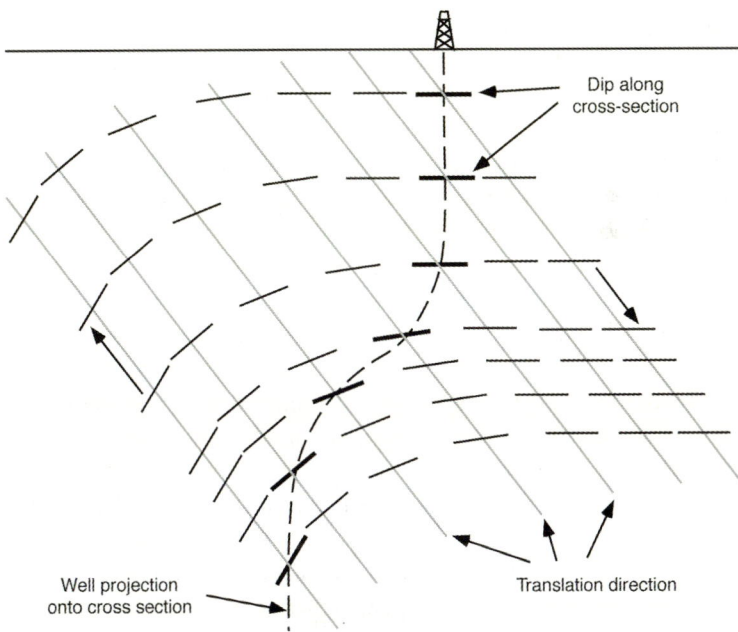

Figure 2.1.21. Translation of dips derived from the dipmeter survey away from the wellbore in a cross-section of arbitrary direction. The similar fold model is used, with arrows indicating the translation direction (redrawn after Etchecopar & Bonnetain, 1992).

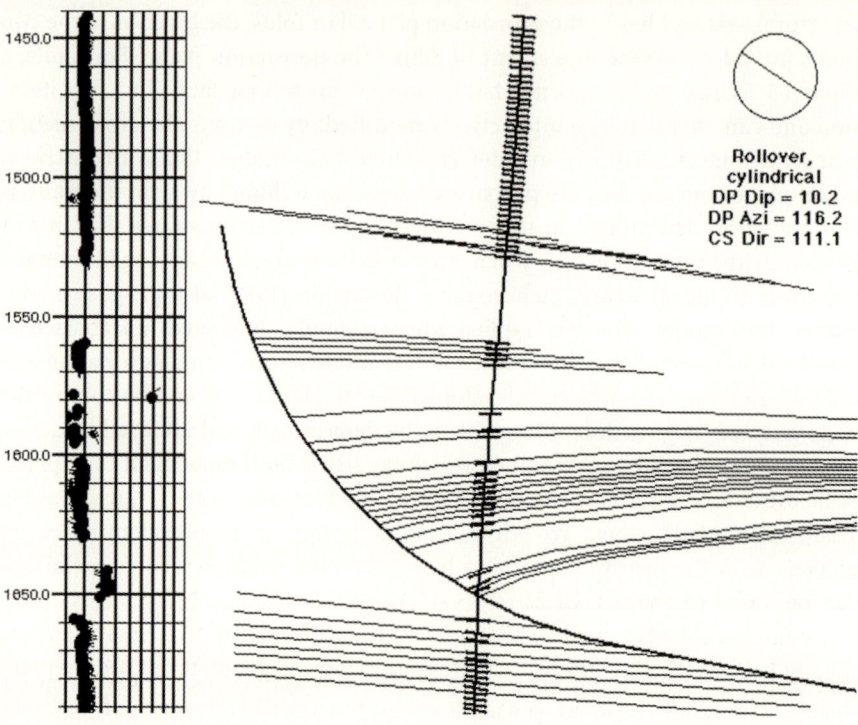

Figure 2.1.22. A cross-section constructed from dipmeter data for which a growth-fault model seems appropriate. Tertiary clastics, Transsylvanian Basin (Romania). Depth is in meters.

structures as there is for tectonic structures. Additionally, dipmeters cannot resolve some of the smaller sedimentary structures such as cross lamination. In the case of cross bedding, Luthi et al. (1990) have shown that dip data can be used to determine the curvature and migration direction of the bedforms as well as the size of the cross-bedded sets. The method is based on the assumption that cross-bedded sets in a reservoir are randomly penetrated, and that the azimuth scatter of the various cross-beds represents the curvature of the bedforms (see figure 2.1.18 for an example of cross bedding). Using Monte Carlo statistical techniques, theoretical dip azimuth distributions are obtained as a function of the curvature and the direction of migration, i.e. the angle at which bedforms migrate with respect to their main orientation. High angles of migration are found in variable flow directions such as eolian dunes in seasonally shifting winds (Rubin & Hunter, 1987). Figure 2.1.23 shows schematic sketches of four situations, low and high curvature bedforms at zero and high angles of migration. The resulting dips and azimuths for a sequence of cross-beds are schematically shown on stereographic plots. Low curvature, straight-crested bedforms are characterized by a narrow cluster, while higher curvature, crescentic bedforms exhibit a wider azimuthal

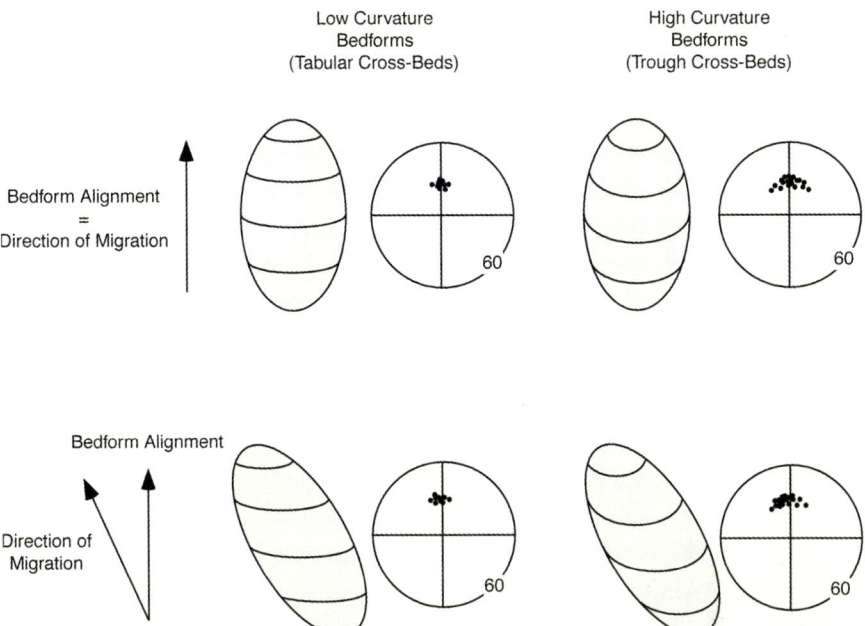

Figure 2.1.23. Theoretical dip and azimuth distribution for cross-bedded sequences as a function of bedform curvature and angle of migration, defined as the angle between bedform alignment and direction of migration. Notice the subtle differences in amount of scatter and asymmetry of the data points on the stereoplots.

scatter. For no angular migration, there is azimuthal symmetry, while for increasingly oblique migration there is increasing asymmetry in the azimuth distribution. For strongly crescentic bedforms, Luthi et al. (1990) found that the azimuthal scatter can show two distinct peaks, corresponding to the two elongate sides of the bedform. This technique is discussed in more detail in chapter 3.1.3 in the context of a case study.

Anxionnaz & Delhomme (1998) have proposed a method to reconstruct cross-beds from dipmeter or borehole image data. The method resembles the structural reconstruction method proposed by Etchecopar & Bonnetain (1992) and will be discussed in more detail in chapter 3.1.3. At present, no other quantitative geometric models for reconstructing sedimentary structures from dipmeter results are found in the literature. This probably has to be attributed to the scarcity of quantitative field data from outcrops.

Statistical Methods on Dip and Curve Characters

Dipmeter curves respond to local changes in resistivity usually caused by differences in porosity and/or clay content. Figure 2.1.24 is a digital scan of a thin section across a layered eolian sandstone with large differences in packing

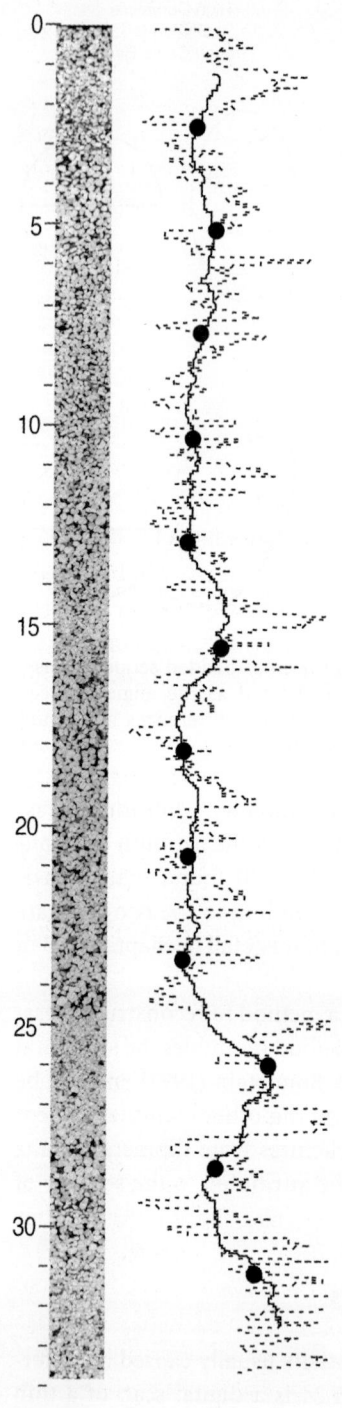

Figure 2.1.24. A digitized thin section from a laminated eolian sandstone (left) is used to calculate a dipmeter response over the width of the image. The dashed line corresponds to a dipmeter curve recorded by a small electrode sampled every 0.01mm. The solid line corresponds to a dipmeter curve recorded with an electrode of the actual size but with the same sampling as the dashed line. The dots correspond to measurements made at the sampling rate of an actual dipmeter tool. Resistivities increase to the right and cover about one order of magnitude. Jurassic Navajo formation, Utah, USA. Vertical scale in millimeters.

and therefore of porosity. The dashed curve on the right is the expected resistivity variation, calculated from the porosity using Archie's law (equation 1.3.1) over a narrow vertical window. The solid line is the resistivity variation that a dipmeter electrode would record if it were logged over the surface of this rock. It is obtained through a convolution of the dashed line with a simple boxcar response for the electrode, but with the same vertical sampling as the dashed line. The dots correspond to locations where a dipmeter tool would record a measurement, i.e. every 2.54 millimeters. This example helps appreciate the scale at which dipmeter measurements take place: while the finer details are lost, the three or four tight laminae still show up quite clearly. Lovell et al. (1997) have come to similar conclusions based on experiments on cores.

Such detailed resistivity measurements are instructive for dipmeter interpretation. They tell us that as long as there are laminations within the resolution of the dipmeter measurements, i.e. one about centimeter, events can be correlated and thus dips can be computed. Laminations below this scale and massive bedding produce no reliable dips. Figure 2.1.25 shows three dipmeter curve characters and their associated bedding types: While there are abundant correlations in the laminated silty shale example, the cross-bedded sands already produce fewer and more scattered dips, and the conglomeratic sandstone produces no coherent dips due a complete lack of layering. Therefore, the number of dips per unit interval and their scatter, as well as the dipmeter curves themselves are characteristic of bedding and facies.

Delhomme and Serra (1984) used these observations to develop a computer program that converts the dipmeter curves into several curves characterizing the nature of bedding. Among the parameters they propose are

- the activity of the curves
- the frequency of peaks or troughs per unit interval
- the average event thickness per curve as determined from a pattern correlation program
- the ratio of peak over trough thickness per curve (the "balance")
- the density of correlations per unit interval, and
- the sharpness of the events

These "synthetic" curves are used in a clustering procedure to subdivide a well into zones of similar dipmeter characteristics reflecting their bedding style. In a later development (Anxionnaz et al, 1990), the dip scatter was calculated as the spherical standard deviation of successive dips, and the number of dipmeter-derived curves was significantly reduced. Bedding interfaces and resistivity trends within a bed were also determined and displayed as a visual help, together with the dip scatter and the bedding zonation (figure 2.1.26). Since dipmeters have a vertical resolution far better than other logs, this approach permits a very accurate determination of bedding thicknesses and interfaces. The dipmeter-derived curves, however, do not possess this original high resolution

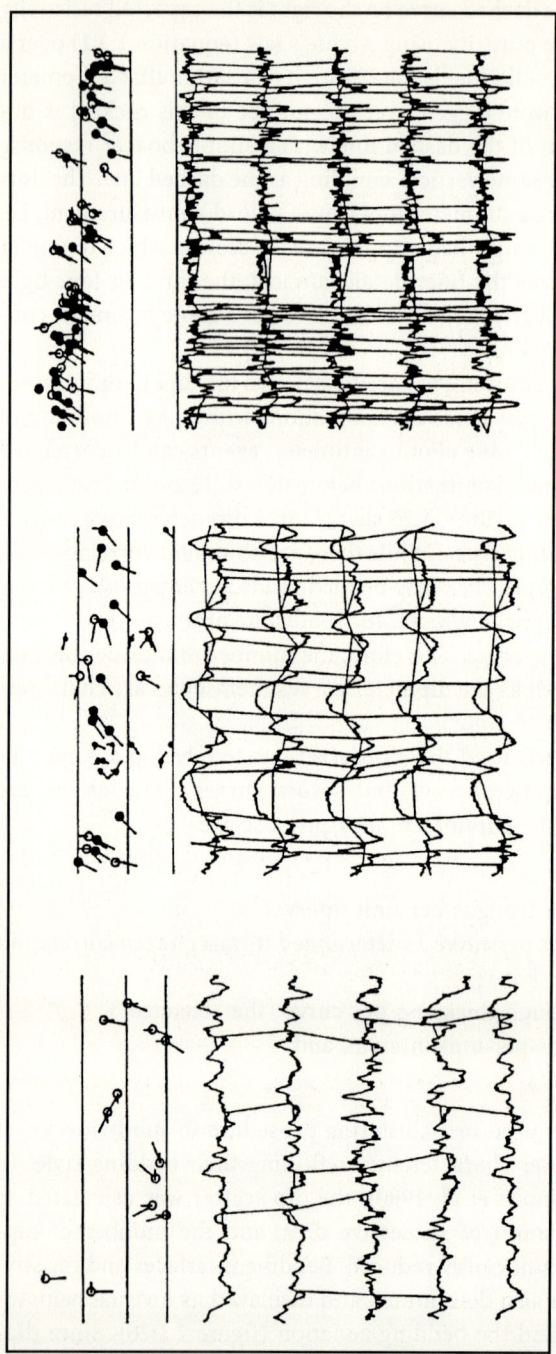

Figure 2.1.25. Dips and dipmeter curves for three lithologies and associated bedding types. From top: Thinly bedded silts and shales; interbedded cross-bedded sands and shales of medium thicknesses; a conglomerate. Each section covers approximately 2 meters.

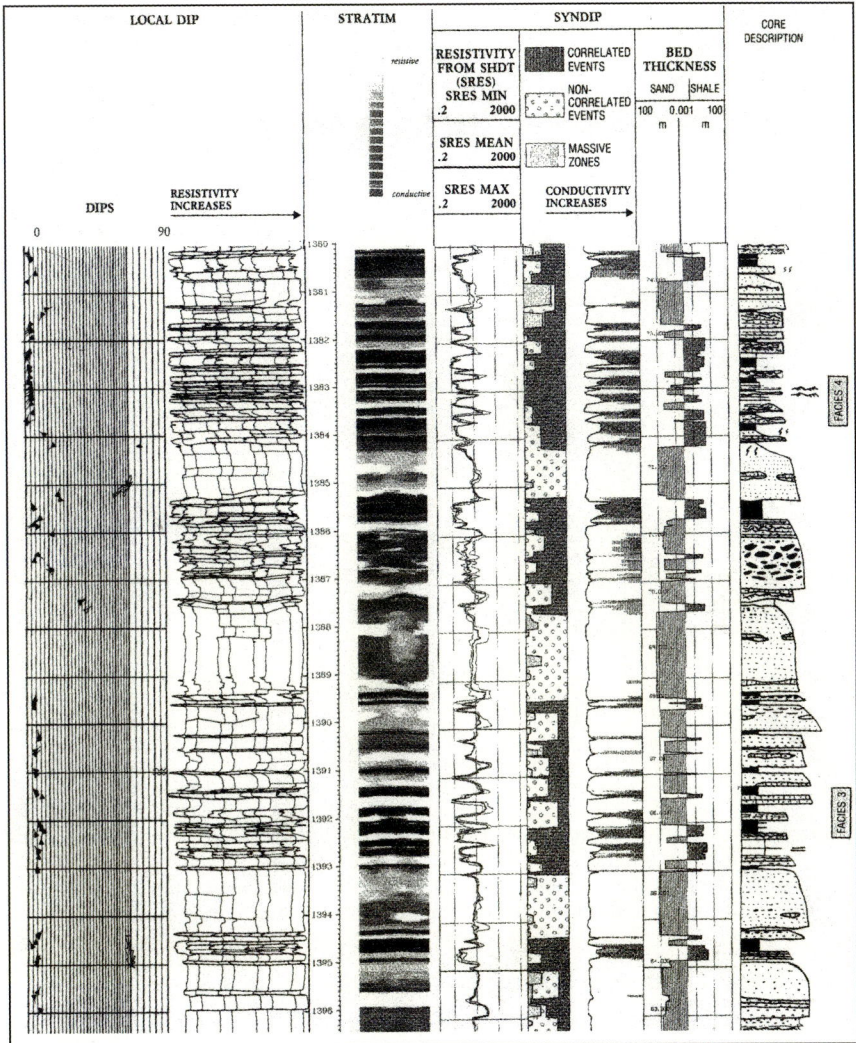

Figure 2.1.26. Example of dipmeter-derived bedding zonation with from left to right: Dips from pattern recognition, an image constructed from the dipmeter curves, recomputed resistivity curves, a zonation based on the frequencies of correlated, non-correlated and massive events, a running average of the bed thickness, and a core description.

any more, and in the clustering procedure Anxionnaz et al. (1990) propose to include other open-hole logs in order to capture the lithological and petrophysical properties. They also developed a knowledge base through which geological significance is conveyed to the data. The procedure is similar to what geologists do when describing cores: They gather all the information over an interval of interest and then interpret the data in terms of depositional environment, geometry of the deposit, diagenetic overprints etc. An example is shown in figure

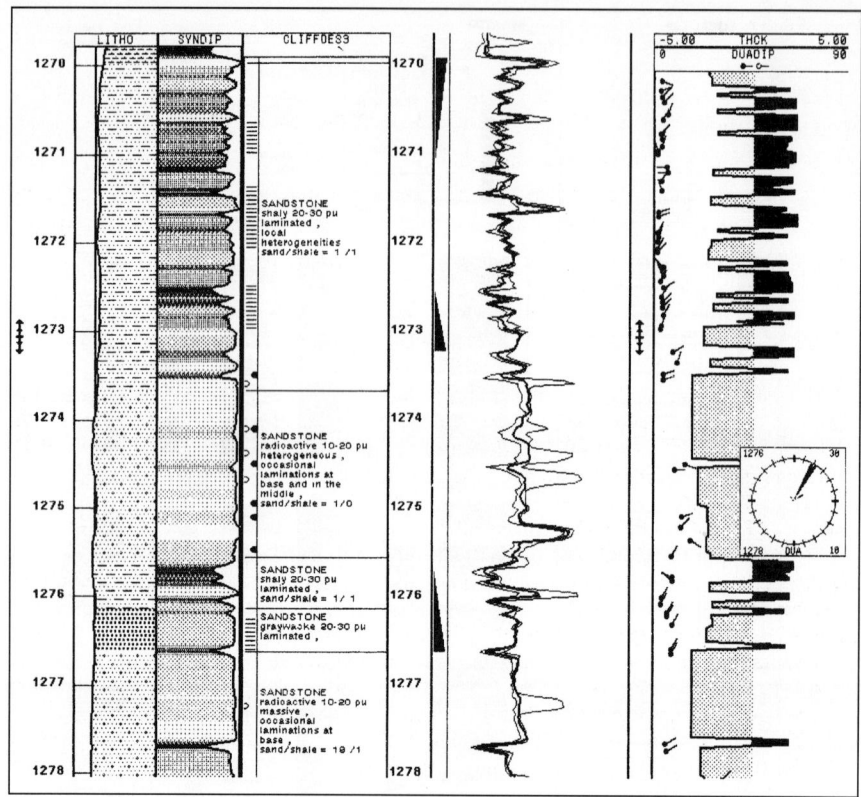

Figure 2.1.27. Example of a core-like description based on dipmeter-derived bedding and open-hole log characteristic. On the left a log-derived and a dipmeter-derived zonation, followed by a geological description from the knowledge base, recomputed resistivity curves, dip results and a bedding thickness column (from Anxionnaz et al., 1990).

2.1.27, together with the proposed geological interpretations. Balossino & Anxionnaz (1996) describe a similar approach applied to a particular reservoir, where by the geological interpretation of the obtained zonation was established from cores and subsequently applied to other, uncored sections of the well.

The high resolution of dipmeter curves cannot only be used for sedimentary interpretations, but also to improve the petrophysical evaluation of reservoirs, which is usually based on logs with a substantially lower resolution. Allen (1984) developed a methodology wherein bed boundaries are determined from high-resolution logs such as the electromagnetic propagation tool (EPT[*]) or the dipmeter tool. For each bed an iterative forward modeling procedure is then used to determine the most likely formation properties. This approach is described in more detail in chapter 3.2.2.

---

[*] Mark of Schlumberger

2.1.5 Summary

Dipmeters are invaluable aids in determining the large-scale tectonic structures as well as smaller-scale sedimentary structures in the vicinity of the wellbore. Other important applications include their use for contour mapping, cross-sections, facies recognition and the analysis of thinly bedded reservoirs. Interactive workstations featuring graphic aids, statistical programs, and three-dimensional geometric reconstruction techniques have greatly facilitate the use of dipmeters.

References

Adams JT, Ayodele JK, Bedford J, Kaars-Sijpestijn CH, Watts NL (1992) Application of dipmeter data in structural interpretation, Niger delta. In: Hurst A, Griffiths CM, Worthington PF (eds) Geological Application of Wireline Logs II. Geol Soc Spec Publ 65: 247–264.
Allaud LA, Ringot J (1969) The high resolution dipmeter tool. Log Analyst 10, 3: 3–11.
Allaud LA, Martin M (1977) History of a technique. John Wiley, New York.
Allen DJ (1984) Laminated sand analysis. Trans 25th Ann Symp Soc Prof Well Log Analysts: Paper XX.
Antoine JN, Delhomme JP (1990) A method to derive dips from bed boundaries in borehole images. Trans Soc Petrol Eng Ann Meeting: Paper 20540.
Anxionnaz H, Delfiner P, Delhomme JP (1990) Computer-generated corelike description from open-hole logs. Am Ass Petrol Geol Bull 74: 375–393.
Anxionnaz H, Delhomme JP (1998) Near-wellbore 3D reconstruction of sedimentary bodies from borehole electrical images. Trans 39th Symp Soc Prof Well Log Analysts: Paper N.
Balossino P, Anxionnaz H (1996) Deriving textural and geometric information from dipmeter data as a help to define subsurface geological models. Trans Europ Assoc Geol Eng.
Bengtson CA (1981) Statistical curvature analysis techniques for structural interpretation of dipmeter data. Am Ass Petrol Geol Bulletin 65: 312–332.
Bigelow EL (1973) High-resolution dipmeter uses in Michigan's Niagaran reef. Oil & Gas J 71: 78–88.
Billings MP (1972) Structural geology. Prentice Hall, Engelwood Cliffs.
Bricaud JM, Poupon A (1959) Continuous dipmeter survey: The poteclinometer and the microfocussed devices. Trans 5th World Petrol Congress: 225–239.
Brown LF, Fisher WL (1977) Seismic-stratigraphic interpretation of depositional systems: Examples from Brazilian rift and pull-apart basins. In: Payton EC (ed) Seismic stratigraphy – applications to hydrocarbon exploration. Am Ass Petrol Geol Mem 26: 213–248.
Cameron GIF (1992) Analysis of dipmeter data for sedimentary orientation. In: Hurst A, Griffiths CM, Worthington PF (eds) Geological Applications of Wireline Logs II. Geol Soc Spec Publ 65: 141–154.
Campbell, RL (1968) Stratigraphic applications of dipmeter data in Mid-Continent. Am Ass Petrol Geol Bull 52: 1700–1719.
Chauvel Y, Seeburger D, Orjuela A (1984) Applications of the SHDT stratigraphic high-resolution dipmeter to the study of depositional environments. Trans 25th Ann Symp Soc Prof Well Log Analysts: Paper G.
Cheung PSY, Heliot D (1990) Workstation-based fracture evaluation using borehole images and wireline logs. 65th Ann Conf Soc Petrol Eng: Paper 20573.
Cox JW (1968) Interpretation of dipmeter data in the Devonian carbonates and evaporites of the Rainbow and Zama areas. Petrol Soc Canad Inst Min Paper 6820.

Davis R, Austin H, Carlbom I, Frawley B, Pruchnik P, Sneiderman R, Gilreath AJ (1981) The dipmeter advisor – interpretation of geological signals. Proc 7th Int Joint Conf Artificial Intelligence, Vancouver: 846–849.

de Chambrier P (1953) The microlog continuous dipmeter. Geophysics 18: 929–951.

Delhomme JP, Serra O (1984) Dipmeter derived logs for sedimentological analysis. Trans 9th European Formation Evaluation Symp Soc Prof Well Log Analysts: Paper 50.

Devilliers MC, Werner P (1990) Example of fault identification using dipmeter data. In: Hurst A, Lovell MA, Morton A (eds) Geological Applications of Wireline Logs. Geol Soc Spec Publ 48: 287–295.

de Witte AJ (1956) A graphical method of dipmeter interpretation. J Petrol Technol 8:192–199.

Doll HG (1943) The S.P. dipmeter. J Petrol Techn Paper 1547.

Doveton JH (1994) Geologic log analysis using computer methods. Am Assoc Petrol Geol, Computer Applications in Geology No 2, Tulsa.

Dumont A, Kubacsi M, Chardac JL (1987) The oil-based mud dipmeter tool. Trans 28th Symp Soc Prof Well Log Analysts: Paper LL.

Ekstrom MP, Dahan CA, Chen MY, Lloyd PM, Rossi DJ (1987) Formation imaging with microelectrical scanning arrays. Log Analyst 28: 294–306.

Etchecopar A, Bonnetain JL (1992) Cross sections from dipmeter data. Am Assoc Petrol Geol Bull 76: 621–637.

Faivre O, Catala G (1995) Dip estimation from azimuthal laterolog tools. Trans 36th Symp Soc Prof Well Log Analysts: Paper CC.

Gilreath JA, Maricelli JJ (1964) Detailed stratigraphic control through dip computations. Am Ass Petrol Geol Bull 48: 1902–1909.

Hepp V, Dumestre AC (1975) Cluster – a method for selecting the most probable dip results from dipmeter surveys. Ann Conf Soc Petrol Eng: Paper 5543.

Höcker, CFW, Eastwood, KM, Herweijer, JC, Adams JT (1990) Use of dipmeter data in clastic sedimentological studies. Am Ass Petrol Geol Bull 74: 105–118.

Kleinberg RL, Chew WC, Chow EY, Clark B, Griffin DD (1987) Microinduction sensor for the oil-based mud dipmeter. 62nd Ann Conf Soc Petrol Eng: Paper 16761.

Luthi SM, Banavar JR (1988) Application of borehole images to three-dimensional geometric modeling of eolian sandstone reservoirs, Permian Rotliegende, North Sea. Am Ass Petrol Geol Bull 72: 1074–1089.

Luthi SM, Banavar JR, Bayer U (1990) Models to interpret bedform geometries from cross-bed data. J of Geol 98: 171–187.

Lovell MA, Harvey PK, Williams CG, Jackson PD, Flint RC, Gunn DA (1997) Electrical resistivity core imaging: A petrophysical link to borehole images. Log Analyst Nov-Dec, 45–53.

Mitchum RM, Vail PR (1977) Seismic stratigraphy and global changes of sea level, part 7: Seismic stratigraphic interpretation procedure. In: Payton EC (ed) Seismic stratigraphy – applications to hydrocarbon exploration. Am Ass Petrol Geol Mem 26, pp 135–143.

Morrison R, Thibodaux J (1984) The six-arm dipmeter, a new concept by Geosource. Trans 25th Symp Soc Prof Well Log Analysts: Paper MMM.

Nurmi, RD (1984) Geological evaluation of high-resolution dipmeter data. Trans 25th Symp Soc Prof Well Log Analysts: Paper M.

Payre X, Serra O (1979) A case history: turbidites recognized through dipmeter. Trans 6th European Formation Evaluation Symp, Soc Prof Well Log Analysts: Paper K.

Phillips S (1987) Dipmeter interpretation of turbidite channel reservoir sandstones, Indian Draw field, New Mexico. In: Tillman RW, Weber KJ (eds) Reservoir sedimentology. Soc Econ Paleontologists Mineralogists Spec Pub 40, pp 113–128.

Ramsay JG, Huber MI (1987) Techniques of modern structural geology, volume 2: Folds and fractures. Academic, New York.

Rider M (1996) The geological interpretation of well logs. Second edition. Gulf Publ. Co, Houston.

Rønningsland TM (1990) Structual interpretation of dipmeter results in the Gullfaks field. Hurst A, Lovell MA, Morton A (eds) Geological Applications of Wireline Logs. Geol Soc Spec Publ 48: 273–286.

Rosthal RA, Young RA, Lovell JR, Buffington L, Arceneaux CL (1995) Formation evaluation and geological interpretation from the resistivity-at-the-bit tool. 70th Ann Conf Soc Petrol Eng: Paper 30550.

Rubin DM, Hunter RE (1987) Bedform alignment in directionally varying flows. Science 237: 276–278.

Schlumberger C and Doll HG (1933) The electromagnetic teleclinometer and dipmeter. Proc. First World Petroleum Congress, London: 424–430.

Schlumberger C, Schlumberger M, Doll HG (1935) The electromagnetic dipmeter and determination of the direction of dip of the sedimentary strata cross-cut by drilling. Internat. Congress Mines Metallurgy and Geological Technology, Paris, October 1935.

Schlumberger M, Doll HG (1937) Discussion of results obtained with the electrical dipmeter on a known geologic structure. Second World Petroleum Congress, Paris, June 1937.

Schlumberger (1981) Dipmeter interpretation. Volume 1 – Fundamentals. Schlumberger, New York.

Serra O (1984) Fundamentals of well-log interpretation. 1. The acquisition of logging data. Elsevier, Amsterdam.

Serra O (1986) Fundamentals of well-log interpretation. 2. The interpretation of logging data. Elsevier, Amsterdam.

Vincent P, Gartner JE, Attali G (1977) Geodip: An approach to detailed dip determination using correlation by pattern recognition. 52nd Ann Conf Soc Petrol Eng: Paper 6823.

Ye SJ, Rabiller P, Keskes N (1997) Automatic high resolution sedimentary dip detection on borehole imagery. Trans 38th Symp Soc Prof Well Log Analysts: Paper O.

2.2 Electrical Borehole Imaging

2.2.1 Tool Principles

The development of dipmeters with a larger number of electrodes has laid the foundation for electrical imaging. This is perhaps be illustrated by an example such as figure 2.2.1, which is from a Middle Eastern carbonate reservoir: The dipmeter curves from the SHDT* (Chauvel et al., 1984) are seen to carry a strong bedding signal, and the dip correlation programs have consequently found numerous dips. In addition to these correlations, however, conductive spikes are seen which correlate between the curves but show significant changes in their shapes. They are highlighted on the plot for better clarity and are seen to intersect the bedding. From their steep dips, their conductive nature, and the oil production in this well, these features are interpreted as fractures. Because fractures and beds intersect each other, most dip calculation programs cannot obtain dips for both features. In this case, all dips are from bedding, although fracturing is more important for the production of this reservoir. Figure 2.2.1, in fact, can be viewed as an image with a very poor lateral resolution. In order to obtain a more complete image, more electrodes have to be put on each pad, resulting in an electrical image of the borehole wall on which bedding and fracturing can be easily distinguished.

Currently, three electrical borehole-imaging methods exist:

- *Micro-electrical imaging*, a wireline logging method based on the dipmeter principle and producing a high-resolution electrical current map of the borehole wall;
- *Azimuthal resistivity imaging*, a wireline method producing a calibrated resistivity image with a laterolog technique at relatively low resolution, and
- *LWD resistivity imaging*, an electrical imaging method with electrodes in the rotating bottomhole assembly, producing a resistivity image immediately behind the drill bit with a resolution intermediate to the first two methods

Common to these three imaging techniques is that they operate in conductive borehole muds only. In all of them the azimuthal position of the tool has to be continuously recorded. This is commonly done with an inclinometry unit containing a solid-state triaxial fluxgate magnetometer. No electrical borehole imaging method is currently commercially available for oil-based mud, but acoustic and density imaging tools can operate in oil-based mud (chapters 2.3 and 2.4). Prensky (1999) gives a comprehensive list of papers on borehole imaging tool techniques, primarily the electrical ones, and cites numerous papers containing applications and case studies.

---

* Mark of Schlumberger

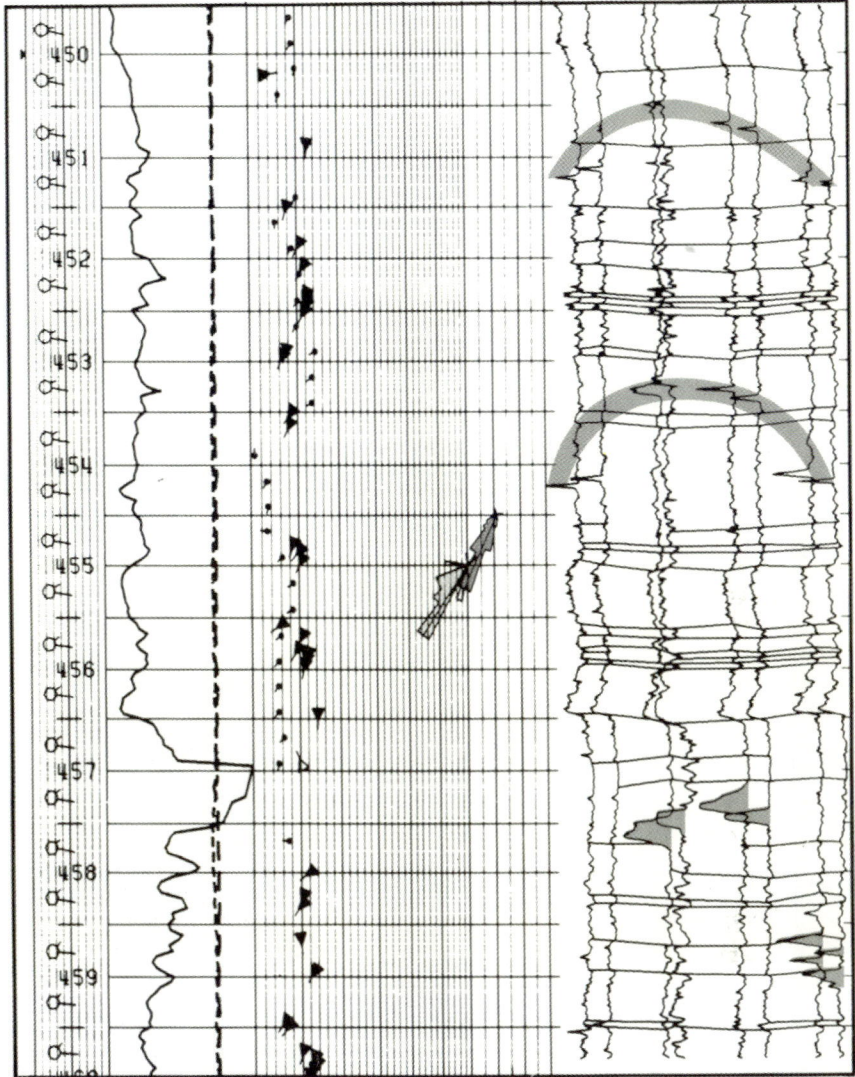

Figure 2.2.1. Dipmeter results over a carbonate reservoir in the Middle East. Both bedding and fracturing (highlighted) can clearly be seen on the dipmeter curves on the right, intersecting each other. The two dip calculation programs only obtain the bedding dips. Survey performed with the SHDT dipmeter tool and processing done with the DualDip software which calculates dips using patterns recognition as well as cross-correlation methods.

Micro-Electrical Imaging

The measurement principle of micro-electrical borehole imaging is practically identical to dipmeters (see chapter 2.1.1). The lower part of the sonde containing the pad-mounted electrodes is kept at a known potential with respect to the upper part of the tool. The two units are separated by an insulated section of the tool. The mud around the pads is more or less at the same potential, thus creating an equipotential line along the borehole surface. This forces the current emitted from the lower part of the tool to penetrate at a right angle into the formation, a condition referred to as passive focusing. The current is a low-frequency alternating current whose physics can be treated with so-called quasi-electrostatics. On the pads are rows of electrodes, with each electrode insulated from the other electrodes and the pad. The current emitted from each

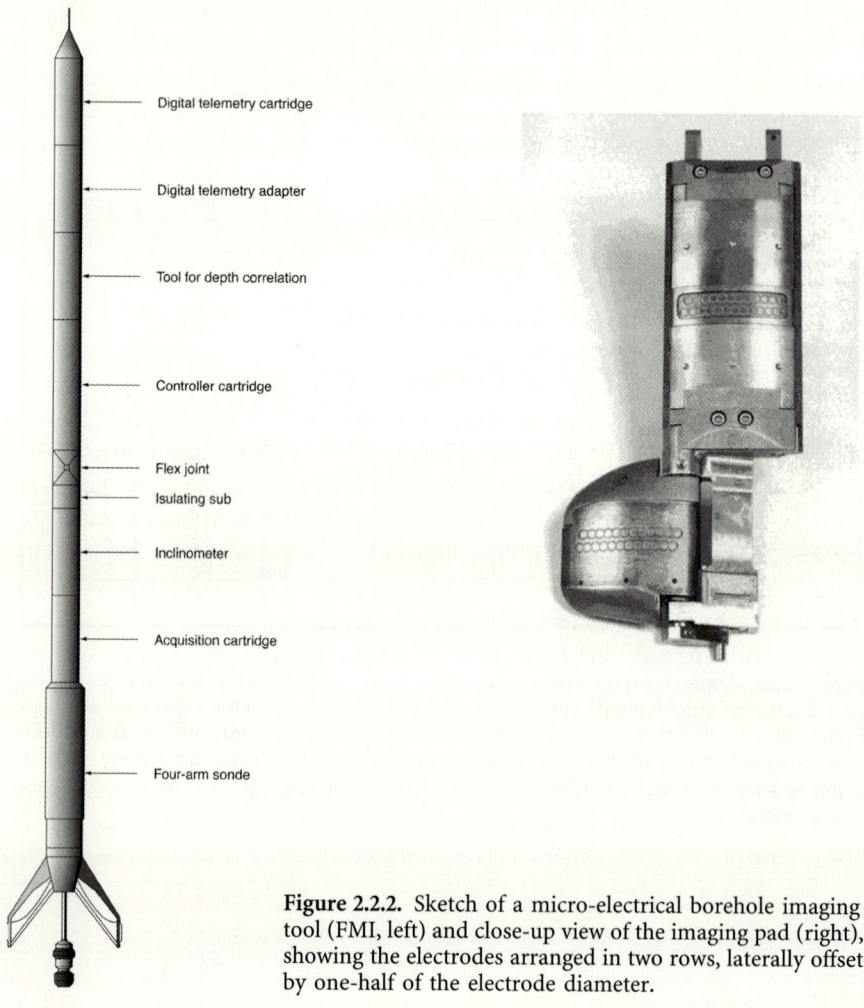

Figure 2.2.2. Sketch of a micro-electrical borehole imaging tool (FMI, left) and close-up view of the imaging pad (right), showing the electrodes arranged in two rows, laterally offset by one-half of the electrode diameter.

electrode is a function of the formation resistivity in front of it and is continually measured. In order to achieve a similar sampling rate in the circumferential (azimuthal) direction as in the vertical direction[1], Ekstrom et al. (1987) concluded that the electrode buttons must be arranged with at least a 50 % overlap in order to avoid undersampling. This led to the development of staggered rows of electrodes on each measurement pad, an example of which is shown in figure 2.2.2 together with a sketch of the tool. Vertical sampling is done versus time, typically at 60 samples per second, hence the term "scanning arrays". After conversion to depth, the vertical sampling distance is 1/10 of an inch, or 2.54 millimeters.

In practice, 27 electrodes of 5 mm diameter and a lateral spacing of 2.5 mm were initially implemented on two pads of the Formation MicroScanner (FMS*) tool (Ekstrom et al., 1987; Lloyd et al., 1986). It produced two 7 cm wide imaging strips at 90° of each other, but it also performed the traditional dipmeter functions, since all four pads were equipped with SHDT button pairs. The lateral coverage, i.e. the percentage of the borehole circumference scanned with this technique, was about 20 % in a standard borehole of 8.5 inches in diameter. For an increased coverage multiple runs had to be made, in the hope that the imaging arrays would be oriented differently in each run.

In subsequent tool generations the imaging arrays were first put on all four pads and then the number of electrodes was increased to improve lateral coverage (Boyeldieu & Jeffreys, 1988; Safinya et al., 1991). In Schlumberger's most recent version, the Fullbore Formation MicroImager (FMI*), each of the four pads contains two rows of twelve electrodes, plus the same number on an attached pad, called a "flap", positioned 15 cm below (Safinya et al., 1991). The tool has therefore effectively eight pads of 24 buttons, or a total of 192 electrodes, each 4 millimeters in diameter and surrounded by a 1-millimeter wide insulator (figure 2.2.2). Lateral and vertical sampling rates are 0.1 inch, or 2.54 mm, and the lateral coverage of the FMI is 72 % in an 8.5-inch borehole. This dense arrangement of electrodes was made possible through multiplexing and the deployment of very large-scale integrated (VLSI) circuits on the pads themselves. The electrode signals are digitized downhole with a precision of 10 bits and transmitted together with auxiliary information such as the total current, voltage, calipers and tool orientation data through the wireline cable at a rate of about 200 kbits/sec.

The signal recorded by these tools consists of two components: A *deep signal* is provided by the total current, which has a similar depth of investigation as a laterolog. A much *shallower signal* is provided by the variations of the button current themselves, which is a measurement equivalent to the microlog. The

---

[1] More correctly in the direction of the borehole axis. However, the terms "vertical resolution" and "vertical sampling" are deeply entrenched in the logging literature.

* Mark of Schlumberger

combined result is that the current of the image tracks the overall trend of a deep resistivity measurement such as the laterolog, while the high-frequency spikes and troughs track a shallow resistivity measurement and originate from within a few centimeters beyond the borehole wall.

Similar tools were developed by Western-Atlas (STAR**) and Halliburton (EMI***) under a licensing agreement with Schlumberger but based on the six-arm dipmeter design. These tools feature six pads, each with an imaging array containing 25 electrodes of the same size as the FMI and the swivel joints already present in the six-arm dipmeters. They produce images with a lateral coverage of 56 % in an 8.5-inch diameter borehole (Seiler et al., 1994).

Azimuthal Resistivity Imaging

This wireline method is based on an idea by Mosnier (1982, 1987) and represents a hybrid between the dual laterolog technique (Suau et al, 1972) and micro-electrical imaging (Ekstrom et al., 1987). It is a measurement using the laterolog principle, which scans the azimuthal distribution of formation resistivities around the borehole. Schlumberger's azimuthal resistivity imager (ARI*) is a multipurpose

---

* Mark of Schlumberger

** Mark of Western-Atlas

*** Mark of Halliburton

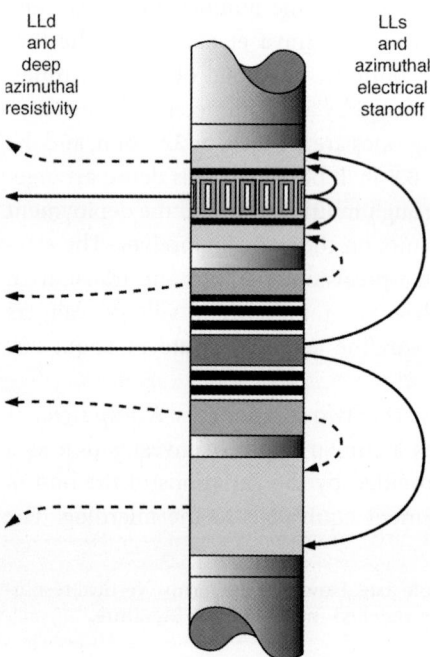

Figure 2.2.3. Sketch of the azimuthal resistivity imager (ARI) tool. In the measurement mode on the left, the tool records a deep laterolog resistivity with the central electrode (A0) and an azimuthal image with the twelve electrodes (A2) shown in the upper part of the tool. In the measurement mode on the right, a shallow laterolog resistivity is measured with the central electrode and an electrical standoff with the electrode array in the upper part of the tool. The latter is used to correct the azimuthal image. Focusing currents are dashed, measurement currents are solid lines.

tool performing resistivity imaging as well as the standard shallow and deep resistivity measurements (Davies et al., 1992; Smits et al., 1995). Borehole imaging is done with an array of twelve azimuthal resistivity electrodes. They are mounted within the so-called A2-electrode, which in the standard laterolog is the focusing electrode for the deep resistivity measurement, and the current return for the shallow resistivity measurement (figure 2.2.3). Currents with a frequency of 35 Hz are emitted from the electrodes and continuously measured. The electrode voltages are held at the same potential as two nearby monitoring ring electrodes using a feedback circuit, thereby providing active focusing and placing the mud at the same potential. Azimuthal focusing is passive through the neighboring electrodes. Knowing the geometric factor k' of this arrangement, the resistivity R of electrode i is then obtained as

$$R_i = k' \cdot V_m / I_i \qquad (2.2.1)$$

where V_m is the voltage of the monitoring electrode and I the current emitted by the electrode. An auxiliary measurement, made at a slightly different frequency, serves to estimate the standoff of the tool and to correct for uneven centering with respect to the borehole axis. The electrode currents are sampled every 1.27 cm (0.5 inch) in imaging mode, providing a vertical resolution of the resistivity image of about 20 centimeter (8 inches). This is three to four times higher than the conventional laterolog, mainly because of the different focusing method. The azimuthal resolution is 60°, and therefore both vertical and lateral resolutions are

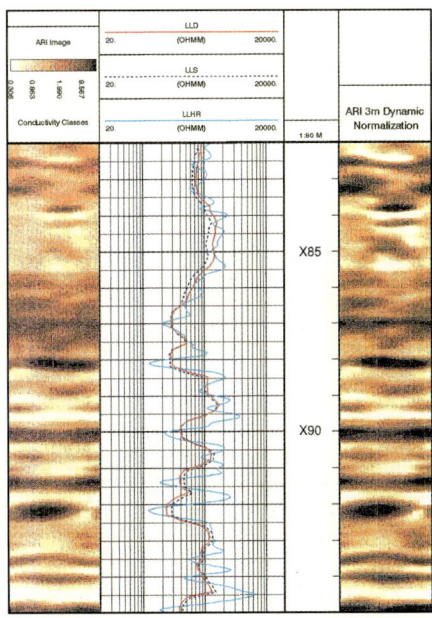

Figure 2.2.4. A resistivity image recorded by the ARI showing well-defined layering in the lower two-thirds of the image, and more massive, partly heterogeneous bedding in the upper part. The curves in the middle represent the shallow and deep laterolog resistivity (LLS and LLD) as well as an averaged resistivity from the azimuthal electrodes (LLHR). The latter shows excellent vertical resolution compared to the standard resistivities. The image on the left has a global ("equalized") color scale, while the colors in the image on the right are dynamically adjusted ("normalized") to provide good contrast at all levels.

approximately a factor of 20 larger than for the micro-electrical imaging method. A similar tool is the ELIAS (Straub et al., 1995) which is available for slim holes and has found acceptance in geothermal wells.

The principal use of this tool is for the petrophysical evaluation of reservoirs, particularly in cases where a higher vertical resolution than the conventional laterolog is required (figure 2.2.4). It is also used to get an idea of the overall bedding style and to spot fractures. Because of the tool's relatively low resolution, only structural dips can be obtained.

LWD Resistivity Imaging

This is the latest development in electrical borehole imaging and is the first LWD borehole imaging technique. The tool borrows from the two methods described above, although it is adapted to the particular conditions during drilling by taking advantage of the fact that the bottom-hole assembly rotates when no downhole motor is used. A single measurement electrode is mounted on the bottom hole assembly and makes an azimuthal scan while the tool rotates.

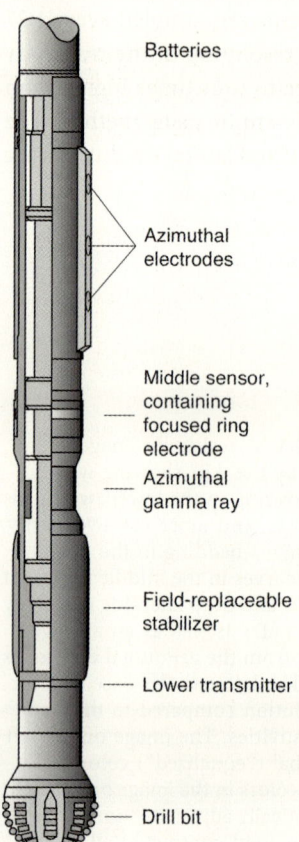

Figure 2.2.5. Sketch of the RAB (resistivity-at-the-bit) tool, showing the azimuthal electrodes in the uppermost part, which rotate with the entire bottom-hole assembly, scanning the borehole wall and recording an azimuthal image directly behind the drill bit. The tool can be mounted directly after the drill bit, or behind a downhole motor.

As the drill bit progresses, the electrode follows it, describing a spiral path down the borehole.

The Resistivity-At-the Bit tool (RAB*, figure 2.2.5), like the wireline azimuthal resistivity imaging, performs in fact other measurements too (Bonner et al, 1994; Rosthal, 1995; Lovell et al., 1995). Its name comes from a measurement made with the drill bit serving as an electrode through which a current is fed from toroids placed higher up in the tool. This resistivity measurement is sensitive to a volume of the formation in the immediate vicinity of the drill bit, and it allows drillers and geologists to make rapid decisions on when to set casing or where to start coring. Known as "geostopping" (see chapter 3.5.2), this procedure represents one of the significant benefits LWD data contribute to reservoir development. The close vicinity of this resistivity measurement to the drill bit also allows it to be used in "geosteering", whereby the resistivity response is compared to a layered formation model and deviations are used to correct the drilling direction (see chapter 3.5.2).

In another mode, the RAB tool performs resistivity measurements at various depth of investigation using a cylindrical focusing technique and three measurement electrodes as well as a ring electrode (Bonner et al, 1994). These measurements have an excellent vertical resolution and little sensitivity to "shoulder beds", i.e. to over- or underlying layers of different resistivities. The purpose of these measurements of variable depths of investigation is to perform a radial resistivity profiling for a better evaluation of the fluid invasion from the wellbore into the formation.

These same electrodes are also used to produce quantitative resistivity images such as shown in figure 2.2.6. Since the tool rotates, the button currents – usually of the electrode with the deepest depth of investigation – are recorded as a func-

---

* Mark of Schlumberger

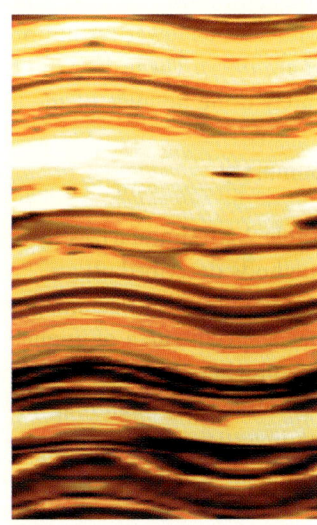

Figure 2.2.6. A RAB image covering approximately ten meters in a well layered sequence of beds in a deviated well. Notice that beds down to about ten centimeters can be resolved.. In the middle of the image a fracture is seen to cross cut the layers. Its apparent dip is lower than the beds, but because of the hole deviation its true dip is actually higher.

tion of the tool azimuth. The buttons with shallow and medium depth of investigation can also be recorded and may produce very useful information when displayed alongside each other (see figure 3.5.10). Together with the borehole deviation data, which are standard in measurement-while-drilling, the resistivity images can be oriented with respect to north, or, in the case of horizontal wells, with respect to the top of the hole. Typically, the RAB tool records data every ten seconds. In imaging mode, a rotation speed of 30 rotations per minute or higher ensures a full azimuthal scan, consisting typically of 56 azimuthal samples. The vertical resolution can be as high as 2.5 cm (1 inch), but is more typically around 5 cm (2 inches) and decreases at bit penetration rates above 60 ft/hr. The azimuthal resolution is in the order of ten degrees. Both depth and azimuthal resolution are thus about 5 times lower than in micro-electrical imaging. Compared to the two previous techniques there is no azimuthal focusing in this method, and the presence of conductive features such as fractures may compromise the azimuthal resolution additionally. In intervals where a downhole motor is activated to turn the drill bit, the tool does not rotate (it will "slide"; see chapter 3.5.1) and no azimuthal image is obtained.

Currently only the resistivity measurements of the RAB tool can be transmitted in real time to the surface using mud pulses, while the images have to be stored in downhole memory and are retrieved later for processing and interpretation. A method has been proposed to compute dips downhole and to transmit them in real time to the surface (Rosthal et al., 1995). This allows monitoring layer dips instantaneously together with the well trajectory. Real-time transmission of resistivity images may soon be available with new developments in downhole image compression, or improvements in data transmission.

The gamma ray sensor on this tool is eccentered such that it can provide an azimuthal measurement as well. Its count rates can be binned into quadrants (top, left, bottom, right) which can be used to produce a crude image, although with a much lower resolution than all the electrical imaging techniques (Rosthal et al., 1995). These data can be transmitted in real time to the surface but this is rarely done because of the low resolution of the technique.

Summary

Table 2.2.1 summarizes the main characteristics of the three electrical imaging methods described above. Prilliman et al. (1997) describe comparisons of LWD and wireline electrical imaging methods. They conclude that each of the methods has a specific area of application, with the wireline methods being particularly useful where high-resolution and accurate depth is required. LWD methods are recommended when real-time information is needed or when the rocks have to be investigated while invasion effects and drilling-induced fracturing are still minimal. All methods are confined to water-based mud systems, but efforts are underway to develop a tool that will also work in oil-based mud.

Table 2.2.1 Comparison of the three current electrical borehole imaging methods (Schlumberger tools only)

	Micro-Electrical Imaging (Wireline)	Azimuthal Resistivity Imaging (Wireline)	LWD Resistivity Imaging
Tool name	FMI	ARI	RAB
Tool type	pad tool	mandrel tool	LWD tool
Azimuthal measurements	192 electrodes	12 electrodes	56 samples
Azimuthal resolution	3 degrees	60 degrees	~15 degrees*
Vertical sampling	2.5 mm	1.3 cm	~ 2.5–5 cm*
Vertical resolution	0,5 cm	20 cm	~ 5-10 cm*

* Approximate values, depending on rate of penetration of the drill bit.

2.2.2 Image Processing

With the emergence of electrical borehole imaging it became necessary for the first time in well logging to apply image processing to logging data. The main goals of image processing are to improve the quality of the images, to provide good image contrasts and eliminate artefacts as much as possible. *Artefacts* are non-geological features encountered on borehole images that are caused either by an unexpected or abnormal borehole condition, or by tool peculiarities or sensor malfunctioning (Bourke, 1989; Cheung, 1999; Lofts & Bourke, 1999). Examples of the first type of artefacts are induced fractures and reduced signals in the presence of breakouts or washouts. These artefacts, as Cheung (1999) pointed out, can in fact be very useful to the interpreter, for example the direction of induced fractures or of breakouts. It is, therefore, not always desirable to remove these effects. The second type of artefacts, however, needs remedy, but this is often difficult since the signal may be completely lost, such as in the case of a dead electrode on the microelectrical array. The approach, therefore, often involves inter- or extrapolation from areas where there is a good geological signal. In general, the deeper-reading imaging tools are less affected by borehole conditions, but they may be as prone to other types of artefacts as the shallower reading tools.

Borehole image processing usually involves several sequential steps, some of which may apply only for a given tool. Figure 2.2.7 depicts a sequential flow chart of the common processing steps.

Borehole images are discretely sampled digital maps of the button currents I_i or resistivities R_i as a function of depth and azimuth. Since there are two independent variables, they are strictly two-dimensional, albeit in polar rather than Cartesian coordinates. The original sampling, however, is made as a function of time and electrode position, and thus the most important step in image proces-

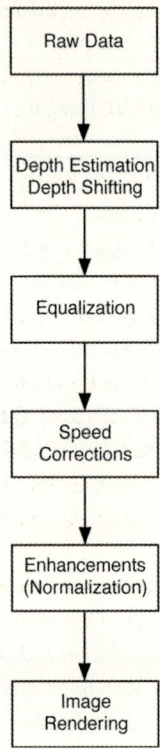

Figure 2.2.7. The processing chain for electrical borehole images

sing consists of placing the samples at their "correct" depth and azimuth. Imaging tools, particularly those featuring pads, can experience very irregular movements downhole, and the term "correct" is therefore somewhat arbitrary and perhaps better replaced by "optimum estimation" of the sample position. The process is, however, very important because of the high vertical and lateral sampling, particular in micro-electrical imaging where only small errors in sample position can degrade the image quality considerably.

Micro-Electrical Imaging. These images are sampled at a rate of 2.5 m both azimuthally and vertically, requiring the samples to be as accurately positioned as possible so that unpleasant effects on the image can be avoided.

The samples are first positioned at a depth corresponding to the time at which they were sampled by taking into account the nominal cable speed and their physical position on the tool with respect to a reference ("tool zero"). The individual buttons are then checked for any systematic offsets compared to other buttons. Such offsets can be caused by slightly different measurement circuits or pad contacts, resulting in the image trace to be darker or brighter than the average. In order to correct this, the average current and standard deviation of each button (the "gain" and the "offset") are calculated over a large sliding window, and corrections are made such that all buttons have the same dynamics. While this *equalization* is not strictly physically correct (one could imagine a button following a conductive feature for a long interval), it is in practice justified as the tools generally rotate while being pulled uphole, and button orientation distributions can be considered random.

The downhole triaxial accelerometer data is then used to estimate the instantaneous downhole tool velocity through a Kalman filter (Chan, 1984). The downhole tool speed may significantly differ from the uphole cable speed because of cable elasticity, tool weight, friction, or jamming of the tool in irregular borehole geometries. The difference to the nominal cable speed is used to apply an additional vertical shift to the sample positions. Since the output is an imaging array sampled at regular spatial intervals, interpolation is necessary in cases where the tool moved fast, and averaging in cases where it moved slow. The azimuthal position of each sample is obtained from the triaxial magnetometer data. As an additional image-processing step, the two subsets of the image from the two rows on each pad can be correlated with each other and matched for a better fit. This step is not based on the tool motion and is somewhat cosmetic, but the vertical shifts with this method are generally small.

Up to this step, the original current measurement of the electrodes is usually preserved in its original, high-precision form, but during normalization, quantization, and rendering the data is significantly reduced[2]. The purpose is to produce an image with color levels and contrasts that make is easy for the interpreter to analyze. *Normalization* of the image over a vertically sliding window of a given length can enhance the contrast in the image. It involves calculating the statistics such as mean and standard deviation of all data points at a particular window position, and assigning the measurements into user-determined bins. Usually, the number of bins is kept low, for example to 16 or 32, corresponding to four and five bits respectively. Such a dynamic normalization produces image contrasts at all levels regardless of the absolute value of the original measurement. The color changes on the image are, therefore, only relative to the adjacent image regions. However, the values in the image array are no longer high-precision real numbers, but integers without a fixed relationship to the original sample values. This makes it impossible to recalculate the original data from such an image.

The *quantization* procedure is very similar to dynamic normalization, but generally the assignments of colors (the "binning") is made using the dynamics of the entire image, not of a sliding window. Figure 2.2.8 illustrates the mapping from image data into color bins. By tradition, colors ranging from black for conductive through brown, reddish, yellow and white for resistive have been used. Dark has been chosen to represent high conductivities because shales are often

---

[2] Quantization is the assignment of a color class to a range of image values. Rendering is the process of physically producing an image, either on a screen or on a print.

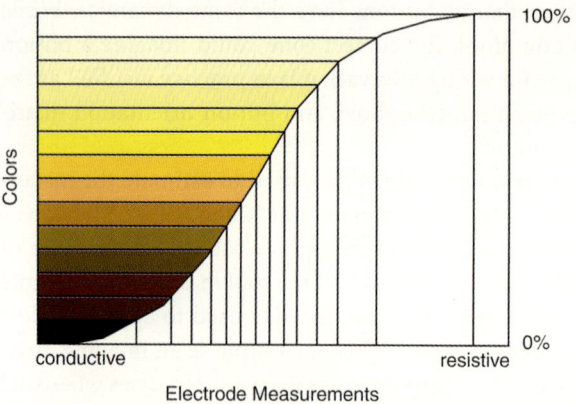

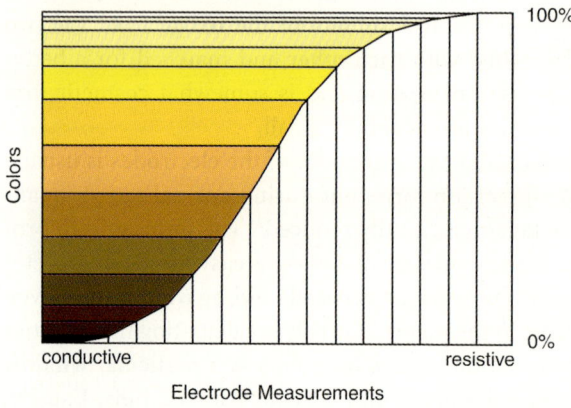

Figure 2.2.8. Color schemes used for borehole images, illustrated with cumulative frequency plot. Top: a histogram-normalized, non-linear color scheme where every color receives the same number of samples. Bottom: a linear color scheme where the electrical data are evenly subdivided and assigned to a color. The number of samples for each color is a function of the distribution of the original data.

more conductive and darker-colored than other lithologies. Different color schemes or just a gray-scale range can of course also be used, however inverting the intensity scales such that conductive becomes white is not advised for the sake of consistency. In the color distribution at the top of figure 2.2.8, every color bin receives an equal number of samples. This is a non-linear process and requires sorting of the samples prior to binning. The result is usually an image with a great degree of contrast that helps spotting image details in areas where sample values differ little from each other. However, these images often appear somewhat harsh in contrast. On the other hand, the color distribution at the bottom of figure 2.2.8 distributes the measurements into equal bins, usually on a logarithmic scale. This often results in an image where intermediate colors abound and few extreme colors such as black or white occur, except for unusual distributions of the input data. The image is often pleasant to the eye but may lack details in certain areas, a shortcoming that can be corrected on an interactive workstation through local image enhancements.

Figure 2.2.9 illustrates these various image processing steps on an example where significant speed corrections had to be done. These were necessary because of a very irregular tool movement, illustrated in figure 2.2.10 as the acceleration measured along the tool axis and the calculated downhole tool

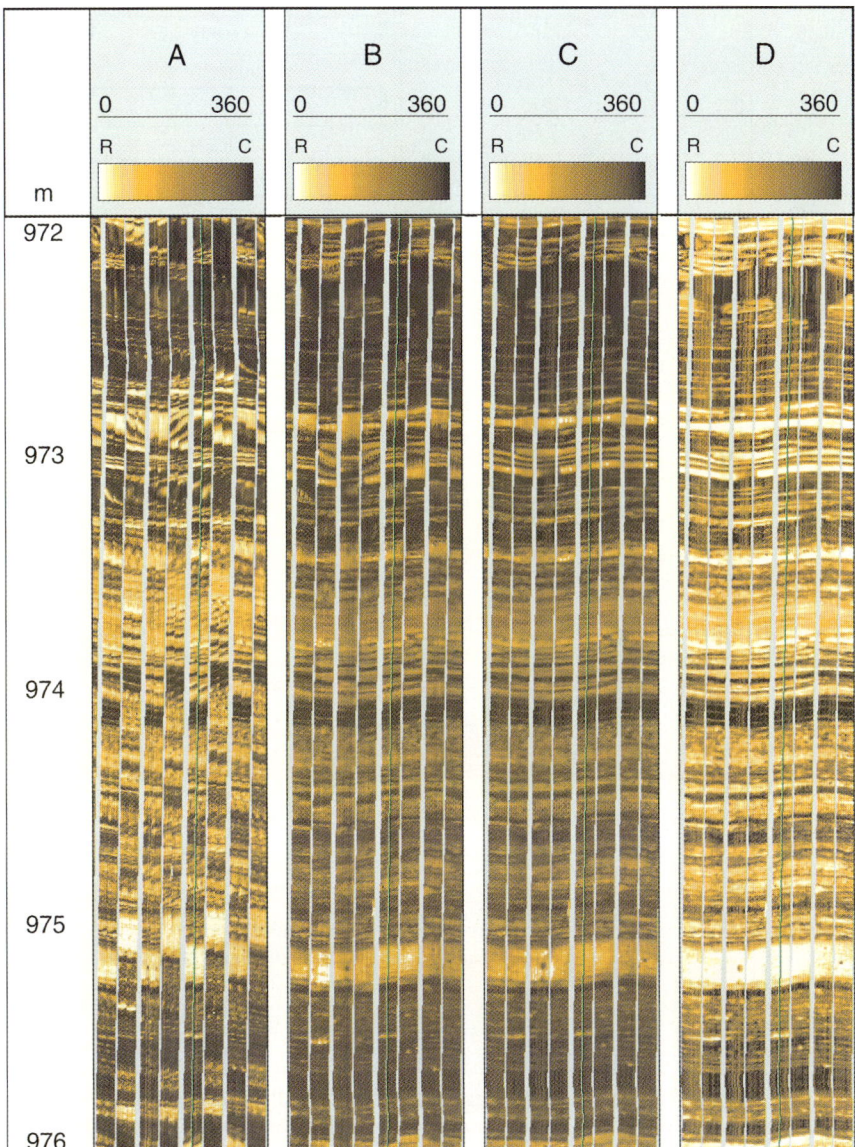

Figure 2.2.9. Images corresponding to four steps in the image processing chain. A: raw image before any processing. B: image after depth shifting. C: image after equalization and accelerometer-based speed correction. D: image after image-based speed correction and dynamic color normalization. FMI images over a Cambrian carbonate interval, West Texas.

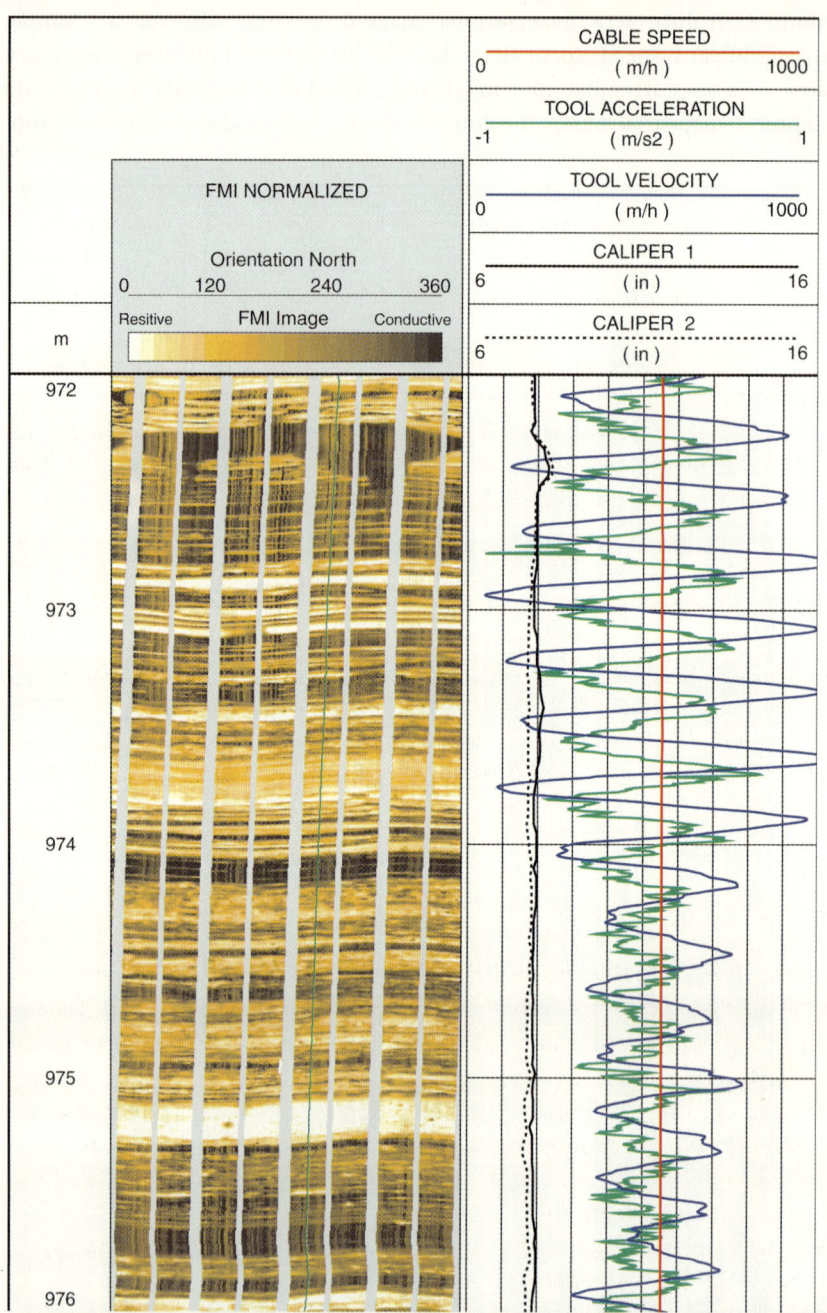

Figure 2.2.10. Vertical tool acceleration (green) of the same interval as figure 2.2.9, together with the calculated downhole tool velocity (blue) and the uphole cable speed (red). The tool speed oscillates like a yo-yo between practically zero and over 1000 m/hr in a regular manner, caused by the cable elasticity. The images have been successfully corrected for this movement with a two-step speed correction.

speed. The latter oscillates between over 1000 m/hr and practically zero, while the uphole cable speed is constant and around 550 m/hr. The good quality of the images on both figures demonstrates that a correct estimation of the downhole tool speed was possible.

Azimuthal Resistivity Imaging. The processing here is simpler because the images have a lower resolution and sample position therefore does not require the same accuracy. Additionally, there are no depth shifts of the electrodes because they are all placed at the same depth. Equalization and accelerometer-based speed corrections are done in the same manner as above, and image enhancements can also be applied. As an optional step, azimuthal sample interpolation can be made at the graphics level to reduce the blocky appearance of the images. Vertical sampling is much higher than the resolution of the tool (see table 2.2.1) for reasons related to the calculation of fracture aperture (see below). Therefore no sample interpolation is necessary in the axial direction.

LWD Resistivity Imaging. Here the most important concern is the correct depth positioning of the samples. Since the drill bit penetration rate can vary considerably, and since the image data are sampled versus time, the spatial distance of the samples can vary greatly. The values for the depth sampling given in table 2.2.1 are thus only approximate. In the image processing chain, typically a resampling by interpolation is made so that imaging arrays with a constant sampling density are obtained. No further speed corrections are made, but equalization, enhancement, and rendering are done as in the previously described methods.

2.2.3 Graphic Representation

The basic graphic representation of borehole images is the same as Zemanek et al. (1970) proposed for ultrasonic borehole televiewer images (chapter 2.3). The vertical axis represents depth, and the horizontal axis azimuth, ranging from north on the left to east, south, west and north again on the right edge. The measurements themselves are displayed either as color or gray-scale intensities, with the convention that higher conductivities are displayed in darker tones. This representation corresponds to unfolding the borehole cylinder onto a plane, causing features with an apparent dip different from zero to take the form of a sinusoid (figure 2.2.11). The steeper the apparent dip is, the higher the amplitude of the sinusoid is for a given vertical scale.

In horizontal wells, the reference frame is often changed to distance along hole on the principal axis, and bottom to top and again to bottom on the azimuth axis. These images are sometimes interpreted by laying them out horizontally, not vertically. Since horizontal wells often follow bedding, the apparent dip can be very high, forming sinusoids with very large amplitudes.

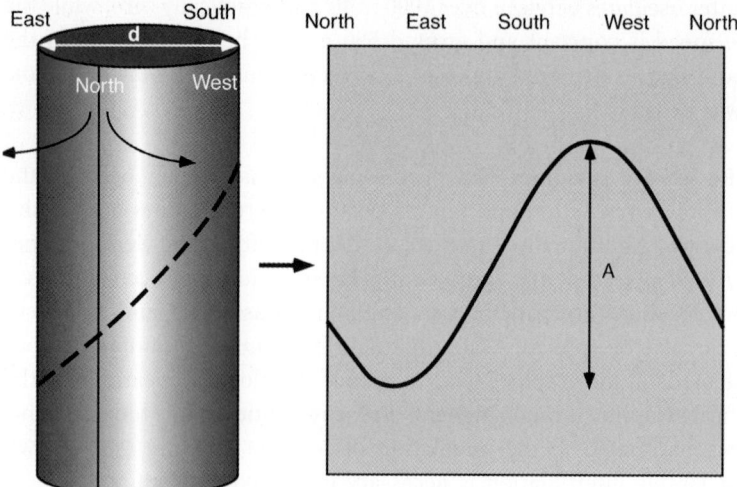

Figure 2.2.11. Projection of borehole images onto a plane with azimuth and depths as coordinates. A dipping plane becomes a sinusoid on this projection. The dip can be calculated from the amplitude A and the borehole diameter d, while the azimuth is read off directly at the trough of the sinusoid.

Conversely, steeply dipping features have low apparent dips and therefore low-amplitude sinusoids.

Borehole images are usually plotted on paper, on the wellsite or in the office, with compressed scales such as 1/200 being used for a quick overview, quality control, and structural interpretations. More expanded scales such as 1/40, 1/20 or 1/10 are used for more detailed sedimentological interpretations. Many

Figure 2.2.12. Interactive workstations are very useful tools for interpreting borehole images.

plotters can only render a limited number of colors, and much detail can get lost. It is therefore preferred to display borehole images on workstations, where greater resolution, image scrolling capabilities, and access to image processing routines are available (figure 2.2.12). Additionally, the images can be projected onto a cylinder representing the borehole and viewed interactively from any angle.

2.2.4 Dip Calculations

There are several ways of calculating dips from borehole images:

- Extracting a subset of the image, to be used as dipmeter curves in the usual dipmeter processing chains
- Interactive picking of features on a workstation by fitting a plane and calculating its dip and azimuth
- Calculating dips automatically and directly from images using layer tracing algorithms
- For the LWD resistivity images a specific approach is required in horizontal wells where high apparent dips prevail

The dipmeter approach has been discussed in detail in chapter 2.1.2. Measurements from electrodes whose positions correspond to those of dipmeter buttons are extracted from the images and processed like dipmeter curves, either with a cross-correlation or a pattern-recognition method. This approach is fast and as reliable as standard dipmeter processing. However, it misses out on the major advantage of borehole imaging, the possibility to classify geological features based on the image and to select only those dips that are of interest for the interpreter. Faivre & Catala (1995) have demonstrated that the dipmeter approach can also be applied to resistivity imaging with the ARI tool.

The interactive approach can be taken on workstations. A feature of interest such as a bed boundary, a fault, or a fracture is chosen and a sinusoid is fitted to it. Generally, the user selects a few points on the feature of interest, and a sinusoid is fitted to it. Alternatively, an adjustable sinusoid template appears on the screen and is manipulated by the user until it fits a feature. Calculating the dip and azimuth is straightforward: The azimuth is directly obtained from the lowest point on the sinusoid, while the dip δ is obtained as $\delta = \arctan(A/d)$, where A is the amplitude and d the borehole diameter (figure 2.2.11). Care has to be taken to rotate this dip by the borehole deviation so that true dips are obtained. In highly deviated or horizontal boreholes, the dip and azimuth seen on such a graphic representation are often quite different from the true dip and azimuth. It is also important to realize that the images are plotted versus depth and *azimuth*, not azimuthal distance. Therefore, the image in an enlarged part of the borehole is exactly as wide as in an in-gauge section, but two beds with the same dip will have different amplitudes, because in the enlarged borehole, the amplitude will be higher than in the narrow hole. For this reason, it is good practice to plot the borehole caliper next to the image, and a reference log such as a gamma ray mag also be helpful. The relatively small diameter of the borehole allows many geological features to be approximated by a sinusoid, even if they are not perfectly planar. If a feature is seen to have a poor fit to a sinusoid, it may either be non-planar, or the borehole is non-circular.

Automatic dip determination from images is a delicate image-processing problem because on borehole images the bedding signal is often affected by other factors such as fracturing. Antoine & Delhomme (1990) use a line-tracing algorithm proposed by Kass & Witkin (1987) to define the bed boundaries and subsequently to calculate a dip. This approach is computationally very complex and slow. Wong et al. (1989) and Torres et al. (1990) use the Hough transform on ultrasonic images to trace sinusoidal events, and Hall et al. (1996) applied the method to electrical images. The procedure consists of two steps: First, the images are filtered such that short line segment representing edges are detected and retained. Second, the three-dimensional sinusoid space of depth-amplitude-phase is scanned to see if there is a good fit of a sufficient number of line segments with a sinusoid. Figure 2.2.13 illustrates this schematically: For a given line segment, there is a large number of possible sinusoids which fit it, although only four are shown. When a second line segment is taken into account, there is only one that fits both. In the actual implementation, a large number of line segments are simultaneously looked at over a given depth interval. All possible sinusoid fits are retained and their frequency is then counted over small cubic blocks within the three-dimensional space of depth/dip/azimuth. The block with the highest count is the winner and represents the best-fit sinusoid. The runner-up can

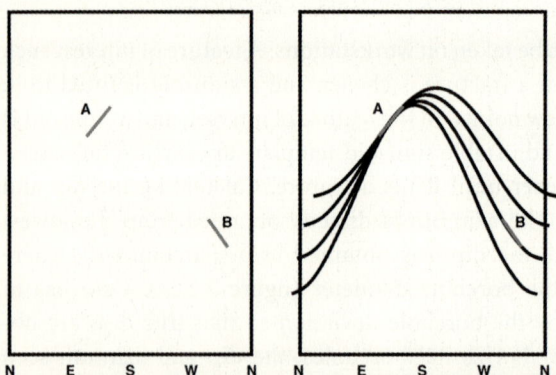

Figure 2.2.13. Principle of the Hough transform. First line segments corresponding to edges are determined on the image (left). Sinusoid fitting is then done to all line segments. While numerous sinusoids fit line segment A, only one fits both A and B (right). It is retained as one of the possible solutions for this interval.

be retained too and may correspond to sinusoids at a different depth, or with different amplitudes (dips) or phases (azimuths). Although computationally heavy, this procedure has the considerable advantage that it can handle intersecting features such as bedding and fracturing.

More recently, Ye et al. (1997) have published a method of dip determination driven by geologic reasoning. Similar to the pattern recognition approach in dipmeter processing, their method is hierarchical: It first determines the major bed

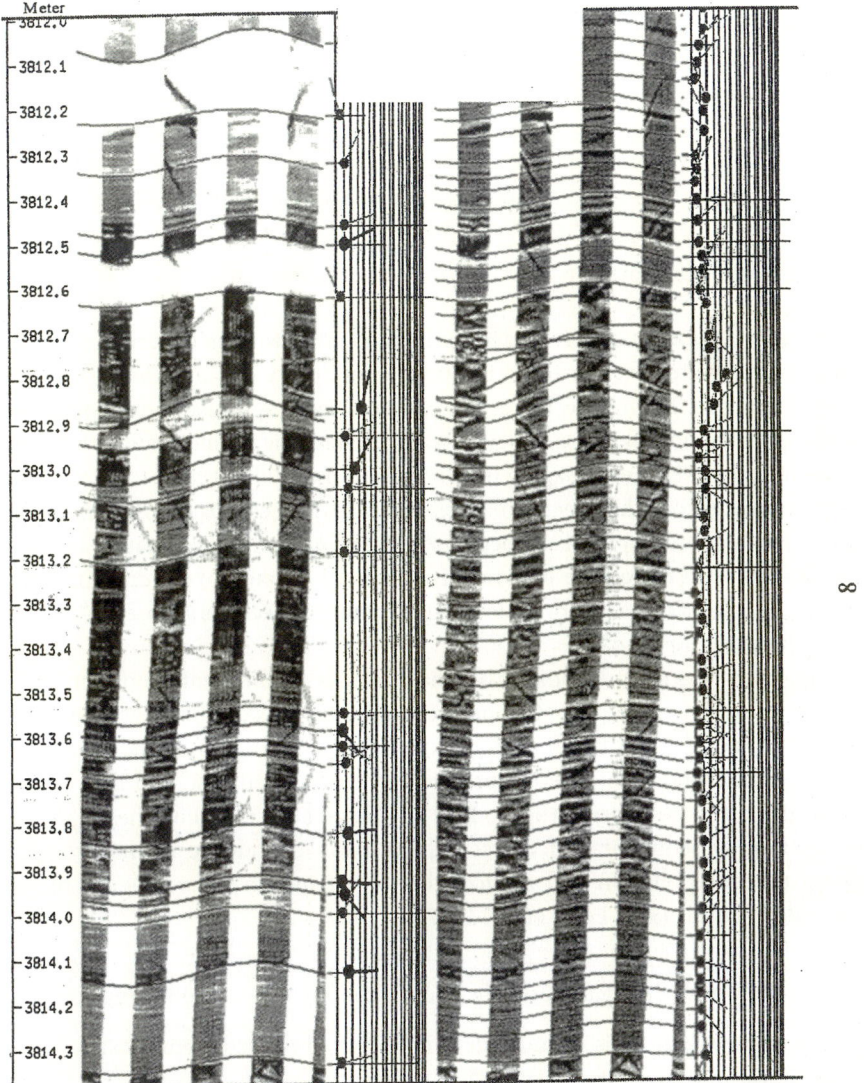

Figure 2.2.14. Automatic dip determination on micro-electrical borehole images using a hierarchical algorithm (from Ye et al., 1997). Both major (left) and minor (right) bed boundaries are identified, and fractures are successfully eliminated.

boundaries, then the internal stratification ("lamination") of the beds. On the microelectrical image of each pad, the principal edges are first detected based on thresholding of the first derivative and lateral connectivity. Thinner beds or layers not extending over the entire pad are eliminated until every image is divided into beds. The bed boundaries are then matched across the different pads on the basis of planarity, contrast, and other factors, and dips are calculated. Before proceeding to the internal stratification analysis, Ye et al. (1997) eliminate conductive heterogeneities such as fractures with an exponential operator on the second derivative of the image, and the erased parts are then reconstructed through interpolation from the neighboring areas. Laminations often do not appear continuously across the entire image, and therefore a local estimation of the bedding orientation is done. It is followed by fitting sinusoids, first over a coarse window, then over a smaller window to fine-tune the fit. A result on an image with fracturing, major and minor bedding surfaces is shown in figure 2.2.14.

2.2.5 Image Interpretation

The first electrical images, recorded in the mid-1980s with an experimental one-pad device, created excitement in the small community who had the privilege of participating in the project. Never before had a logging measurement provided such detailed and obvious geological features. Bedding, faulting and fracturing were, for the first time in logging history, visible and did not have to be inferred from dipmeters or other open-hole well logs. The first comparisons of electrical images with photographs of slabbed cores turned out to be good matches (figure 2.2.15), although there were obvious discrepancies because of the differing physical principles. Electrical borehole imaging comes indeed very close to "electrical coring"(figure 1.2.4).

The interpretation methods for borehole images are different from dipmeters. Bedding surfaces can be directly observed and their dips can be measured selectively for each type of bedding interface (figure 2.2.16). This is the inverse of dipmeter interpretation, where dips have to be measured first, and then geological structures are inferred (figure 2.1.13). The greatest difference lies in the visualization of sedimentary features, which instill confidence in the geologist doing the interpretation and reduces the uncertainty in the geological model he builds (Luthi, 1990; Lofts et al., 1997).

Interpretation of electrical borehole images can include the following steps:
- *Bedding analysis*: This includes determination of the sedimentary structures, zonation into beds of similar bedding style, measurements of relevant dips, and evaluation of the petrophysical properties for each bedding type. A particularly interesting aspect is the quantification of reservoirs with bed thicknesses below the resolution of conventional open-hole well logs.

- *Structural analysis:* Faults, folds, unconformities, fractures and – to a lesser degree – breakouts can be identified and incorporated into a structural geological model. Large-scale structural analysis proceeds in the same manner as for dipmeter analysis, but is helped by the direct determination of major discontinuities (faults, unconformities) from the images.

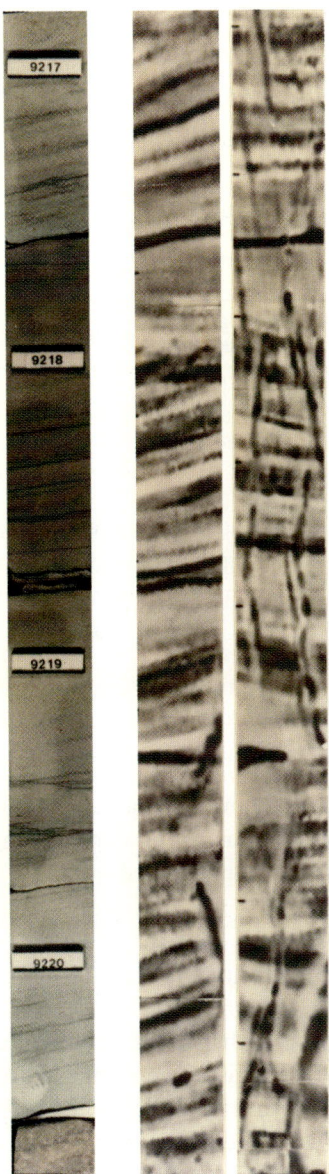

Figure 2.2.15. Comparison of photograph of a slabbed core (left) and with two images recorded with the first experimental electrical imaging tool (right), which contains only one pad with 40 buttons. The comparison of the core with the first borehole image shows an excellent match of the bedding features, notably the mud draping, cross-strata and bounding surfaces in this tidal sandstone. The image of the second run, recorded at a right angle to the first one, shows abundant conductive wisps, which are induced fissures on the borehole wall. Travis Peak formation, Cretaceous, Nagocdoches County, Texas. Depths in feet.

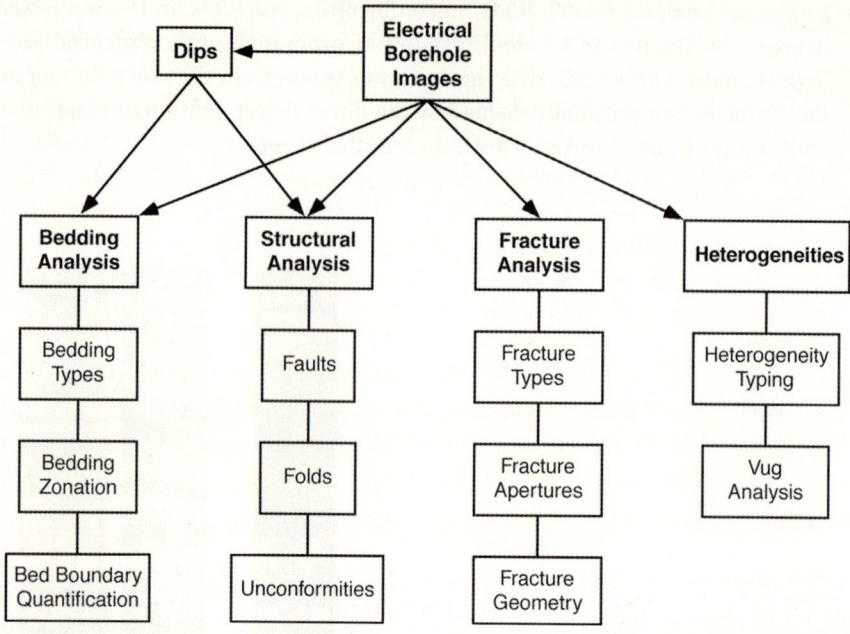

Figure 2.2.16. Borehole image interpretation scheme, corresponding to the approach discussed in the text.

- *Fracture analysis:* Fractures can be characterized for their geometry and probable origin, but also for their apertures, an important parameter in reservoirs producing predominantly from fractures.

- *Heterogeneities:* The two-dimensional nature of borehole images helps assess the degree of heterogeneities in rocks. Examples include vuggy and karstic porosity in carbonates, sandstone lenses, nodules, and diagenetic concretions. Semi-quantitative methods have been proposed for several of these features.

In the following paragraphs, these steps are discussed in more detail. Several case examples involving all of these aspects are found in part 3 of this book as well as in the volume edited by Lovell et al. (1999).

Bedding Analysis

This is perhaps the most important contribution of electrical borehole images to the geologist. Two principal aspects are distinguished: Zonation into bedding types, and quantification of bed boundaries. The nomenclature for bedding and sedimentary structures we use in the subsequent chapters follows the terminology of Collinson & Thompson (1982).

Bedding types are expressed on borehole images through their image texture. While texture in rocks commonly refers to microscopic properties, in image analysis it describes regions with characteristic brightness, colors, slopes, or sizes of the image elements. Textural image segmentation is a process done by the brain at an early stage in the analysis of visual scenes (Hubel, 1988), prior to object recognition and scene analysis. Comparisons of borehole images with slabbed cores show that the image texture correlates with bedding types (figure 2.2.17), although there are differences due to the two different physical processes involved. Therefore, a zonation into image textures effectively produces a *bedding zonation*. This can be done visually on a paper plot or on a workstation, or alternatively through an automated computational procedure. The eye is a powerful image analyst, and the results from a visual, interactive analysis are often of great reliability although they take considerable time, whereas computational approaches are faster but need subsequent validation. Cored sections can serve as control intervals in the validation process, either by guiding the search for specific bedding types in a deterministic manner, or by comparing bedding

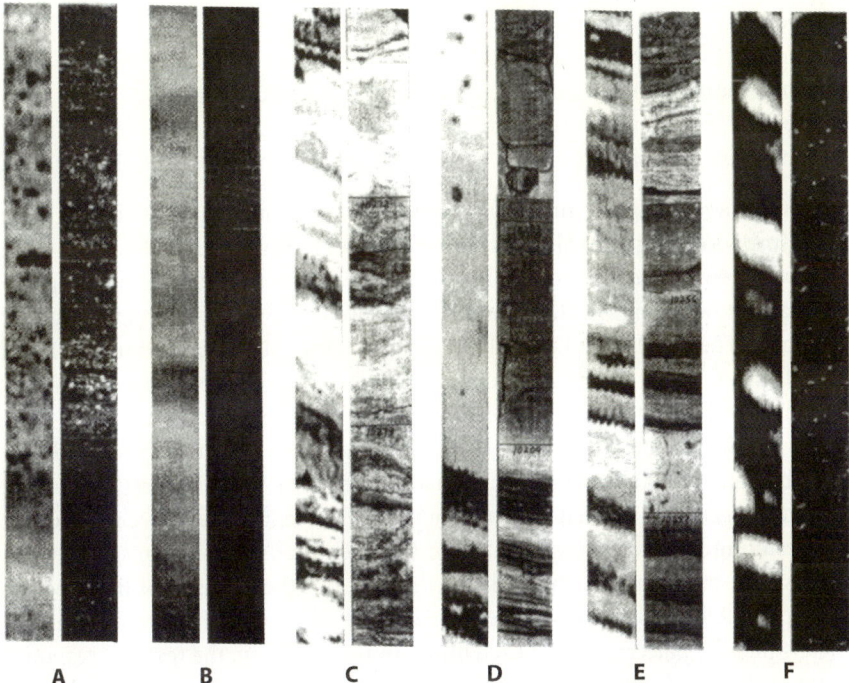

Figure 2.2.17. Comparison of electrical images (left) to photographs of slabbed cores (right) for six different fluvial facies with characteristic bedding types from an Alaskan oil field. A: conglomerate. B: cross-bedded medium-grained sandstone. C: bioturbated argillaceous fine-grained sandstone. D: argillaceous fine-grained sandstone with carbonaceous clasts and bedding type E at bottom. E: interbedded fine-grained sand-, silt- and mudstones. F: mudstone with rootlets (from Luthi, 1994).

types with a heuristic image zonation (a "core calibration"; see several papers in Harvey & Lovell, 1998). Among the depositional bedding types that can be identified visually from the images without much ambiguity are the following:

- Conglomerates, with the individual clasts often showing up as resistive nodules, sometimes with imbrications
- Massive bedding in sandstones, carbonates, evaporites and igneous rocks
- Planar laminations and cross-beds in sandstones and carbonates
- Graded bedding, often expressed by a resistivity gradient
- Interbeds of sandstones and shales, whereby the lower limit of the beds resolved is given by the tool resolution
- Silt/mud laminations, but also foliations in metamorphic rocks, if their spacing is above the tool resolution

The following structures caused by syn- and post-depositional deformation, biogenic and chemical action can also be identified on borehole images:

- Deformed bedding due to dewatering, slumping and other mass flow types
- Bioturbation
- Rootlets
- Mudcracks
- Concretions such as anhydrite nodules, dolomitized zones etc.
- Pressure solution features such as stylolites
- Secondary porosity such as vugs, karstic features, collapse breccias etc.

Features that are aligned mostly parallel to bedding planes, such as certain types of bioturbation, larger fossils, erosional structures (scouring and tools marks) are generally hard to identify on borehole images. Examples of selected bedding features are shown in figures 2.2.17 and 2.2.18 as well as in other figures throughout the book. The articles by Luthi (1990), Harker et al. (1990), and Bourke (1992) also illustrate many of these features and outline procedures on interactive interpretation of electrical borehole images using a workstation.

An automated method to zone borehole images into bedding types was developed by Luthi (1994) through a technique referred to as "texture energy" analysis. The image gets first filtered with a large number of texture-sensitive masks, and the resulting images then get analyzed for their statistical responses to these masks. The masks are mathematically defined, but they consist of lines, edges, and spots. A high response of the image to the filtering with any of the masks means that the patterns match each other. In the final step, the textural response curves get clustered into a user-defined number of classes. Figure 2.2.19 shows a sequence from a Paleozoic fluvio-glacial reservoir in the Middle East. The image zonation shows a good correspondence to the core-derived zonation, particularly when keeping in mind that the texture zonation is based entirely on the image, without any recourse to additional logging measurements. Notice in particular that the oil-bearing laminated sandstone, designated as Sl,la on the core analysis

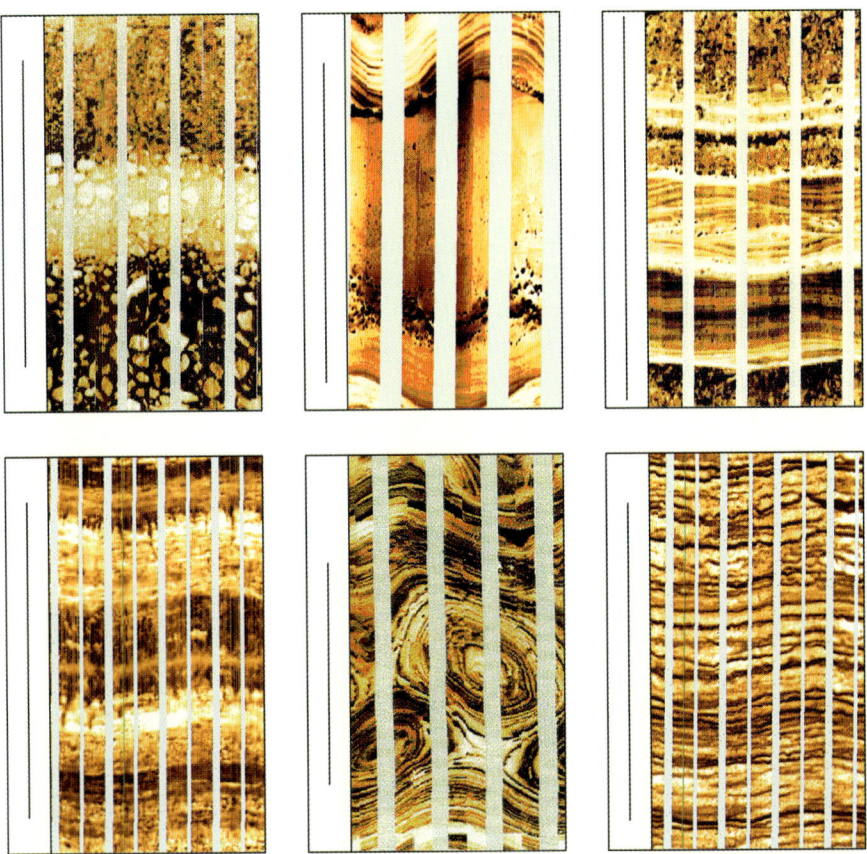

Figure 2.2.18. Various examples of bedding types on electrical borehole images (all FMI). From upper left to lower right: A conglomerate in a debris flow; rip-up clasts in a fluvial channel fill; thin cross-beds in a tidal sandstone; burrowing in a carbonate/clastic mixed shelf deposit; overturned sand layers in a slump; stylolites in a limestone. Bar on left of images corresponds to one meter.

and being the commercially most interesting target, shows a good match with the image-based zonation.

A similar method has been proposed by Hall et al. (1996), who use the concept of "inertia" to derive image texture. Inertia is a measure of the probability of a certain color difference in the image to be present between two pixels a given distance away from each other. A high probability of a large pixel value difference results in a high inertia. The calculation is made in various directions of the image. Thus, a horizontally laminated lithofacies, for example, has a high inertia perpendicular to bedding and a low inertia parallel to it. Zones with similar response in these inertia curves are then grouped using an unsupervised artificial neural network. This step is equivalent to the clustering performed with the texture energy curves in the previous step.

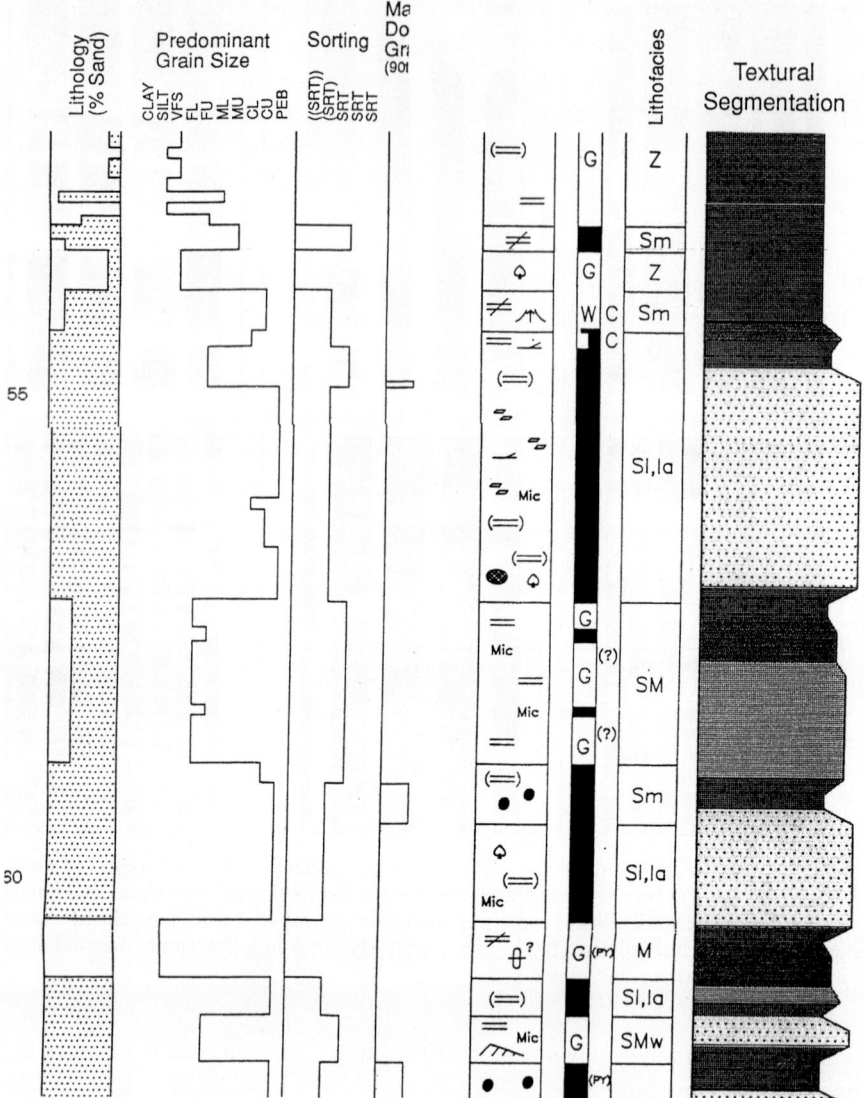

Figure 2.2.19. Zonation according to texture energy in a Paleozoic fluvio-glacial reservoir from the Middle East. The left columns are core descriptions, while the column on the right contains the zonation based on textural analysis of FMS images (from Luthi, 1994). Notice the good match of the laminated sands (Sl, la) with the image texture shown in a stippled pattern.

Kraaijveld & Epping (2000) propose filtering the images with a 2-D Gaussian kernel and its two partial derivatives at a variety of kernel scale. This so-called "scale-space filtering" (Kass & Witkin, 1987) probes the textural properties of the image as a function of scale, and, through the derivatives, the orientation of image elements. From this, they proceed to estimate the local dips in the image and to zone it into image textures using a neural network classifier. The network is trained on intervals where the facies is known, for example where core is available. The scale space technique seems to be particularly useful when data are noisy, since it acts as a blurring filter. These various processing elements are incorporated into a toolbox that can be applied not only to electrical, but to any type of borehole image.

Quantifying bed boundaries is of particular importance and represents a major difference between borehole images and dipmeters. Among the dipmeter programs, only pattern-recognition algorithms have the built-in capability of distinguishing between different bedding surfaces, but the interpreter has often great difficulties in sorting them out within the large number of tadpoles on the plot. Cross-correlation programs lump all bedding surfaces over a correlation interval together and output one average dip. On borehole images, however, the inter-

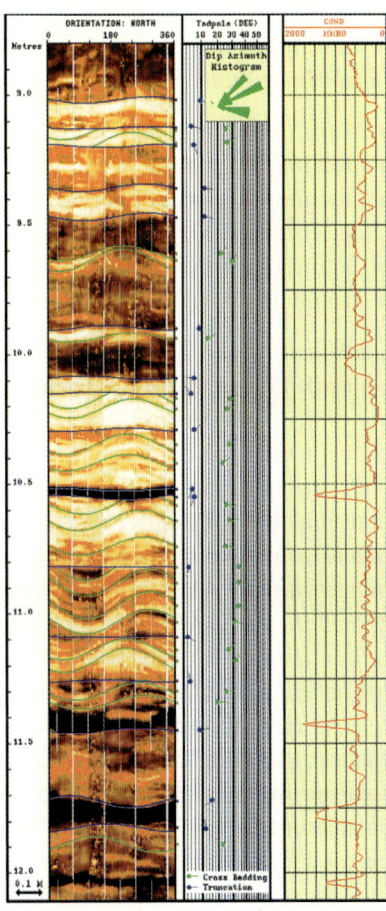

Figure 2.2.20. Example of bedding surfaces on FMI image. Cross-beds (in green) dip towards east and northeast, while bounding surfaces (in blue) show low dips with scattered azimuths. Red curve in right track is conductivity and indicates interdune deposits where it reads low, particularly in the lower half.

preter can easily make the distinction through visual inspection just like on cores. Figure 2.2.20 is an example from a cross-bedded sequence, where set boundaries and cross bedding can be distinguished and grouped into separate classes. They are indicated by sinusoids fitted to them and highlighted in separate colors. Similarly, Luthi & Banavar (1988) have shown in a study of an eolian sequence that first-order bounding surfaces (Brookfield, 1977), i.e. interdune/dune contacts, as well as reactivation surfaces can be distinguished and their dips can be quantified. Such data can lead to substantial improvements of a geological model as discussed by Plumb & Luthi (1989) for an eolian reservoir. More recently, Anxionnaz & Delhomme (1998) propose extrapolating these bedding surfaces away from the borehole in a similar manner as for the construction of cross-section from dipmeter data using the method developed by Etchecopar & Bonnetain (1992) discussed in chapter 2.1.4. Grace et al. (1986) demonstrate the use of selective bedding dips for fluvial reservoir modeling in Texas, while Thompson & Snedden (1996) describe a similar effort in deltaic reservoirs of Nigeria. Slatt et al. (1994) use FMS data from turbidite reservoirs in the Gulf of Mexico in combination with analog outcrop data to improve the reservoir model.

Among the bed boundaries that can be identified with little ambiguity from borehole images are the following, listed in more or less hierarchical order:

- Unconformities
- Boundaries of bedsets and larger bedding units
- Bounding surfaces of various degrees in cross-bedded sets (coset boundaries, set boundaries, reactivation surfaces)
- Bounding surfaces of individual beds, including erosional bases
- Laminations within beds (only if thicker than tool resolution)

Classification and quantification of bed boundaries is mostly done interactively on workstations, but the automatic dip calculation procedures described above are designed to perform this task in a more automated mode and may soon be robust enough to be widely applicable.

A particularly interesting problem in the oil industry is the accurate thickness determination of thinly bedded reservoirs. Microelectrical imaging has a theoretical resolution of one-half centimeter, which is one to two orders of magnitude better than most electrical, nuclear and acoustic open-hole well logs. However, as Trouiller et al. (1989) have shown in a comparison study with cores, there is an estimation bias for beds below five centimeters (figure 2.2.21): Sands tend to be overestimated and, conversely, shales tend to be underestimated. They found the relative error to increase with thinner beds. Trouiller et al. (1989) calibrate the electrical imaging response with a resistivity tool in order to compensate for baseline drifts induced by borehole effects, "shoulder effects" (the influence of adjacent layers) and vertical anisotropy. In this way, a constant cutoff can be used to separate shales from sands. Using core comparisons and forward modeling results of the tool response, they found that for beds thinner than

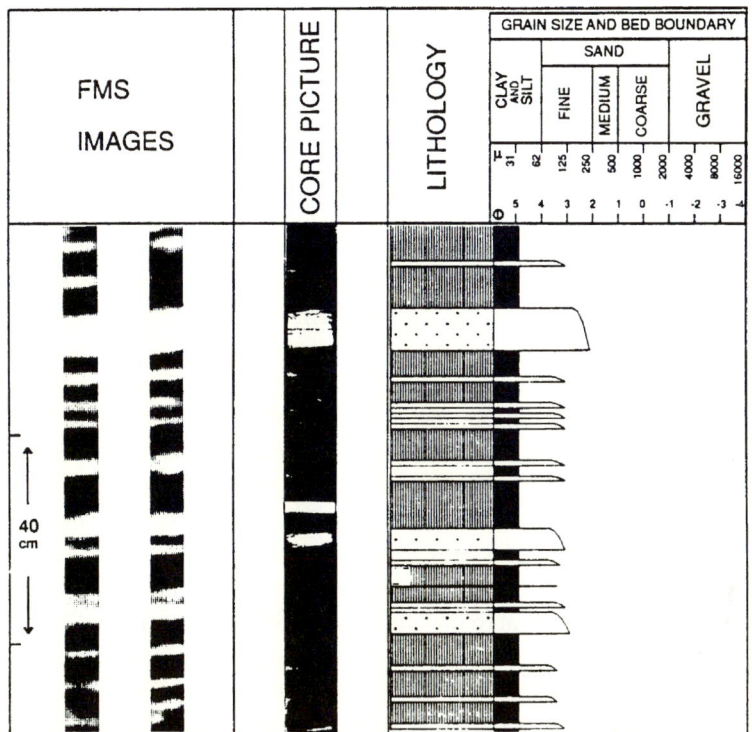

Figure 2.2.21. Comparison of micro-electrical images (left) with core photographs (middle) and core description (right) in a thinly bedded reservoir. Notice scale bar, on the left (from Delhomme et al., 1989).

five centimeters a simple, linear correction could be applied to obtain accurate thickness estimates. Sullivan & Schepel (1995) use such an approach to determine the net sand count in a thinly bedded turbidite reservoir in the Gulf of Mexico. Additionally, they use the resulting bedding zonation to refine the petrophysical evaluation, finding an improvement in the net reservoir volume of 20 to 30 percent when compared to conventional log analysis. This difference is entirely attributed to the limited resolution of standard open-hole well logs. Hackbarth & Tepper (1988) and McGann et al. (1988) report similar studies. Chapter 3.2.2 discusses the evaluation of thinly bedded reservoirs in more detail.

Structural Analysis

In the field of structural analysis, borehole images have also allowed the interpreter to take an important step forward compared to dipmeter analysis (e.g. Gonfalini & Anxionnaz, 1990; Lovell et al., 1995; Onions & Whitworth, 1995). Faults, unconformities, and fractures are easily and unambiguously determined on borehole images, as illustrated by the fault in figure 2.2.22. Large-scale structures, for example folds, still are best analyzed using dipmeter results, but images can be used as guides in the process.

Faults are usually characterized by a steeply dipping fault plane, which in some cases is a complex zone featuring several fault planes with associated fault gouges, breccias and fractures (Koepsell et al., 1989). The most diagnostic feature is the termination of bedding planes on the fault plane (figure 2.2.22). In cases where the throw is small, beds can be correlated from one side to the other of the fault, but this is only possible when the displacement along the fault plane is less than the amplitude of the sinusoid described by the fault plane on the borehole image. Such faults are generally not very significant on a reservoir scale, but they indicate that larger faults dipping in the same direction may be present. Normally, however, the throw of the fault plane cannot be determined from borehole images alone, nor the sense of relative movement along the fault. This means that normal, reverse or strike-slip faults cannot be distinguished based on borehole images alone, although in case of a layer drag along the fault plane, normal and reverse faults may be distinguished from each other. In a normal fault, the layers in its vicinity tend to be tangential to the fault plane, while in a reverse fault, they tend to be deformed in the opposite direction, showing a relative increase of dip relative to the fault plane. Otherwise, such differences should be made from measurements on a larger scale, for example seismic surveys, or based on knowledge of the general tectonic situation in a given area.

Unconformities, i.e. surfaces of erosion or non-deposition, show up on borehole images as an abrupt change in bedding style and, in the case of angular unconformities, in bedding dip. The latter is best seen on dip results derived from the borehole images, but is often also quite obvious from the layer geometry on the images immediately above and below the unconformity. Conglomerates can be telling signs of a disconformity, i.e. an unconformity where the layers above and below are parallel to each other. However, in the absence of a conglomerate, disconformities may be difficult to identify. As in dipmeters, subtle unconformities created by baselaps or toplaps are hard to spot because of the small angular disparities involved, and a careful dip and image analysis coupled with seismic interpretation is required.

Folds can show spectacular patterns on borehole images, particularly when layer dips change rapidly over a short interval. Most folds drilled in the oil industry, however, are relatively large structures which are best analyzed using dipmeter data on a compressed scale (for example 1/200) so that dip changes with depth are better seen. Borehole images, when displayed at such compressed scale, show abundant, seemingly flat-layered bedding due to the strong vertical compression. Large-scale fold analysis, therefore, has to proceed as in dipmeter interpreta-

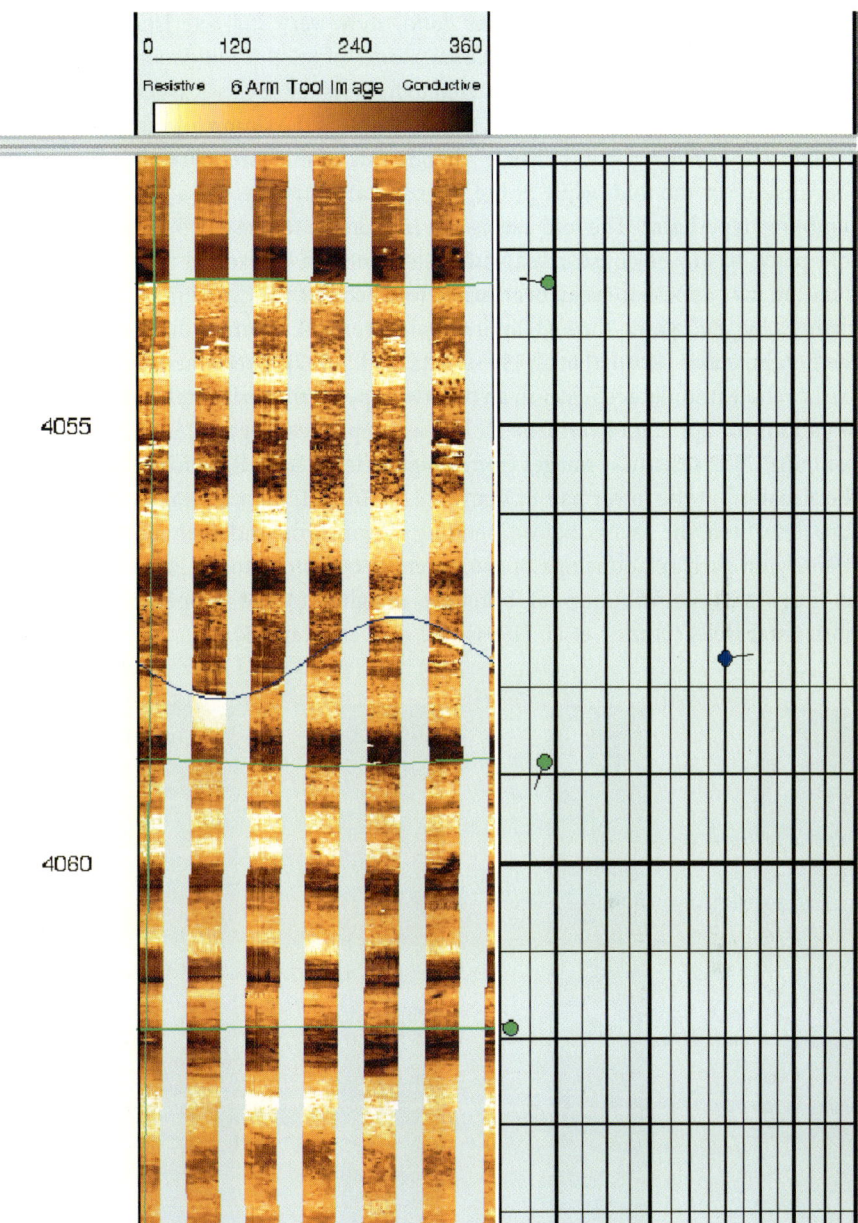

Figure 2.2.22. A fault intersected by the borehole, recorded by the six-arm EMI tool. The throw of this fault cannot be established because layers cannot be correlated from one side of the fault to the other. Fault dips 50° east (blue) while layer dip is below 10°. (green)

tion. Overturned folds, on the other hand, show very characteristic patterns on borehole images. A solid computational model of such a fold is shown in figure 2.2.23, which is a virtual core constructed to help visualize classical geological structures on such cylindrical surfaces. A field example is the slump fold shown in figure 2.2.18e. The characteristic "bullseye" on one side of the borehole wall is located on the "inside" of the fold, while at the opposite azimuth there is a layer pattern resembling hyperbolas. The fold axis is at a right angle to these two features, and the fold plane has to be constructed with a stereonet. In horizontal wells, a layer undulation may appear to be an overturned fold because the dip azimuth of the layers changes by 180°, going through an interval of vertical apparent dip with respect to the borehole axis (Rosthal et al, 1995; Lofts et al., 1997). The same effect happens if the borehole undulates slightly in an otherwise perfectly horizontal layer. These two situations of "apparent overturned folds" are depicted in figure 2.2.24. Figure 2.2.25 contains LWD resistivity images from a highly deviated well, which corresponds to the situation in the lower half of figure 2.2.24. In fact, the well trajectory here first dips less than the layers before bending downwards and then upwards again. Identifying such relationships of the borehole trajectory to the layer geometry is an important contribution of borehole imaging to the proper placement of horizontal wells (Hurley et al. (1994), see also chapter 3.5).

Figure 2.2.23. A solid computational model of an overturned fold as seen on a core surface or a borehole wall. This virtual core has been obtained by folding a layered block and then extracting a cylinder through a Boolean intersection (from Luthi, 1992).

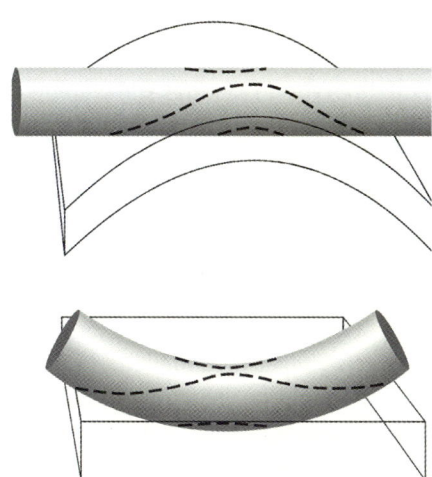

Figure 2.2.24. Schematic depicting an apparent overturned fold in a horizontal well, caused by a slight layer undulation (top), and in a slightly undulating well with horizontal layers (bottom).

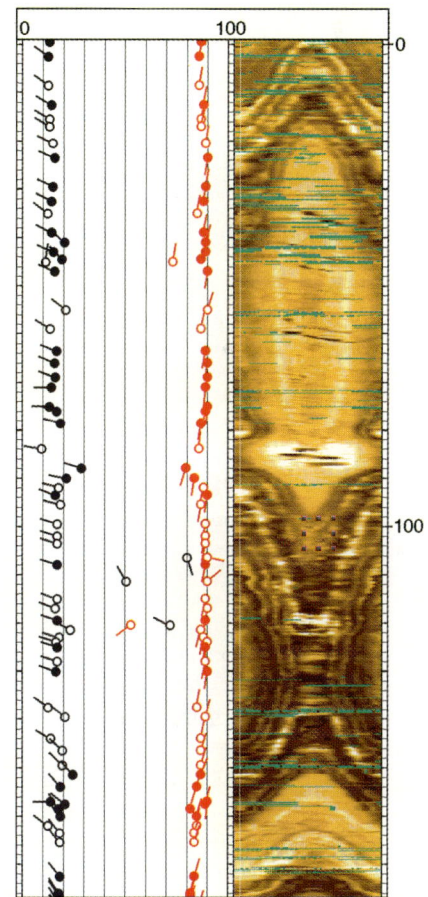

Figure 2.2.25. Images recorded with the RAB tool in a horizontal well. The image is represented with the bottom of the well in the middle, and the top at the edges. The well first dips slightly less than the layers, then bends down and up again with respect to the layers. The apparent dips, shown in red on the arrow plot, illustrate these reversals, while the true dips in black testify to the uniform layer dip of about 15° towards WNW. The lower part corresponds to the sketch at the bottom of figure 2.2.24. Depth is along hole in feet. Green indicates no data.

107

Fracture Analysis

Fracture types. Fractures show up on electrical images because their resistivities often differ greatly from those of the host rock. In the case of open fractures, the drilling fluid typically invades the fracture, creating a thin, conductive sheet. By contrast, a completely cemented fracture forms a thin resistive sheet. In both cases, the current lines are strongly modified compared to a uniform medium.

For a *cemented* dipping fracture, the current lines get squeezed when the tool is just below and in front of the fracture, giving rise to an artificially high resistivity. Conversely, as soon as the tool has passed the fracture, the current lines diverge stronger than they would otherwise, and the apparent resistivity is lower than it should be. The fracture itself is too thin to be measured directly by the tools. The overall effect is thus a change in resistivity from one side to the other of the fracture, which gives rise to a halo-like effect on the images. This effect is mirrored on the opposite side of the borehole, where almost the exact inverse electrical distortions occur (figure 2.2.26).

Open fractures, when filled with conductive mud, are easily identified on electrical borehole images as conductive features cross-cutting the bedding (figure 2.2.27), although in rare cases they may be layer-parallel.

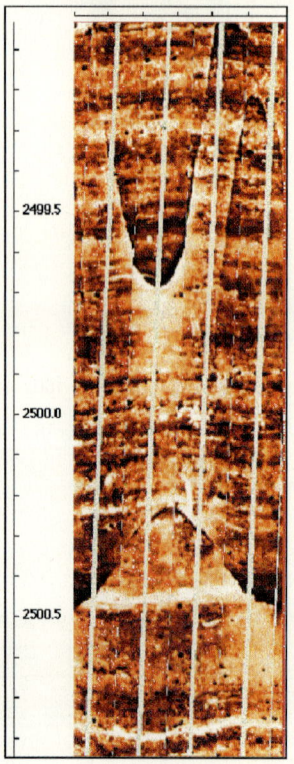

Figure 2.2.26. Micro-electrical image of a cemented fracture at the top, showing characteristic halo effects due to the insulating thin sheet formed by the fracture cement. The lower feature shows the same halo effect, but layers are displaced, suggesting a fault. Depth is in meters.

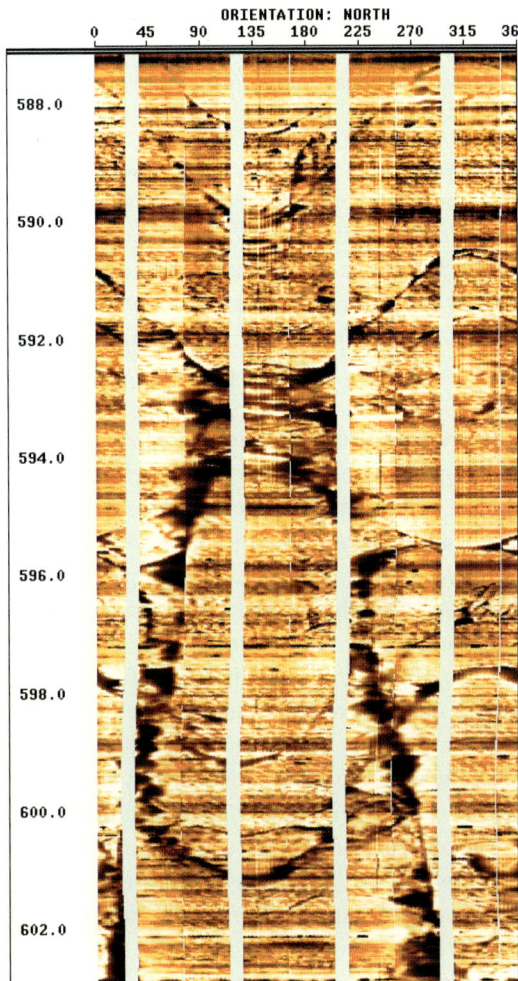

Figure 2.2.27. FMI images over a layered carbonate sequence with at least two distinct planar fracture sets, on dipping towards SSW, the other dipping steeply towards NNE.

Based on their image morphology, several types of fractures can be distinguished:

- Highly irregular fracture networks associated with brecciation. These can be caused either during deposition, as collapse breccias during karstification, or during tectonic movements in fault zones.

- Single, usually planar fractures, either caused by shearing or tension during tectonic movements. They can be relatively extensive or confined to individual beds.
- Solution-enlarged fractures of either of the previous types, often locally enlarged into vuggy to cavernous pores.
- Thin localized joints, terminating within the borehole. Some may be natural fractures caused by tectonism, others induced by thermal stressing or borehole-related stress amplification in the rock.

- Induced fractures, caused entirely by the drilling of the borehole. Two principal types of induced fractures occur, tensile fractures in the direction of maximum stress, and shear joints in the direction of minimum stress leading to borehole breakouts.

The case studies in chapter 3.3 illustrate some of these fracture types. Lehne (1990) applied a morphological classification from electrical images and other logs to characterize the fracture system in the chalk of the Tommeliten Gamma field in the North Sea. Other field examples are reported in Heliot et al. (1990), Hornby & Luthi (1992), Mendoza (1996), Haller & Porturas (1998) and Ayadi et al., (1998). Standen (1991) offers guidelines for the analysis of fractures from electrical images, largely based on experience in Middle Eastern reservoirs.

Fracture aperture calculation. The calculation of fracture aperture from electrical images is based on forward modeling of the tool response to a fracture, after which a computational inversion of is performed to obtain the fracture aperture. This is an important quantity because it controls the flow from reservoirs where production is mostly from fractures. In an ideal situation, the permeability should scale with the square of the fracture aperture, while the flow scales with the cube of the aperture. Thus, flows from two planar, infinite fractures of 10 to 100 microns width will differ by a factor of 1000, if all other factors are the same. In reality, fractures are rugose, partially cemented and either laterally confined or connected with other fractures, all of which may significantly affect the flow. However, it is still very useful to have an idea of how wide a fracture is, even if it is only relative to other fractures.

The presence of a fracture in the borehole results in a distortion of the current lines towards the fracture, even before the electrical sensor is in front of it. Much of the emitted current flows in the fracture itself and with increasing distance away from the wellbore gradually radiates outwards into the surrounding rock (figure 2.2.8 top). As the sensor approaches the fracture, the total emitted current increases until it peaks when the sensor is exactly in front of it (figure 2.2.28 bottom). Luthi & Souhaité (1990) and Faivre (1993) demonstrated for the FMI and ARI respectively that the sum of the extra current A emitted by the tool due to the fracture is proportional to the fracture aperture W, following a relationship of the form

$$W = a \cdot A \cdot R_m^b \cdot R_{xo}^{1-b} \qquad (2.2.2)$$

Here R_m and R_{xo} are the mud and shallow rock resistivity, a is a tool constant, and b is a value slightly smaller than 1. Thus, the fracture width can be calculated as the extra current integrated along the trace of an electrode passing the fracture. This inversion is mostly sensitive to an accurate calculation of A and to the mud resistivity. Fracture dip and tool standoff from the borehole wall do not affect the value of A significantly and hence do not have to be taken

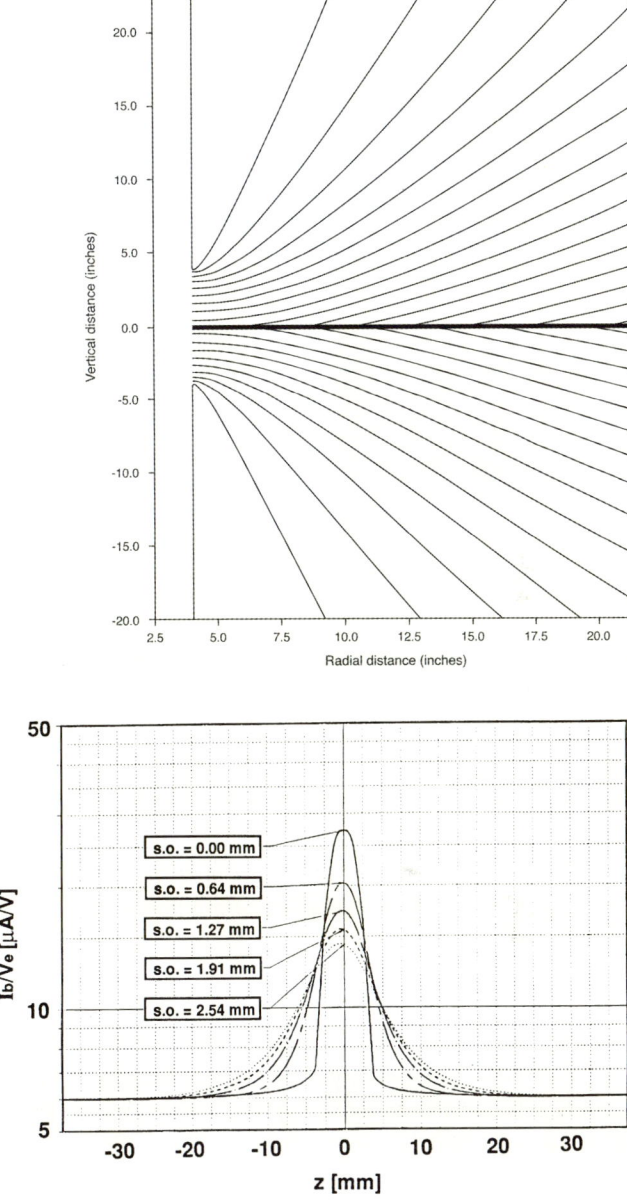

Figure 2.2.28. Top: Current lines of from a pad in front of a fracture, calculated with a finite element modeling code. Rock resistivity is 10 Ωm, mud resistivity 0.1 Ωm, pad is 20 cm long, electrode is 5 mm in diameter, and fracture is horizontal with an aperture of 2.5 mm. Radial and vertical distances in inches (courtesy J. Lovell). Bottom: Response of a FMI electrode when passing a fracture, as a function of standoff, i.e. the distance to the borehole wall. Notice how the current increases before the button is in physical contact with the fracture. The total area of the current is the same for all standoffs (from Luthi & Souhaité, 1990).

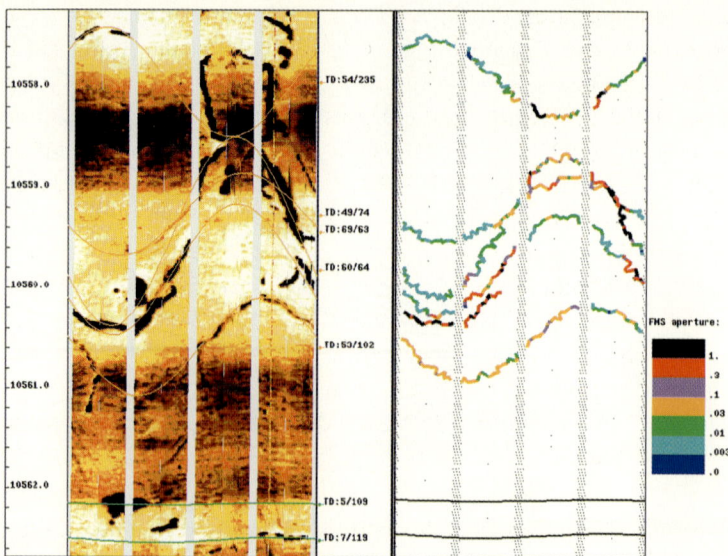

Figure 2.2.29. Fracture apertures from FMI images. Sinuosoids traced at left are used to calculate dips and azimuths. The traces on the right follow the maximum current values with their colors indicating apertures in millimeters. Notice that the fracture dipping at 60°/63° shows apertures between 100 microns and one millimeter, while all others are less than 30 microns wide. Depth is in feet.

into account in the inversion. Since the calculation can be done on any of the sensors crossing the fracture, aperture values along the entire fracture trace can be obtained. Figure 2.2.29 shows such a calculation in a case where the fracture has first been interactively selected by the geologist, then automatically traced by a program and finally analyzed for apertures along its trace. Fracture apertures between 1 micron and about one millimeter can be quantified with micro-electrical imaging using the FMI. This range can is extended upwards with the ARI resistivity imaging tool to about ten millimeters.

It may seem strange, at first, that such tiny features can be quantified with a sensor of much larger size and a considerable area of investigation. It is possible because of the large resistivity contrast between fracture and matrix, which often amounts to several orders of magnitude. This causes a measurable increase in button current even for a very thin, open fracture. Its inversion into fracture apertures is possible because of a simple relationship between a small number of relevant parameters. The detection threshold for a fracture is estimated to be around one micron, perhaps even less, while the resolution threshold, i.e. the distance at which two fractures can be separated from each other, is much larger and corresponds to the values mentioned in table 2.2.1 for the various tools. However, it is important to keep in mind that fractures can be modified in the vicinity of the borehole wall due to mud and drill bit influence, and often a local increase in fracture apertures can be expected.

In addition to the apertures, the dip and azimuth of a fracture can be measured from borehole images. This results in a fairly complete characterization of the fractures *in situ*. Cheung & Heliot (1990) describe a workstation-based implementation of the procedure, while Dennis et al. (1987), Laubach et al. (1988), Casarta et al. (1989), Plumb & Luthi (1989), Mercadier & Mäkel (1991), Mercadier & Livera (1993) and Sullivan & Schepel (1995) all describe case studies of fractured reservoirs where electrical borehole images have been successfully utilized. Similar fracture studies in research wells are found in Pezard & Luthi (1988) and MacLeod et al. (1992). Hornby & Luthi (1992) integrate microelectrical images with deep resistivity and Stoneley wave reflection measurements for a more complete analysis of fractured reservoirs. Kato et al. (1995) combine FMI logs and cores for analyzing a geothermal well in Japan, while Genter et al. (1991) combine electrical and acoustic borehole images in a geothermal well in France.

Induced fractures. Several factors can induce fracturing of the rocks in a borehole, with induced hydraulic fractures and breakouts being most commonly observed. A good summary of the borehole failure mechanisms leading to induced fractures can be found in Aadnøy and Bell (1998), who also illustrate the typical fracture morphologies seen on electrical borehole images.

When a borehole is drilled, the stress regime is strongly modified due to the removal of rock material. Kirsch (1898, in Bell, 1990) have formulated the basic relationships when a hole is drilled into a plate under stress. The principal far-field stress gets concentrated as tangential stresses in the two sections of the hole where the wall is parallel to the maximum stress direction, while at 90° to these directions the tangential stresses can be substantially reduced. The larger the difference in far-field stress magnitudes, the stronger these effects will be. This situation is depicted in figure 2.2.30, with the resulting tangential stresses indicated as contour lines. An isotropic pressure contribution from the mud in the borehole is also taken into account. Similar calculations can be done for radial stresses and shear stresses using the equations given in Bell (1990). If tensile failure occurs, it happens in the section where tangential stresses are minimal, i.e. in the azimuth of maximum horizontal stress (Dart & Zoback, 1989). Alternatively, there can be shear failure in the section of the borehole where tangential stresses are maximal, i.e. in the azimuth of minimum horizontal stress. According to Zoback et al. (1985), one can predict the region on the wellbore where breakout failure will occur if the horizontal stresses, the shear strength and internal friction of the rock are known. The first mode of failure is loosely referred to an *induced or hydraulic fracturing*, and is quite commonly observed on electrical images in many boreholes, often without the drillers and engineers being aware of its existence. The second mode of failure, termed *breakouts*, results in a spalling or caving of the borehole until the extra stress concentration is sufficiently released and the borehole remains stable (Bell, 1990).

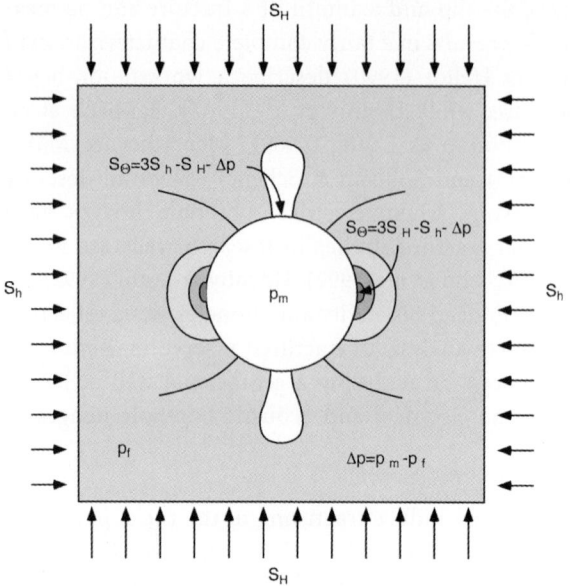

Figure 2.2.30. Stress modification around a borehole in a rock with principal horizontal stress S_H, minimum horizontal stress S_h, pore fluid pressure p_f and with mud pressure p_m in the borehole. Contours of tangential stresses are shown in a schematic manner with darker colors representing higher stresses. Notice the high stress concentration in immediate vicinity of the borehole in the direction of minimum stress, and the stress reduction at a right angle to it. Depending on rock strength and fluid pressure, borehole failure may occur in either of these regions.

Figure 2.2.31 shows *induced fractures* formed after a small hydraulic fracturing operation (a "mini-frac") on micro-electrical images recorded before and after the procedure. The two induced fractures are almost exactly vertical, at 180° from each other and indicating a principal horizontal stress direction from east to west. Fracture aperture calculations on such fractures are not very meaningful, since the fracture is tapering out towards its limit. However, the strike of induced fractures is important to know as it will be the same for large-scale hydraulic fracturing, and it will therefore dictate the drainage direction within the reservoir. Among the principal criteria to diagnose induced fracturing is the shape, which is often quite irregular, branched and discontinuous, and the azimuth, which tends to be very constant and in the direction of principal horizontal stress. In deviated holes, the relationship of the locally modified stress regime with the far-field regional stress is more complicated. One observes very often an induced fracture pattern in which short, branched fractures are seen on opposite sides of the hole, ending abruptly with a small downward hook in the direction of the far-field regional stress, or at a bed boundary. Figure 2.2.32 shows an example where natural fractures may have been additionally enlarged and extended in the direction of the borehole when the hole was drilled. As a rule of thumb, an induced fracture always shows a certain geometric relationship to the borehole, while a natural fracture does not and often cuts across the entire borehole without changing its appearance.

Breakouts are a geometric modification of the borehole and grow until stresses are sufficiently reduced and fracturing ceases. Microelectrical imaging methods

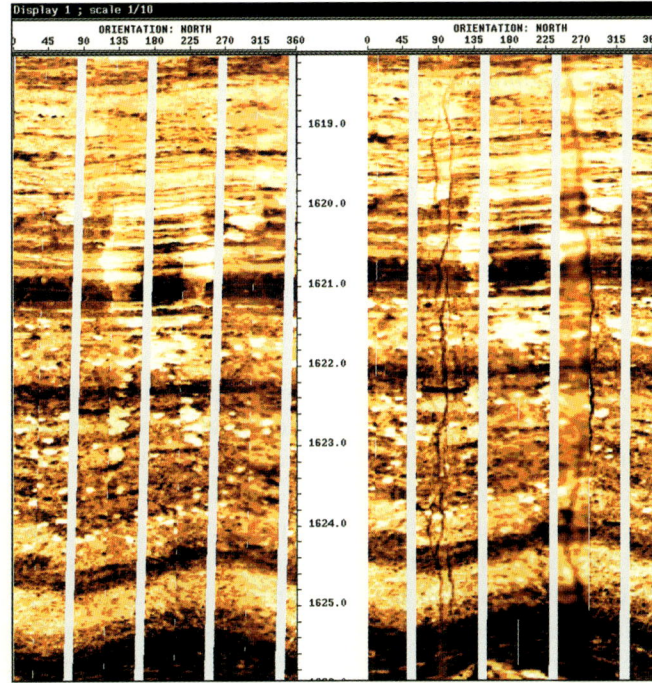

Figure 2.2.31. Borehole images in a clastic sequence before (left) and after (right) hydraulic fracturing. The failure occurs in the direction of maximum stress, which in this case is E-W.

Figure 2.2.32. Natural fractures in a deviated well, showing deflections in the direction of the borehole axis. This may indicate a fracture enlargement or modification during the drilling process.

115

are not very sensitive to them, since the pads will just follow the caved-out section, but the caliper readings can be characteristic. Resistivity imaging in wireline mode corrects for the sensor standoff, and therefore no large effect is seen either. In LWD mode, breakouts may be present when the images are recorded, but they take time to develop to their full size (Figure 3.5.11). Ultrasonic imaging, on the other hand, is very sensitive to the geometry of the borehole, and numerous examples of borehole breakouts have been documented in the literature (see chapter 2.3).

Heterogeneities

A principal feature of borehole imaging is its two-dimensionality, distinguishing it from conventional, one-dimensional logging. Borehole imaging can therefore capture lateral variations in the rock, usually referred to as heterogeneities. Here we consider cases where there are large electrical differences within a bed caused by heterogeneous porosity and saturation distributions. Badr & Ayoub (1989) and Nurmi et al. (1990) observed on microelectrical borehole images that many Middle Eastern carbonate reservoir are very heterogeneous despite their reputation of being rather, homogeneous packets. Roca-Ramisa (1994) found the same to be true in Far Eastern reservoirs. These heterogeneities are largely attributed to the distribution of porosity, which can range from uniform to layered,

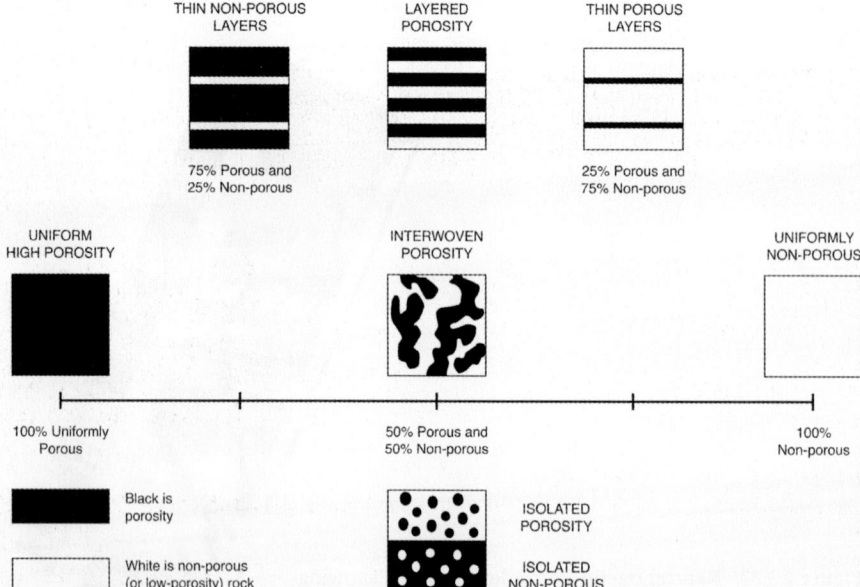

Figure 2.2.33. Geometric classification of heterogeneities according to fabric in carbonate rocks (from Nurmi et al., 1990)

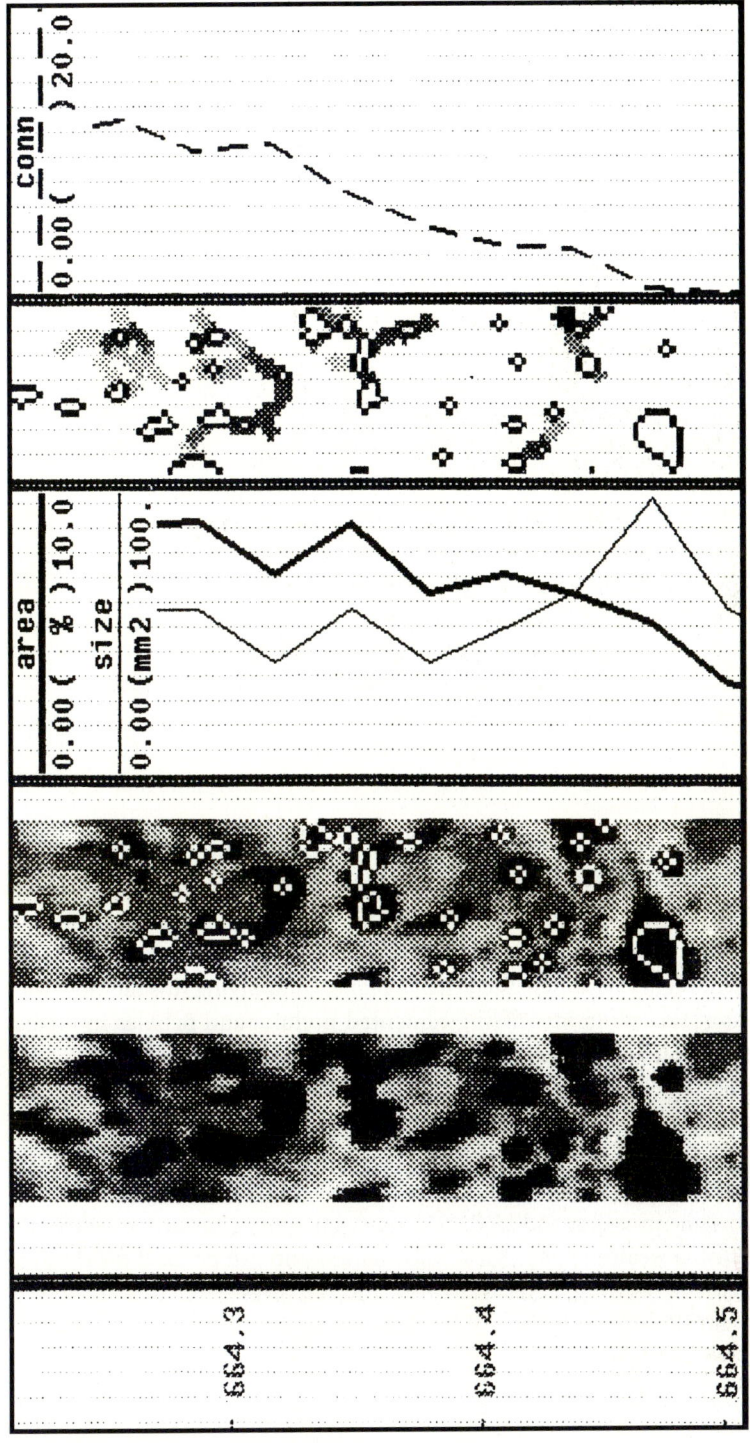

Figure 2.2.34. Heterogeneity analysis in a vuggy limestone. From left: Raw micro-electrical image; vug contours; vug area and median vug size calculated over a 0.5 ft sliding window; vug contours and conductive paths; vug connectedness (from Delhomme, 1992).

patchy, interwoven, and isolated. They propose a classification to distinguish these various types of heterogeneities, which is reproduced in figure 2.2.33. In order to quantify these heterogeneities, several automated methods have been developed. Delhomme (1992) uses methods from mathematical morphology to extract meaningful measures of heterogeneities. He first marks and outlines objects of interests using the marker-controlled segmentation method proposed by Meyer & Beucher (1990) and applied to electrical borehole images by Rivest et al. (1992). These can be conductive or resistive spots, lenses, or lines. Delhomme (1992) then computes object attributes such as area percentage, size, contrast, and orientation. In a third step, he calculates statistics and summarizes the results as logs. This can be applied to the extraction vugs represented on the images as conductive spots, and the summary log then records the percentage of the image containing such features (a property related to the porosity contribution of the vugs) and their average sizes. Figure 2.2.34 shows the results of such a calculation, including the connectedness of the vugs. The latter is derived from the conductive areas connecting the vug outlines and should be related to the cementation exponent m (equation 1.3.3). An estimation of the cementation exponent can be obtained from simultaneous analysis of a shallow resistivity and an electromagnetic propagation measurement, in which the first is sensitive to the tortuosity of the electrical path, while the second is not (Focke & Munn, 1987). Delhomme (1992) found that a high value of m indeed corresponds to a low connectedness of the vugs, while his vug porosity calculation tracks the standard total porosity from open hole logs very well. Similar methods of quantifying heterogeneities are discussed in Sovich & Newberry (1993), and Standen et al. (1993).

Heterogeneity analysis can be performed in a similar fashion for resistive features on the borehole images. Hall et al. (1996) propose applying the Hough transform not only to find sinusoids, but also to identify ellipsoidal objects. The procedure is similar but computationally more demanding because an ellipsoid is much higher parameterized than a plane. By applying this method, they found a certain success in identifying cemented nodules and pebbles in images from a fluvial sequence in the Nile delta.

2.2.6. Summary

Electrical borehole images are available in three modes, micro-electrical imaging, resistivity wireline imaging, and LWD resistivity imaging. They are robust methods that furnish a wealth of geological information and can be used for the analysis of structures, bedding, heterogeneities, and fractures. They can only be recorded in water-based mud, but at high logging speeds.

References

Aadnøy BS, Bell JS (1998) Classification of drilling induced fractures and their relationship to in-situ stress directions. The Log Analyst, 39, n. 6: 27–42.

Antoine JN, Delhomme JP (1990) A method to derive dips from bed boundaries in borehole images. 65th Ann Conf Soc Petrol Eng: Paper 20540.

Anxionnaz H, Delhomme JP (1998) Near-wellbore reconstruction of sedimentary bodies from borehole electrical images. Trans 39th Symp Soc Prof Well Log Analysts: Paper N.

Ayadi M, Pezard PA, Bronner G, Tartarotti P, Laverne C (1998) Multi-scalar structure at DSDP/ODP Site 504, Costa Rica Rift, III: faulting and fluid circulation. Constraints from integration of FMS images, geophysical logs and core data. In: Harvey PK, Lovell MA (eds) Core-Log Integration. Geol Soc Spec Publ 136, pp 311–326.

Badr AR, Ayoub MR (1989) Study of complex carbonate reservoir using the formation MicroScanner (FMS) tool. Middle East Oil Tech Conf and Exhibition, Soc Petrol Eng: Paper 17977.

Bell, JS (1990) Investigating stress regimes in sedimentary basins using information from oil industry wireline logs and drilling records. In: Hurst A, Lovell MA, Morton A (eds) Geological Applications of Wireline Logs. Geol Soc Spec Publ 48, pp 305–325.

Bonner S, Bagersh A, Clark B, Dajee M, Dennison M, Fredette M, Grogan O, Hall JS, Jundt J, Kwok E, Lovell JR, Rosthal RA, Allen DA (1994) A new generation of electrode resistivity measurements for formation evaluation while drilling. Trans 35th Symp Soc Prof Well Log Analysts: Paper OO.

Boyeldieu C, Jeffreys P (1988) Formation microscanner – new developments. Trans 11th European Form Eval Symp Soc Prof Well Log Analysts: Paper X.

Bourke L (1989) Recognizing artifact images of the Formation Microscanner. Trans 30th Symp Soc Prof Well Log Analysts: Paper WW.

Bourke L (1992) Sedimentological borehole image analysis in clastic rocks: a systematic approach to interpretation. In: Hurst A, Griffiths CM, Worthington PF (eds) Geological Applications of Wireline Logs II. Geol Soc Spec Publ 65, pp 31–42.

Brookfield ME (1977) The origin of bounding surfaces in ancient aeolian sandstones. Sedimentology 24: 303–332.

Casarta LJ, McNaughton DA, Bornemann E, Bettis FE (1989) Fracture identification using a new wellbore imaging device in the Lisburne carbonate, Prudhoe Bay, Alaska. Trans 30th Symp Soc Prof Well Log Analysts: Paper XX.

Chan DSK (1984) Accurate depth determination in well logging. IEEE Trans Acoust Speech and Signal Proc, ASSP–32: 42–28.

Chauvel Y, Seeburger D, Orjuela A (1984) Application of the SHDT Stratigraphic Dipmeter to the Study of Depositional Environments. Trans 25th Symp Soc Prof Well Log Analysts: Paper G.

Cheung PSY, Heliot D (1990) Workstation-based fracture evaluation using borehole images and wireline logs. 65th Ann Conf Soc Petrol Eng: Paper 20573.

Cheung PS (1999) Microresistivity and ultrasonic imagers: tool operations and processing principles with reference to commonly encountered image artefacts. In: Lovell MA, Williamson G, Harvey PKJ (eds) Borehole images: applications and case histories. Geol Soc Spec Publ 159, pp. 45.57.

Collinson JD, Thompson DB (1982) Sedimentary structures. George Allen & Unwin, London.

Dart RL, Zoback ML (1989) Wellbore breakout stress analysis within the central and eastern continental United States. The Log Analyst 30: 12–25.

Davies DH, Faivre O, Gounot MT, Seeman B, Trouiller JC, Benimeli D, Ferreira AE, Pittman DJ, Smits JW, Randrianavony M, Anderson BI, Lovell J (1992) Azimuthal resistivity imaging: a new generation laterolog. 67th Ann Conf Soc Petrol Eng: Paper 24676.

Delhomme, JP (1992) A quantitative characterization of formation heterogeneities based on borehole image analysis. Trans 33rd Symp Soc Prof Well Log Analysts: Paper T.

Dennis B, Standen E, Georgi DT, Callow GO (1987) Fracture identification and productivity predictions in a carbonate reef complex. 62nd Ann Conf Soc Petrol Eng: Paper 16808.

Ekstrom MP, Dahan CA, Chen MY, Lloyd PM, Rossi DJ (1987) Formation imaging with microelectrical scanning arrays. Log Analyst 28: 294–306.

Etchecopar A, Bonnetain JL (1992) Cross sections from dipmeter data. Am Assoc Petrol Geol Bull 76: 621–637.

Faivre O, Catala G (1995) Dip estimation from azimuthal laterolog tools. Trans 36th Symp Soc Prof Well Log Analysts: Paper CC.

Focke JW, Munn D (1987) Cementation exponents in Middle Eastern carbonate reservoirs. Soc Petr Eng, Formation Evaluation: 155–167.

Genter A, Martin P, Montaggioni P (1991) Application of FMS and BHTV tools for evaluation of natural fractures in the Soultz geothermal borehole GPK1. Geothermal Science and Technology 4, 189–214.

Gonfalini M, Anxionnaz H (1991) A complete use of structural information from borehole imaging techniques (FMS); a case history for a deep carbonate reservoir. Trans 31st Symp Soc Prof Well Log Analysts: Paper J.

Grace LM, Luthi SM, Pirie RG (1986) Stratigraphic interpretation using formation imaging and dipmeter analysis. 61St Ann Conf Soc Petrol Eng: Paper 15611.

Hackbarth CJ, Tepper BJ (1988) Examination of BHTV, FMS and SHDT images in very thinly bedded sands and shales. 63rd Ann Conf Soc Petrol Eng: Paper 18118.

Hall J, Ponzi M, Gonfalini M, Maletti G (1996) Automatic extraction and characterization of geological features and textures from borehole images and core photographs. Trans 37th Symp Soc Prof Well Log Analysts: Paper CCC.

Haller D, Porturas F (1998) How to characterize fractures in reservoirs using borehole and core images: case studies. In: Harvey PK, Lovell MA (eds) Core-Log Integration. Geol Soc Spec Publ 136, pp 249–259.

Harker SD, McGann GJ, Bourke LT, Adams JT (1990) Methodology of formation microscanner tool image interpretation in Claymore and Scapa fields (North Sea). In: Hurst A, Lovell MA, Morton A (eds) Geological Applications of Wireline Logs. Geol Soc Spec Publ 48, pp 11–25.

Harvey PK, Lovell MA (1998) Core-Log Integration. Geol Soc Spec Publ 136.

Heliot D, Etchecopar A, Cheung P (1990) New developments in fracture characterization from logs. In: Maury V, Fourmaintraux D(eds) Rocks at Great Depth. Balkema, Rotterdam, 1471–1478.

Hornby BE, Luthi SM (1992) An integrated interpretation of fracture apertures computed from electrical borehole scans and reflected Stoneley waves. In: Hurst A, Griffiths CM, Worthington PF (eds) Geological Applications of Wireline Logs II. Geol Soc Spec Publ 65, pp 185–198.

Hubel DH (1988) Eye, brain and vision. Scientific American Library.

Hurley N, Thorn DR, Carlson JI, Eichelberger SLW (1994) Using borehole images for target-zone evaluation in horizontal wells. Am Ass Petrol Geol Bull 78: 238–246.

Kass M, Witkin A (1987) Analyzing oriented patterns. Computer Vision Graphical Models and Image Processing: 37: 362–285.

Kato O, Doi N, Akazawa T, Sakagawa Y, Yagi Y, Muraoka H (1995) Characteristics of fractures based on FMI logs and cores in well WD-1 in the Kakkonda geothermal field, Japan. Geothermal Resources Council Transactions 19: 317–322.

Kirsch G (1898) Die Theorie der Elastizität und die Bedürfnisse der Festigkeitslehre. Zeitschrift des Vereines Deutscher Ingenieure: 42, 797–807.

Koepsell RJ, Jenson FE, Langley RL (1989) Gulf Coast fault orientation determined by formation imaging techniques. Trans 30th Symp Soc Prof Well Log Analysts: Paper VV.

Kraaijveld MA, Epping WJM (2000) Harnessing advanced image analysis technology for quantitative core and borehole image interpretation. Trans 41st Symp Soc Prof Well Log Analysts: Paper VV.

Laubach SE, Baumgardner RW, Monson ER, Hunt E, Meador KJ (1988) Fracture detection in low-permeability reservoir sandstone – comparison of BHTV and FMS logs to core. 63rd Ann Conf Soc Petrol Eng: Paper 18119.

Lehne KA (1990) Fracture detection from logs in the North Sea chalk. In: Hurst A, Lovell MA, Morton A (eds) Geological Applications of Wireline Logs. Geol Soc Spec Publ 48, pp 263–271.

Lloyd PM, Dahan C, Hutin R (1986) Formation imaging with microelectrical scanning arrays – a new generation of high resolution dipmeter tool. Trans 10th Europ Form Eval Symp Soc Prof Well Log Analysts: Paper L.

Lofts JC, Bedford J, Boulton H, van Doorn J, van der Groen P, Jeffreys P (1997) Feature recognition and the interpretation of images acquired from horizontal wellbores. In: Lovell MA & Harvey PK (eds) Developments in Petrophysics. Geol Soc Spec Publ 122, pp 345–366.

Lofts JC, Bourke LT (1999) The recognition of artefacts from acoustic and resistivity borehole imaging devices. In: Lovell MA, Williamson G, Harvey PKJ (eds) Borehole images: applications and case histories. Geol Soc Spec Publ 159, pp. 59–76.

Lovell MA, Williamson G, Harvey PKJ (1999) Borehole images: applications and case histories. Geol Soc Spec Publ 159.

Lovell JR, Rosthal RA, Arceneaux CL, Young RA, Buffington L (1995) Structural interpretation of resistivity-at-the-bit images. Trans 36th Symp Soc Prof Well Log Analysts: Paper TT.

Luthi SM, Banavar JR (1988) Application of borehole images to three-dimensional geometric modeling of eolian sandstone reservoirs, Permian Rotliegende, North Sea. Bull Am Ass Petrol Geol 72: 1074–1089.

Luthi SM (1990) Sedimentary structures of clastic rocks identified from electrical borehole images. In: Hurst A, Lovell MA, Morton A (eds) Geological Applications of Wireline Logs. Geol Soc Spec Publ 48: 3–10.

Luthi SM, Souhaité P (1990) Fracture apertures from electrical borehole scans. Geophysics 55: 821–833.

Luthi SM (1992) Solid computational models of geological structures in boreholes. In: Pflug R, Harbaugh JF (eds) Computational graphics in geology: Three-dimensional modeling of geologic structures and simulating geologic processes. Lecture Notes in Earth Science, Springer, pp 51–61.

Luthi SM (1994) Textural segmentation of digital rock images into bedding units using texture energy and cluster labels. Math Geol 26: 181–196.

Mc Gann GJ, Riches HA, Renoult DC (1988) Formation evaluation in a thinly bedded reservoir – a case history – Scap field, North Sea. Trans 29th Symp Soc Prof Well Log Analysts: Paper V.

MacLeod CJ, Parson LM, Sager WW & ODP Scientific Party (1992) Identification of tectonic rotations in boreholes by the integration of core information with Formation MicroScanner and Borehole Televiewer images. In: Hurst A, Griffiths CM, Worthington PF (eds) Geological Applications of Wireline Logs II. Geol Soc Spec Publ 65, pp 235–246.

Mendoza JR (1996) The contribution of wellbore imaging to interval selection in naturally fractured reservoirs. Soc Petrol Eng Int Petrol Conf: Paper 35292.

Mercadier CG, Mäkel GH (1991) Fracture patterns of Natih formation outcrops and their implications for the reservoir modelling of the Natih field. 66th Ann Conf Soc Petrol Eng: Paper 21377.

Mercadier CGL, Livera SE (1993) Application of the Formation MicroScanner to modelling of paleozoic reservoirs in Oman. In: Flint SS, Bryant ID (eds) The geological modelling of hydrocarbon reservoirs and outcrop analogues. Spec Publ Int Ass Sedim 15: 125–142.

Meyer F, Beucher S (1990) Morphological segmentation. J Vis Comm and Image Representation 1: 21–46.

Mosnier J (1982) Détection électrique de fractures naturelles et artificielles dans un forage. Annales de géophysique 38: 637–640.

Mosnier J (1987) Méthode électrique de détection des fractures sur parois d'un forage. Bull Soc Geol France 3: 1049–1054.

Nurmi R, Charara M, Waterhouse M, Park R (1990) Heterogeneities in carbonate reservoirs: detection and analysis using borehole electrical imagery. In: Hurst A, Lovell MA, Morton A (eds) Geological Applications of Wireline Logs. Geol Soc Spec Publ 48, pp 95–111.

Onions KR, Whitworth KR (1995) Applications of electrical borehole imaging to mining design. Scientific Drilling 5: 69–76.

Pezard PA, Luthi SM (1988) Borehole electrical images of the Cajon Pass scientific drillhole, California: Fracture identification and tectonic implications. Geophys Res Let 15: 1017–1020.

Plumb RA, Luthi SM (1989) Analysis of borehole images and their application to geologic modeling of an eolian reservoir. Soc Petr Eng Form Eval: 505–514.

Prensky S (1999) Advances in borehole imaging technology and applications. In: Lovell MA, Williamson G, Harvey PKJ (eds) Borehole images: applications and case histories. Geol Soc Spec Publ 159, pp. 1–43.

Prilliman J, Bean CL, Hashem M, Bratton T, Fredette MA, Lovel JR (1997) A comparison of wireline and LWD resistivity images in the Gulf of Mexico. Trans 38th Symp Soc Prof Well Log Analysts: Paper DDD.

Rivest JF, Beucher S, Delhomme JP (1992) Marker-controlled segmentation: an application to electrical borehole imaging. J Electron Imaging 1(2): 136–142.

Roca-Ramisa L (1994) Carbonate characterization and classification from in-situ wellbore images. 23rd Ann Convention Indonesia Petrol Ass 1: 181–188.

Rosthal RA, Young RA, Lovell JR, Buffington L, Arceneaux CL (1995) Formation evaluation and geological interpretation from the resistivity-at-the-bit tool. 70th Ann Conf Soc Petrol Eng: Paper 30550.

Safinya KA, Le Lan P, Villegas M, Cheung PS (1991) Improved formation imaging with extended microelectrcial arrays. 66th Ann Conf Soc Petrol Eng: Paper 22726.

Seiler D, King G, Eubanks D (1994) Field test results of a six arm microresistivity borehole imaging tool. Trans 35th Symp Soc Prof Well Log Analysts: Paper UUU.

Slatt RM, Jordan DM, Davis RJ (1994) Interpreting Formation Microscanner log images of Gulf of Mexico Pliocene turbidites by comparison with Pennsylvanian turbidite outcrops, Arkansas. In: Weimer P, Bouma AH, Perkins BF (eds) Submarine Fans and turbidite systems. Proc 15th Gulf Coast Section Soc Sedimentary Geol: 335–348.

Smits JW, Benimeli D, Dubourg I, Faivre O, Hoyle D, Tourillon V, Trouiller JC, Anderson BI (1995) High resolution from a new laterolog with azimuthal imaging. 70th Ann Conf Soc Petrol Eng: Paper 30584.

Sovich JP, Newberry B (1993) Quantitative applications of borehole imaging. Trans 34th Symp Soc Prof Well Log Analysts: Paper FFF.

Standen E (1991) Tips for analyzing fractures on electrical wellbore images. World Oil 212: 99–117.

Standen E, Nurmi R, El-Wazeer F, Ozkanali M (1993) Quantitative applications of wellbore images to reservoir analysis. 34th Symp Soc Prof Well Log Analysts: Paper EEE.

Straub A, Sano K, Imamura S, Ohhashi T (1995) Development of a borehole imaging system, the ELIAS probe. 1st Ann Symp Soc Prof Well Log Analysts, Japan Chapter: Paper L.

Suau J, Grimaldi P, Poupon A, Souhaité P (1972) The dual laterolog – Rxo tool. 47th Ann Conf Soc Petrol Eng: Paper 6823.

Sullivan KB, Schepel KJ (1995) Borehole image logs: Applications in fractured and thinly bedded reservoirs. Trans 36th Symp Soc Prof Well Log Analysts: Paper T.

Thompson LB, Snedden JW (1996) Geology and reservoir description of 1Y1 reservoir, Oso field, Nigeria, using FMS and dipmeter. In: Pacht JA, Sheriff RE, Perkins BF (eds) Stratigraphic analysis utilizing advanced borehole technology for petroleum exploration and production. Proc 17th Gulf Coast Section Soc Sedimentary Geol: 315–328.

Torres D, Strickland R, Gianzero M (1990) A new approach to determining dip and strike using borehole images. Trans 31st Symp Soc Prof Well Log Analysts: Paper K.

Trouiller, JC, Delhomme, JP, Carlin S, Anxionnaz H (1989) Thin-bed reservoir analysis from electrical borehole images. 64th Ann Conf Soc Petrol Eng: Paper 19578.

Wong SA, Startzman RA, Kuo TB (1989) Enhancing borehole image data on a high-resolution PC. Petroleum Computer Conference, Soc Petr Eng: Paper 19124.

Ye SJ, Rabiller P, Keskes N (1997) Automatic high resolution sedimentary dip detection on borehole imagery. Trans 38th Symp Soc Prof Well Log Analysts: Paper O.

Zemanek J, Glenn EE, Norton LJ, Caldwell RL (1970) Formation evaluation by inspection with the borehole televiewer. Geophysics 35: 254–269.

Zoback, MD, Moss D, Mastin L, Anderson RN (1985) Wellbore breakouts and in-situ stress. J Geophys Res: 90: 5523–5530.

2.3 Acoustic Borehole Imaging

2.3.1 Tool Principles

The "borehole televiewer" (BHTV) developed by Mobil in the late 1960s was the first borehole imaging device that could be run in typical wells of the oil industry (Zemanek et al., 1969, 1970). Optical imaging devices (chapter 2.5) had been developed earlier but were only applicable in holes with clear visibility, i.e. containing either clear water or air. Additionally, they only recorded stationary images, while the borehole televiewer allowed a continuous recording along the borehole wall. The borehole televiewer operated as an ultrasonic reflection scanner of the borehole wall, and its early images showed interesting features on the borehole wall such as fractures, breakouts and major lithological contacts, and in cased hole perforations as well as casing joints. Subsequent development of this measurement technique had been undertaken by Amoco (Wiley, 1980; Broding, 1981), Shell (Rambow, 1984) and Arco (Pasternak & Goodwill, 1983). Today all major oil service companies offer an ultrasonic borehole imaging measurement (table 2.3.1). They all operate in reflection mode although experiments with refraction measurements were also made but no commercial service resulted from this. Most elements of the original borehole televiewer are still found in these newer tools, but the term "televiewer" has been replaced by "ultrasonic imaging" and "scanning" (Faraguna et al., 1989; Hayman et al., 1998).

The central piece is an ultrasonic transducer that also acts as a receiver (figure 2.3.1). It is driven by a motor to rotate around the tool's axis several times per second, during which it emits short bursts of ultrasonic pulses. These travel through the borehole mud, get reflected at the borehole surface and then travel back to the transducer (figure 2.3.2). Two quantities are measured

- The *interval transit time* from the transducer to the borehole wall and back
- The *reflected amplitude* of the signal received at the transmitter

Initially, transducer frequencies of around 2 MHz were used, but in newer tools this was lowered to a few hundred kHz. Azimuthal sampling in the early tools was over 600 per rotation, resulting in an azimuthal spacing of less than one degree. Modern tools have a lower sampling rate of around 180 per rotation, or one sample every two degrees. The azimuthal orientation of the transmitter is obtained by marking each of its passages with respect to a tool reference mark ("tool zero"), whose orientation is obtained from a three-axis accelerometer and magnetometer instrument in the tool.

The ultrasonic scan follows a spiral path up the borehole (or downward, depending on which way the tool is logged), whereby the spacing of the spiral depends on the logging speed: The faster the tool is logged the wider the spiral path will be (figure 2.3.3). Sampling theory determines the optimal logging speed: Ideally, two adjacent sample areas on the borehole wall (the "spot size") should

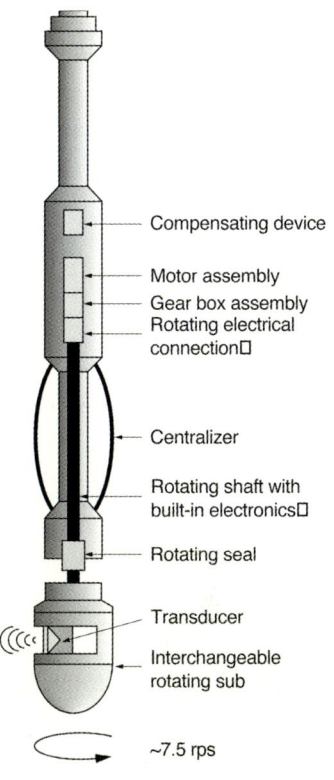

Figure 2.3.1. Sketch of the ultrasonic borehole imager (UBI*) and photograph of its interchangeable rotating assembly (the "sub") mounted at the bottom of the tool.

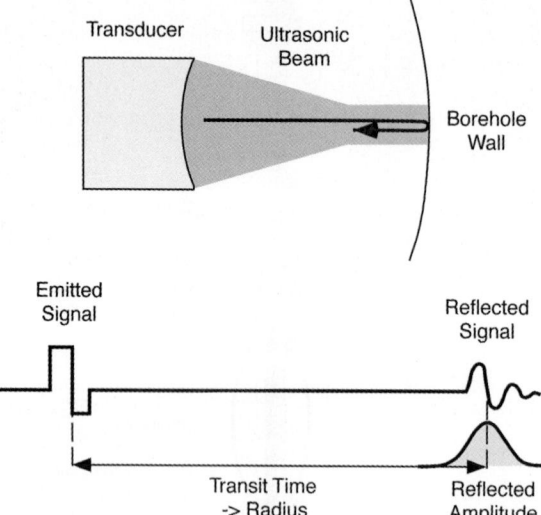

Figure 2.3.2. The ultrasonic borehole imaging principle. The two basic measurements, transit time and reflected amplitude, are obtained from downhole processing (after Hayman et al., 1998).

overlap by 50 % in order to avoid aliasing, i.e. the creation of artificial trends by undersampling. Thus, if the transmitter spins at 8 rotations per second and the spot size is 9 millimeters, the measurement should proceed at 8×9/2 mm/s, corresponding to a logging speed of 130 m/hr. These constraints determine some characteristics of ultrasonic borehole imaging. First, the logging speeds are necessarily slow, considerably slower than for electrical borehole imaging which typically proceeds at 600 m/hr. Secondly, the spot size for the example above results realistically in a resolution of about two centimeters, which is a factor of four more than micro-electrical imaging. In order to improve the resolution, one would have to decrease the spot size and "tighten" the spiral described by the scanning of the ultrasonic beam. However, this would additionally reduce the logging speed. And third, in order to have the same azimuthal and vertical resolution, the number of samples collected during each rotation should be equal to the borehole perimeter divided by half the spot size. For the case mentioned above, this comes to about 150 samples in a 8.5 inch-borehole. Thus 1200 measurements have to be made per second, leaving 0.83 milliseconds for each measurement. At a typical mud velocity of 1500 m/s for the compressional sound wave, and a distance of 7.5 cm between the transducer and the borehole wall, the ultrasonic burst only needs 0.1 millisecond for its flight to the borehole wall and back. This is considerably less than the time allocated from sampling considerations, leaving enough time for downhole processing and preparing the next sample measurements ("dead time"). Thus, the slow logging speed is primarily dictated by the vertical sampling constraints. A higher rotation rate of the transducer could theoretically increase the logging speed, but mechanical con-

Figure 2.3.3. Spiral path followed by an ultrasonic scan of the borehole wall for a high (left) and a low scan rate per foot (right). Samples have to be interpolated to obtain a rectangular image matrix. High scan rates improve the vertical resolution.

siderations and the accuracy of the azimuthal orientation seem to limit this.

In the early borehole televiewer an analog signal was sent uphole, where it was processed and fed to an oscilloscope, which was photographed to produce an image. Modern ultrasonic borehole imaging tools contain a downhole processors that digitize and filter the reflected signals immediately upon arrival (Hinz & Schepers, 1983; Pasternack & Goodwill, 1983; Faraguna et al., 1989; Hayman et al., 1998). Then they determine the reflection peaks and calculate the transit times as well as the amplitudes, the latter by fitting an envelope around the peak signals (figure 2.3.2). The processed data is then transmitted to the surface where the data is put onto an oriented image that can be processed further.

An important improvement in the new tools (table 2.3.1) is the introduction of focused transducers. These have a concave surface that focuses the beam towards the expected target into a smaller area than the size of the transducer itself. In this way, spatial resolution is improved and eccentering effects as well as the sensitivity to rugosity are reduced. The presently used transducers operate in the frequency range from 200 to 500 kHz, which is significantly lower than what was used in earlier tools. These frequencies have been found to be an optimum compromise of high resolution, which increases with higher frequencies due to the shorter wavelengths, and mud penetration, which decreases with increasing frequencies. Mud attenuation, however, is not only a function of frequency, but also of the mud additives and the base fluid. Oil has a higher attenuation than water, and heavy additives will additionally increase the attenuation and thus reduce the signal/noise ratio. The choice of the transducer frequency is therefore dependent on the mud composition and the desired resolution. In Schlumberger's ultrasonic borehole imager (UBI*), there is a single focused transducers which can operate at either 250 kHz with a spot size of 9 millimeters, or at 500 kHz with a spot size of about 5 mm (Hayman et al., 1998). The transducer is mounted on a

Table 2.3.1 Ultrasonic borehole imaging tools provided by three service companies

	Ultrasonic Borehole Imager (UBI)	Circumferential Borehole Imaging Log (CBIL)	Circumferential Acoustic Scanning Tool (CAST)
Scanner Assembly	Rotating heads, with 4 sub sizes	Rotating sensor in oil-filled chamber	Rotating heads, interchangeable
Transducer frequency	focused 250 kHz, 500 kHz	focused (two) 250 kHz	focused 280 kHz, 450 kHz
Scan rate	7.5 rps	6 rps	6–18 rps
Samples/scan	180	125–250	100–200
Scans/foot	12–72	16–64	40–120
Minimum hole size	5.5 in	5 in	5 in

rotating unit at the bottom of the tool (the "sub"), whose size can be adapted to the borehole diameter in order to ascertain optimum focusing. It is therefore in direct contact with the mud. By contrast, the circumferential borehole imaging tool (CBIL**) and the circumferential acoustic scanning tool (CAST***) contain an oil-filled window of an acoustically transparent material in which the transducer rotates (McDougall & Howard, 1989; Seiler et al., 1990; Maki et al., 1991).

Although ultrasonic imaging has the considerable advantage over electrical imaging that it can be done in water- as well as oil-based muds, there is an upper mud weight limit in oil-based mud where the attenuation becomes too high. This limit depends on the mud particles and various other factors, but is generally around 1.8 g/cc, or 15 lbs/gal.

Ultrasonic imaging methods are also used in cased holes for cement evaluation and corrosion monitoring (Broding, 1984; Hayman et al., 1991; Bettis et al., 1993). For cement evaluation usually planar transducers of 200–700 kHz and a relatively large spot size of around 3 cm are used. These emit sufficient energy so that not only the mud/casing reflection, but also the casing/cement and the cement/formation reflection (the "third interface") can be analyzed on the reflected wave train. For corrosion monitoring, high frequencies of several MHz and a very small spot size of 3 mm are used. This allows the detection of small corrosive pits on the inside and outside of the casing.

---

* Mark of Schlumberger
** Mark of Western-Atlas
*** Mark of Halliburton

2.3.2 Tool Response

Reflected Amplitude. All ultrasonic borehole-imaging methods depend on the amount of reflected energy at the interface of interest. This reflection coefficient R is expressed as a fraction of the incident particle velocity, or particle pressure. It is a function of the angle of incidence as well as the density ρ and the compressional wave velocity v. For normal beam incidence, it is found to be

$$R = \frac{\rho_2 v_2 - \rho_1 v_1}{\rho_2 v_2 + \rho_1 v_1} \qquad (2.3.1)$$

where 1 and 2 refer to the two mediums on each side of the interface, i.e. the borehole mud and the formation (Ellis, 1987). The product ρv is called *acoustic impedance* and is expressed in units of Mrayl (1 Rayl = 1 kg·m^{-2}·s^{-1}). The reflection coefficient is directly dependent on the contrast in acoustic impedance at the interface. Table 2.3.2 lists impedance values of some common materials encountered in ultrasonic imaging. It can be seen that for many rock types and for common drilling fluids the value of R is above 0.5, meaning that the reflected energy is higher than the transmitted energy. However, in situations where unfavorable conditions combine, for example when heavy drilling muds are used in relatively uncompacted shales, R can be as low as 0.2. A more important factor of consideration is the change in R from one lithology to another, since it gives an idea of the measurement's sensitivity to lithological changes. Inspection of table 2.3.2 reveals that for most common lithological sequences the variation in R is relatively small, typically in the order of 10 % to 20 %. By contrast, electrical imaging is sensitive to changes in the rock resistivities that typically range over more than one order of magnitude in the same well.

Table 2.3.2 Acoustic properties of some materials (modified from Hayman et al., 1991)

Material	Density (kg·m^{-3})	Acoustic velocity (m·s^{-2})	Acoustic impedance (MRayl)
Air (1-100 bar)	1.3 – 130	330	0.0004 – 0.04
Water	1000	1500	1.5
Drilling fluids	1000 – 2000	1300 – 1800	1.5 – 3.0
Shales	1900 – 2450	2300 – 3400	4.5 – 8.3
Sandstone (20 %)	2320	3900	9.0
Limestone (20 %)	2370	4200	10.0
Dolomite (20 %)	2520	4950	12.5
Anhydrite	2960	6100	18

Ultrasonic imaging has to cope with artefacts caused by uneven borehole conditions (Cheung, 1999; Lofts & Bourke, 1999). Typical boreholes are not perfectly round, but often irregular in shape, both in the azimuthal as well as the axial direction. The tool may also be eccentered due to the borehole deviating from the vertical. Both conditions cause oblique incidence of the acoustic beam onto the borehole surface, at least over a section of each scan. And finally, there may be small-scale rugosity caused by the drill bit that leads to scattering of the incident acoustic beam. All these effects will impact the reflected signal, reducing or altogether concealing the lithological information (figure 2.3.4). In practice, however, it has been found that washouts and rugosity are lithology-dependent, and that therefore *indirect* information on bedding can obtained from ultrasonic images. Shales, for example, can be preferentially washed out, and the resulting signal loss of the ultrasonic image may therefore indicate their

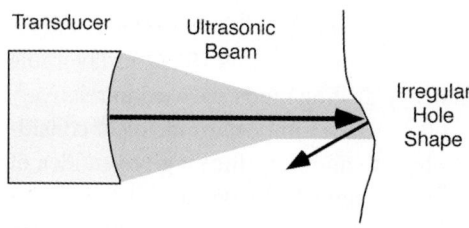

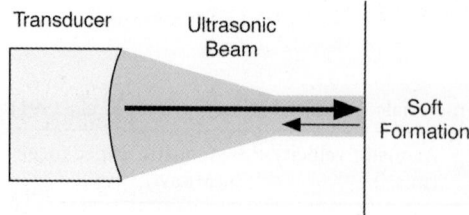

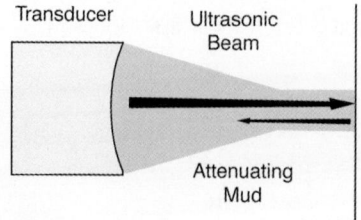

Figure 2.3.4. Factors negatively affecting ultrasonic image quality. From top to bottom: Irregular borehole producing non-orthogonal reflection; soft formation resulting in a weak reflected signal which may be difficult to detect; strongly attenuating borehole mud leading also to a weak signal.

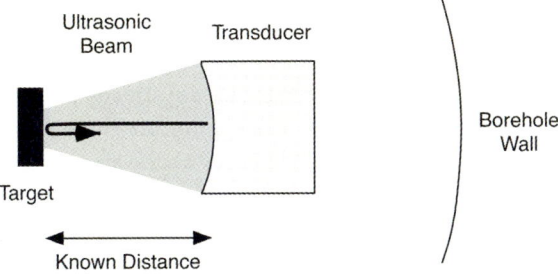

Figure 2.3.5. Measuring the mud velocity with the UBI transducer in the inverted position, firing at a known target on the rotating assembly through the mud.

locations. Similarly, fractures crossing the borehole may show chipping and slight widening at the borehole surface. Together with the usually large contrasts in acoustic impedance between the fluid-filled fracture and the surrounding rock (table 2.3.2), this local damage facilitates the detection of fractures with ultrasonic borehole images. An additional factor in making fractures well visible on ultrasonic borehole images is wave conversion at the fracture surface: Part of the incoming compressional wave is converted into shear waves at the surface of the fracture, thus further reducing the reflected compressional signal.

Transit Time. The interval transit time depends largely on the proper peak detection of the reflected signal. Even in boreholes with irregular shapes, where the reflected signal is strongly reduced compared to an ideal situation, it may still be possible to obtain a correct transit time. Proper positioning of the samples, however, may be more difficult (Georgi, 1985): Just like in seismic with strongly dipping layers, the actual reflection point in ultrasonic borehole imaging may be shifted when beam incidence in non-orthogonal. The corrections needed resemble migration techniques in seismic and can be done using the transit time information (Priest, 1997). They are not yet common practice, perhaps because modern focused transducers are less susceptible to this effect.

The transit time is strictly a function of the distance between the transducer and the borehole wall, and the compressional velocity of the mud. The distance to the borehole wall is more useful to the interpreter than the transit time, since it gives information about borehole elongations, breakouts and washouts (Lysne, 1986). To obtain the transit time, the mud velocity can be measured as the tool is lowered into the well by firing the transducer towards a target at a known distance (figure 2.3.5). With the CBIL, this is done with a separate transducer while logging uphole or downhole. Both techniques give a continuous survey of the mud velocity v_m with depth, allowing the borehole radius r to be obtained as a function of depth z and azimuth a using

$$r(a, z) = \frac{\Delta t(a, z) \cdot v_m(z)}{2} + d \qquad (2.3.2)$$

where d is the distance of the transducer surface from the tool axis.

2.3.3 Processing and Graphic Representations

Much of the processing of ultrasonic images is similar to electric images (figure 2.3.6). The data sent to the surface are already preprocessed downhole, since peak detection and amplitude calculation are done downhole and in real time. The surface processing usually focuses on the amplitude images because they contain more lithological information, while the transit time images reflect the borehole shape. The first step is to orientate the data points with respect to north using the zero passing of the transducer; this may include a correction for irregular borehole shape discussed in the previous paragraph. The data are then shifted to the correct depth using the downhole accelerometer data. Since ultrasonic imaging tools do not have pads like micro-electrical imaging tools, their movements are often smoother and the speed corrections are less severe. When the tool is logged in very heavy mud, the transducer rotations may become slow at times, and the tool speed may have to be adapted continuously. In these cases, proper positioning of the samples can be very difficult. The same is true when the tool is logged on the drill string ("tough logging conditions", or TLC), where the tool stops every time a new stand of pipe is added.

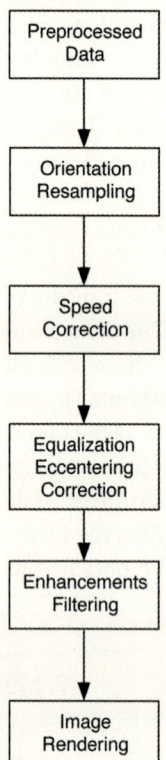

Figure 2.3.6. The various processing steps performed for ultrasonic images.

An aspect of particular importance and unique to ultrasonic borehole images is the correction for *eccentering*. Its goal is to reestablish the reflected amplitude in cases of oblique incidence or larger time of flight due to borehole elongation or tool eccentering (Menger, 1994). The images usually show one or more broad bands of low amplitudes at the azimuths where the beam incidence is not normal (figure 2.3.7). One possible correction is purely statistical: For each azimuthal scan, or for several adjacent scans, the amplitudes are plotted versus azimuth. In the case of borehole ovality or breakouts, there will be an azimuthal dependence of the azimuth, marked by several troughs and peaks. A sinusoid can be fitted to these data, either with one wavelength or two wavelengths (the first- and second-order harmonics). This sinusoid can then be flattened to a straight line with the same amplitude mean, and the actual amplitude measurement can be moved by the same amount. The procedure normalizes the amplitude response in the azimuthal direction and therefore resembles the equalization done on electrical images (chapter 2.2.2). Another correction approach is model-based, wherein the angle of incidence is calculated from the transit time data, and the measured amplitude is corrected using a physical model that relates amplitude reflection versus angle of incidence.

As a final processing step, the ultrasonic images are often noise-filtered in order to remove artefacts caused by small-scale rugosity and normalized in order to enhance the image contrast at all levels. Ultrasonic borehole images have full azimuthal coverage and therefore – unlike microelectrical images – provide a complete two-dimensional array (figure 2.3.8). Many image processing routines are therefore easier to apply than for electrical images, where gaps exist between the images from the different pads (e.g. Wong et al., 1989).

The basic graphic representation is essentially the same as for electrical images, with the depth/azimuth plot initially proposed by Zemanek et al. (1969, 1970) as the most common display. The color schemes are also the same as for electrical images, with brighter colors used for higher reflected amplitudes and lower borehole radii.

Three-dimensional projections of ultrasonic images were already made by Broding (1982) and Lysne (1986), but have only recently become computationally feasible on a larger scale. The borehole shape obtained from the transit time information is used to create a three-dimensional surface, onto which the reflected amplitude image can be mapped. This is a display specific to these images (figure 2.3.9). Since it is a true three-dimensional representation – unlike the core-like projections of electrical images such as figure 2.2.32 – specialized graphic computers are best used for efficient visualization. These representations may become more common in the near future with the availability of more powerful computers.

Another useful display is the borehole cross-section at a given depth. It can be a powerful tool to spot irregularities in the borehole caused by tectonic forces or by the drilling process. An extension of this plot is the "spiral plot", whereby a

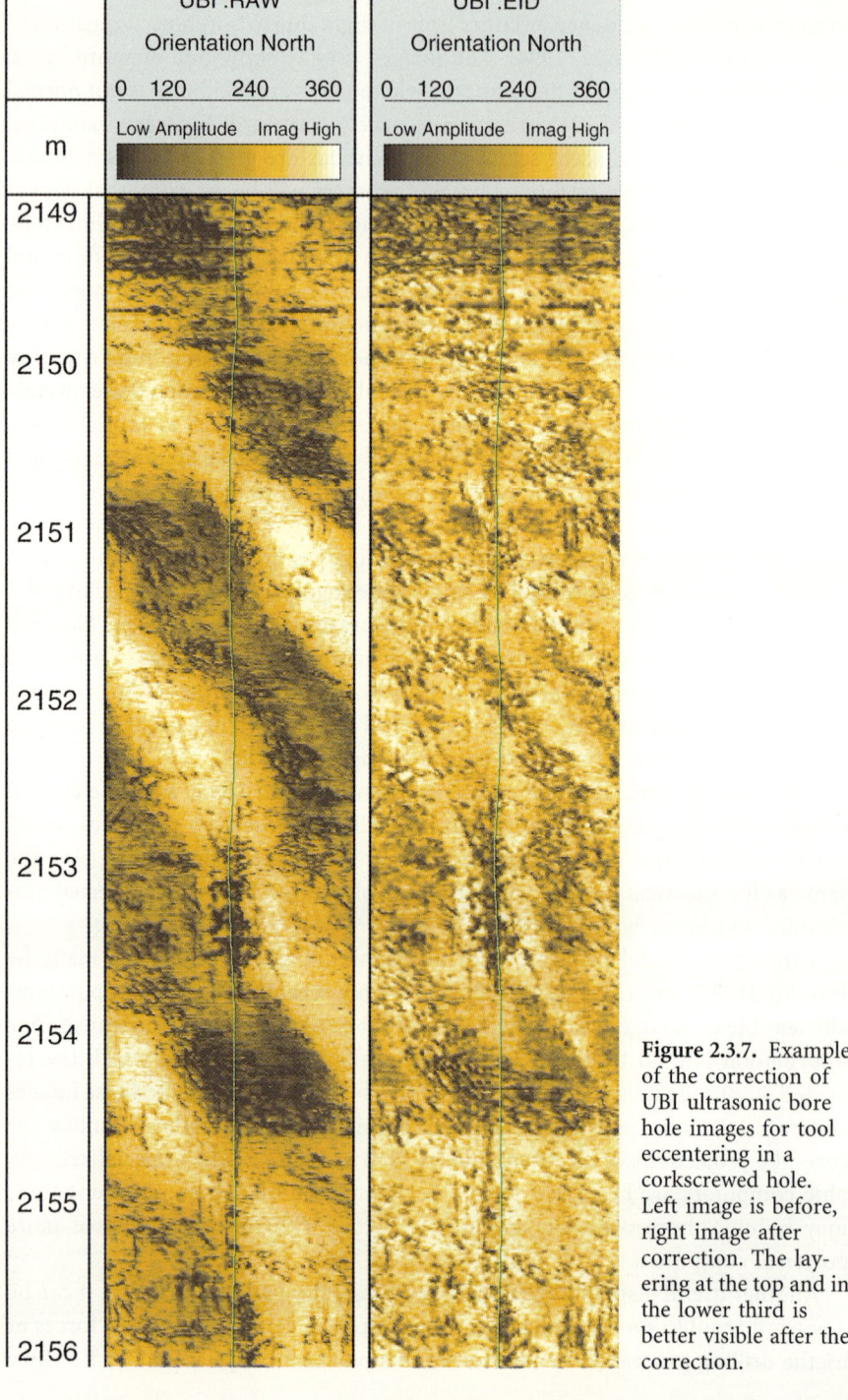

Figure 2.3.7. Example of the correction of UBI ultrasonic bore hole images for tool eccentering in a corkscrewed hole. Left image is before, right image after correction. The layering at the top and in the lower third is better visible after the correction.

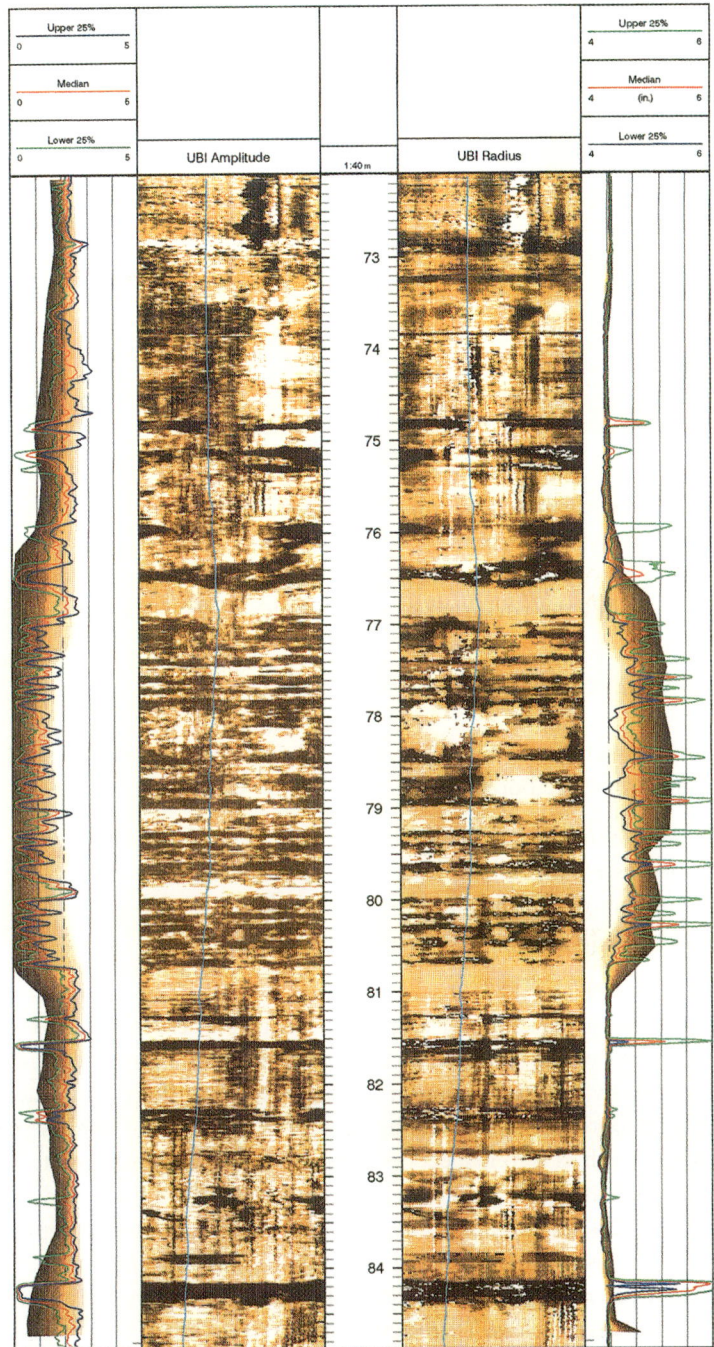

Figure 2.3.8. Graphic display of ultrasonic borehole images with reflected amplitude (left) and transit time (right). The left and right outer tracks contain information on the dynamic color normalization as a function of depth (from Hayman et al., 1998)

Figure 2.3.9. A three-dimensional projection of the borehole shape obtained from the transit time image of the ultrasonic borehole imager, showing a washed out section at the bottom, and a possible breakout above. Interval covers 80 cm corresponding to about 30,000 nodes on the borehole surface (courtesy H. Barrow).

series of successive borehole radius scans are plotted along the spiral path along which the data acquisition proceeded, in a perspective view which serves the same purpose as the previous plot. Some of these applications will be discussed in the following paragraphs and examples of these displays are shown.

2.3.4. Dip Calculations

Similar to electrical borehole imaging, several methods can be used to obtain dips from ultrasonic borehole images:

- Automatic dip calculation by extracting pseudo-dipmeter curves that are then processed with dipmeter processing software
- Interactive dip picking on a workstation
- Automatic dip calculations from the images

Faraguna et al. (1989) discuss the first method. They extract amplitude traces along four given azimuths and subsequently process them with the standard dipmeter processing software, obtaining very similar dips as with the dipmeter. This is currently the most common approach, but is suffers from a lack of dip results in poor borehole conditions because of the high noise level. A preferable albeit slower method is to inspect the images on a workstation, and to pick dips of the obvious features in an interactive manner. This has the added advantage that a reliable classification of the dips is possible, for example by separating bedding-related dips from fracture dips. Automatic dip calculation from the images as discussed in chapter 2.2 is difficult because of the noisy nature of ultrasonic images. Nevertheless, Torres et al. (1990) have shown that the Hough transform can be used to extract dips from ultrasonic images. Newer methods such as the one proposed by Ye et al. (1997) may produce reliable dips where fractures are prevalent, but no results are available to date.

2.3.5. Image Interpretation

In discussing image interpretation of ultrasonic borehole images, we will emphasize those areas where these images have proven particularly useful, and we will restrict ourselves to open hole applications. Among these are mostly structural applications such as

- Fracture analysis;
- Hole shape analysis: breakouts, slippage, ovalization; and
- Bedding analysis

Fractures

Fractures are among the most prominent features on ultrasonic borehole images and have been described from the earliest works onwards (Zemanek et al. (1969, 1970; Keys 1980; Paillet et al., 1985; Lau et al., 1987; Barton & Moos, 1988; Taylor, 1991). While dips and azimuths of fractures can be reliably estimated (except in ovalized boreholes; see above), fracture apertures cannot be obtained. The short wavelength used in ultrasonic borehole imaging causes the reflection to take place practically at the wellbore surface, and slight chipping or widening of the fractures can cause them to appear larger. In an ideal situation, Lincecum (1993) has estimated the fracture detection threshold to be in the order of 25 microns, higher than the corresponding value for microelectrical borehole images of about one micron. In practice, however, Dudley (1993) and Laubach

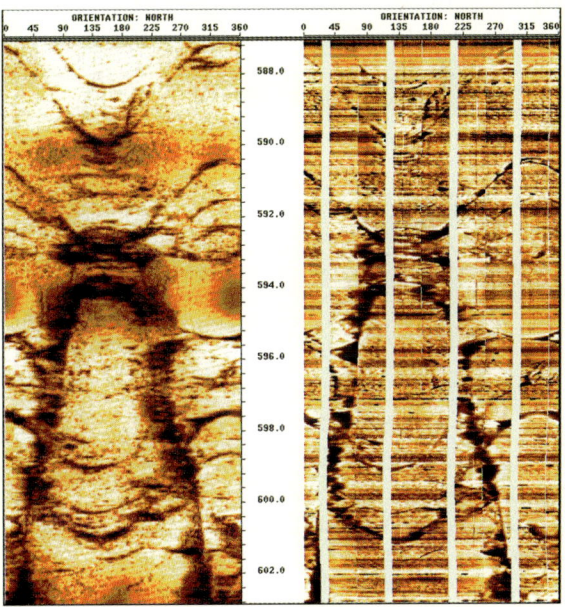

Figure 2.3.10. Ultrasonic (left) and microelectrical (right) borehole images over a layered dolomite interval showing abundant fracturing. For discussion see text.

et al. (1988) found the detection threshold to be considerably higher. Figure 2.3.10 shows an example of the fractured carbonate where at least two fracture sets are present, with one appearing wider than the other. However, due to their steep dips these "wider" fractures are more prone to chipping, and their widths might actually be smaller than those of the other set. The figure also shows that the bedding is better visible on the electrical image than on the ultrasonic image.

Fractures cannot only be quantified for their dips and azimuths, but also classified according to their morphology (Cheung & Heliot, 1990). For identifying induced fractures, the same criteria apply as with electrical images.

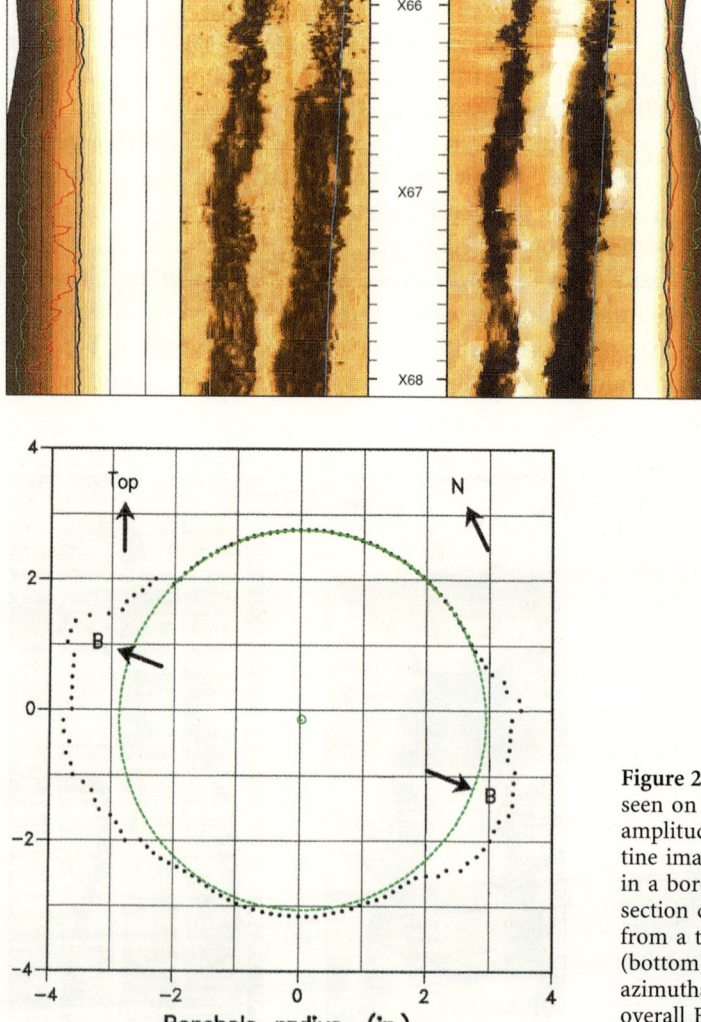

Figure 2.3.11. Breakouts seen on the reflected amplitude and transit tine images (top) and in a borehole cross-section constructed from a transit time scan (bottom). Notice slight azimuthal rotation with overall E-W direction of the breakouts.

Hole Shape Analysis

The transit time images and, to a lesser degree the reflected amplitude image, contains valuable information on the borehole geometry at very high resolution. Of particular interest are

- Borehole breakouts;
- Borehole elongations, caused by a variety of factors; and
- Borehole displacements due to slippage.

Breakouts are caused by stress concentrations in an anisotropic stress regime, leading to shear failure until the stress has been sufficiently relieved so that the borehole becomes stable (see chapter 2.2.5). This results in two damaged and enlarged areas on opposite sides of the borehole. They can be seen on the reflected amplitude image as low-reflectivity zones, while the transit time image shows them as local enlargements. Figure 2.3.11 shows breakouts on the usual depth/azimuth plot and in a borehole cross-section. The azimuth seems to rotate slightly over the interval shown, but also areal extent as well as amount of enlargement of the breakouts can vary greatly (see discussion by Bell, 1990). Considerable work has been done in breakout analysis with ultrasonic borehole images, sometimes combined with data from dipmeter calipers and usually for the purpose of determining regional stress directions (Bell & Gough, 1979; Cox, 1983; Hickman et al., 1985; Plumb & Hickman, 1985; Zoback et al., 1985; Paillet & Kim, 1987; Suter, 1987; Barton et al., 1988; Mastin, 1988; Dart & Zoback, 1989; Cowgill et al., 1992; Hillis & Williams, 1992; Yassir & Dusseault, 1992). The main result of these workers is that breakouts tend to show consistent patterns in a given geographic area, with their alignments oriented in the direction of minimum horizontal stress. In the petroleum industry, this can be used to help establishing the local stress regime.

Borehole elongations are commonly observed on ultrasonic borehole images, and care has to be taken not to confuse them with breakouts. The borehole cross-section plots are helpful in this task. Borehole elongations can be caused by a number of factors:

- The weight of the drill pipe in a deviated well can lead to a characteristic enlargement of the lower part of the hole ("key seat");
- The vibration of the drill string, particularly when downhole motors are used, can cause a spiraling borehole ("corkscrew") which at any given depth will have an elliptical cross-section;
- Anisotropic stresses or rock properties can lead to elliptical enlargements of the borehole ("ovalization").

From the drilling method and the geological context it is often clear which one of the above factors applies. There is in general little useful information gleaned from borehole elongations, perhaps with the exception of the last type, and eccentering corrections applied in the processing chain usually eliminate them quite well on the reflected amplitude images (figure 2.3.7). An example of a key seat is shown in figure 2.3.12, as two ultrasonic image together with a cross-section. As the cross-section suggests, the key seat is caused by the drill pipe wearing out the lower part of the borehole.

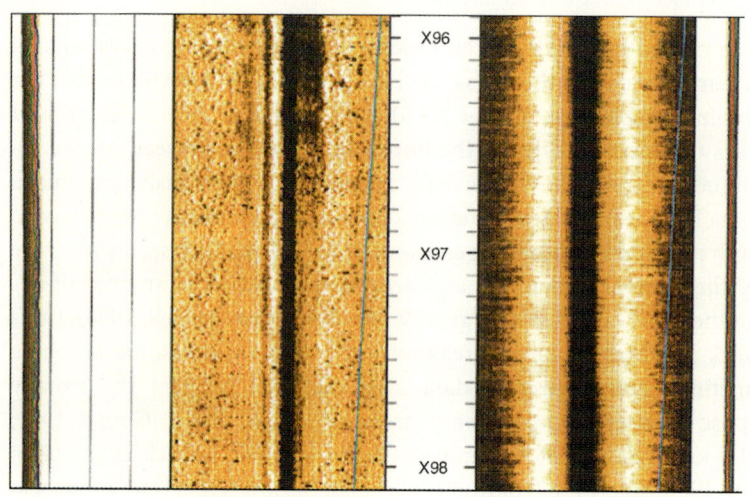

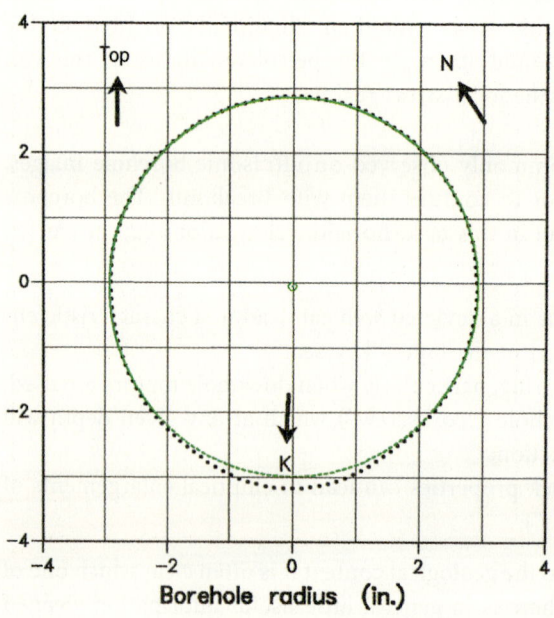

Figure 2.3.12. A borehole elongation due to key seating seen on the reflected amplitude and transit time images (top) and in a borehole cross-section constructed from the transit time image (bottom). The elongation is at the bottom of the hole and probably caused by drill pipe wear on the rock.

Borehole displacement due to slippage is a phenomenon that was discovered only recently using ultrasonic borehole images (Cornet et al., 1997). Figure 2.3.13 show a cross-section image of a well crossing a fault together with a spiral plot of the borehole. Closer inspection of the borehole cross-section plot reveals

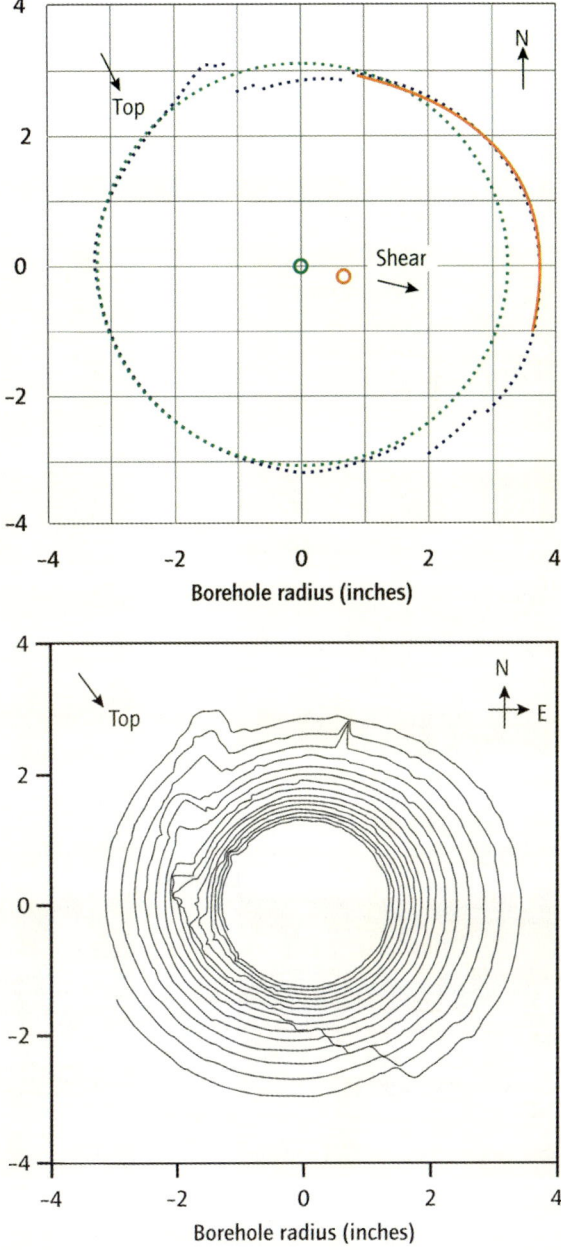

Figure 2.3.13. A borehole affected by strike-slip movement during drilling. The cross-section shows that two circles of equal radius but with offset centers can be fitted. This offset corresponds to the amount of slippage in the horizontal direction (after Hayman et al., 1998).

141

that there are two parts to the borehole, both perfectly in gauge as indicated by the fits to circles corresponding in size to the drill bit diameter. The centers of these two circles are shifted by several centimeters. This is therefore not one of the borehole elongation types mentioned above. Rather, the borehole has moved along the fault, probably because the drill mud has invaded the fault and reduced its friction causing the fault to become reactivated. The shape of the borehole indicates a strike-slip movement, with the displacement of the two circle centers indicating the amount of the movement.

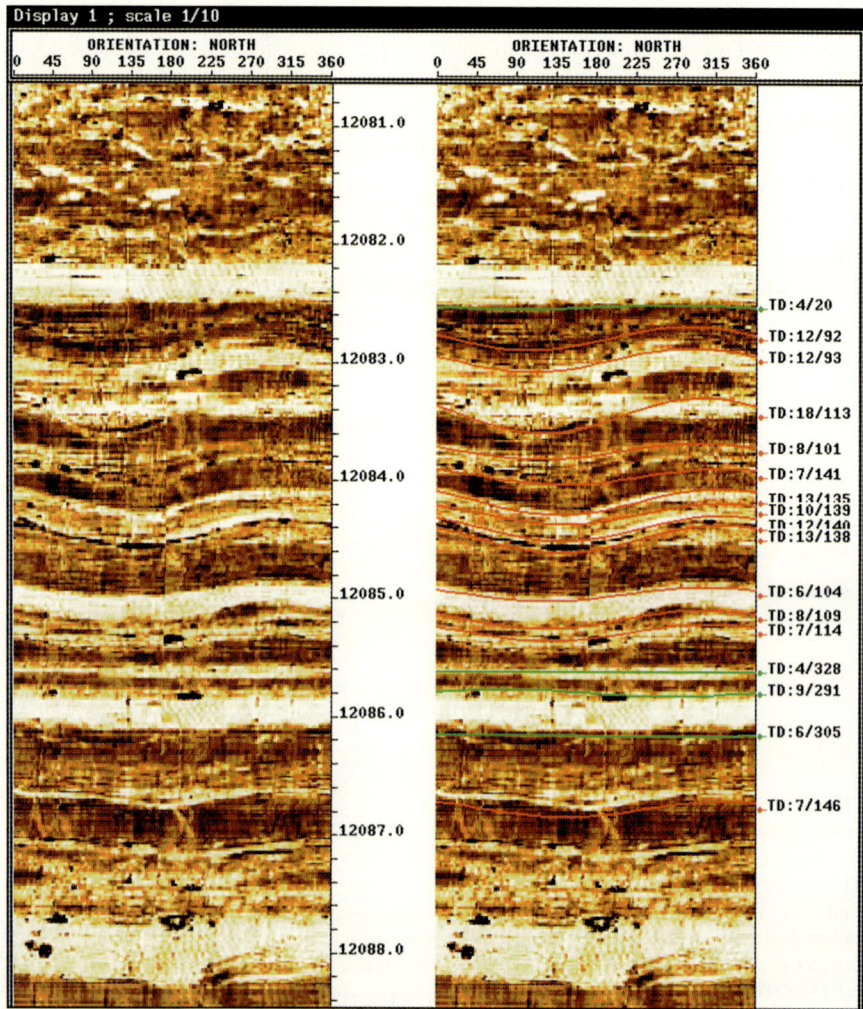

Figure 2.3.14. Ultrasonic images from a cross-bedded sandstone sequence showing bounding surfaces (green sinusoids) and cross-bedded strata (red sinusoids). Depths in feet.

Bedding Analysis

The discussion of acoustic impedances in chapter 2.3.2 led to the conclusion that lithological changes produce only small contrasts on ultrasonic images. Although some success is reported in using the technique for reservoir zonation and sedimentological interpretations (e.g. Hackbarth et al., 1988; Luthi & Banavar, 1988; Plumb & Luthi, 1989; Verdur et al., 1991), it is not as widely used as electrical borehole imaging. The fractured carbonate shown in figure 2.3.10 demonstrates this difference: While the fracture information from the two pictures is equivalent, the bedding is considerably clearer on the electrical image. Under favorable conditions, notably when the borehole wall is smooth and significant lithological changes occur, the ultrasonic images can contain good bedding information. Figure 2.3.14 is an example from a cross-bedded sandstone sequence that in many aspects compares with the electrical image shown in figure 2.2.20. First order bounding surfaces as well as cross-bed strata can easily be distinguished and their dips and azimuths can be measured for further sedimentary interpretation. Figure 2.3.15 shows an overturned fold in a sequence affected by slumping; it is again comparable in quality to electrical images such as the one shown in figure 2.2.18.

Bedding analysis from ultrasonic images is often difficult due to the low signal/noise ratio. It is therefore useful to summarize those extrinsic factors that affect image quality the most:

- Borehole rugosity: The smoother, the better
- Mud fluid: Water is better than oil
- Mud weight: The lighter, the better

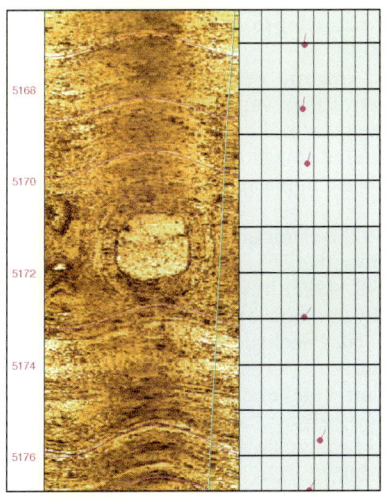

Figure 2.3.15. Azimuthal presentation and core projection of an overturned fold probably caused by slumping in a clastic sequence. Depths in feet.

- Transducers: Higher frequencies give better resolution but weaker signal
- Logging: Slower speeds give better resolution
- Processing: Eccentering corrections and noise filtering improve images

2.3.6 Summary

Ultrasonic borehole images contain useful structural information, such as breakouts and borehole elongations on transit time images, and faults and fractures on reflected amplitude images. Their bedding information is generally not as rich as that obtained from electrical images, but can be sufficient if proper acquisition guidelines are followed. They have the advantage that they can be recorded in oil- as well as water-based mud, although at modest logging speeds.

References

Barton CA, Moos D (1988) Analysis of macroscopic fractures in the Cajon Pass scientific drillhole – over the interval 1829–2115 meters. Geophys Res Let 15: 1013–1016.

Barton CA, Zoback MD, Burns KL (1988) In-situ stress orientation and magnitude at the Fenton geothermal site, New Mexico, determined from wellbore breakouts. Geophys Res Let 15: 467–470.

Bell, JS (1990) Investigating stress regimes in sedimentary basins using information from oil industry wireline logs and drilling records. In: Hurst A, Lovell MA, Morton A (eds) Geological Applications of Wireline Logs. Geol Soc Spec Publ 48, pp 305–325.

Bell JS, Gough DI (1979) Northeast-southwest compressive stress in Alberta: evidence from oil wells. Earth and Planetary Sci Let 45: 475–482.

Bettis FE, Crane LR, Schwanitz BJ, Cook MR (1993) Ultrasound logging in cased boreholes pipe wear. Ann Conf Soc Petrol Eng: Paper 26318.

Broding R (1981) Volumetric scanning well logging. Trans 22nd Symp Soc Prof Well Log Analysts: Paper B.

Broding R (1981) Volumetric scanning well logging. The Log Analyst 23: 14–19.

Broding R (1984) Application of the zonic volumetric scan log to cement evaluation. Trans 25th Symp Soc Prof Well Log Analysts: Paper JJ.

Cheung PSY, Heliot D (1990) Workstation-based fracture evaluation using borehole images and wireline logs. 65th Ann Conf Soc Petrol Eng: Paper 20573.

Cheung PS (1999) Microresistivity and ultrasonic imagers: tool operations and processing principles with reference to commonly encountered image artefacts. In: Lovell MA, Williamson G, Harvey PKJ (eds) Borehole images: applications and case histories. Geol Soc Spec Publ 159, pp. 45.57.

Cornet FH, Helm J, Poitrenaud H, Etchecopar A (1997) Seismic and aseismic slips induced by large-scale fluid injections. Pure Appl Geophys 150: 563–583.

Cowgill SM, Meredith PG, Murrell SAF, Brereton NR (1992) In situ stress orientations in the Witch Ground Graben, North Sea, revealed by borehole breakouts: preliminary results. In: Hurst A, Griffiths CM, Worthington PF (eds) Geological Applications of Wireline Logs II. Geol Soc Spec Publ 65, pp 179–184.

Cox, JW (1983) Long axis orientation in elongated boreholes and its correlation with rock stress data. Trans 24th Symp Soc Prof Well Log Analysts: Paper A.

Dart RL, Zoback ML (1989) Wellbore breakout stress analysis within the central and eastern continental United States. The Log Analyst 30: 12–25.

Dudley JW II (1993) Quantitative fracture identification with the borehole televiewer. 34th Symp Soc Prof Well Log Analysts: Paper LL.

Ellis DV (1987) Well logging for earth scientists. Elsevier, Amsterdam.

Faraguna JK, Chace DM, Schmidt MG (1989) An improved borehole televiewer system – image acquisition, analysis and integration. 30th Symp Soc Prof Well Log Analysts: Paper UU.
Georgi DT (1985) Geometrical aspects of borehole televiewer images. Trans 26th Symp Soc Prof Well Log Analysts: Paper O.
Hackbarth CJ, Tepper BJ (1988) Examination of BHTV, FMS and SHDT images in very thinly bedded sands and shales. Ann Conf Soc Petrol Eng: Paper 18118.
Hayman AJ, Parent P, Cheung P, Verges P (1998) Improved borehole imaging by ultrasonics. Soc Petr Eng Production & Facilities, February 1998, 5–13.
Hayman AJ, Hutin R, Wright PV (1991) High-resolution cementation and corrosion imaging by ultrasound. Trans 32nd Symp Soc Prof Well Log Analysts: Paper KK.
Hickman SH, Healy JH, Zoback MD (1985) In-situ stress, netural fracture distribution and borehole elongation in the Auburn geothermal well. J Geophy Res 90: 5497–5512.
Hillis RR, Williams AF (1992) Borehole breakouts and stress analysis in the Timor Sea. In: Hurst A, Griffiths CM, Worthington PF (eds) Geological Applications of Wireline Logs II. Geol Soc Spec Publ 65, pp 157–168.
Hinz K, Schepers R (1983) SABIS – The digital version of the borehole televiewer. 8th Eur Form Eval Symp, Soc Prof Well Log Analysts: Paper E.
Keys WS (1980) Application of the acoustic televiewer to the characterization of hydraulic fractures in geothermal wells. Proc Geothermal Reservoir Stimulation Symposium: 176–202.
Lau JSO, Auger LF, Bisson JG (1987) Subsurface fracture surveys using a borehole television camera and acoustic televiewer. Can Geotech J 24: 499–508.
Laubach SE, Baumgardner RW, Monson ER, Hunt E, Meadoor KJ (1988) Fracture detection in low permeability sandstone: a comparison of BHTV and FMS to logs and core. 63rd Ann Conf Soc Petrol Eng: Paper 18119.
Lincecum TA, Reinmiller R, Mattner J (1993) Natural and induced fracture classification using image analysis. Trans 34th Symp Soc Prof Well Log Analysts: Paper J.
Lofts JC, Bourke LT (1999) The recognition of artefacts from acoustic and resistivity borehole imaging devices. In: Lovell MA, Williamson G, Harvey PKJ (eds) Borehole images: applications and case histories. Geol Soc Spec Publ 159, pp. 59–76.
Luthi SM, Banavar JR (1988) Application of borehole images to three-dimensional geometric modeling of eolian sandstone reservoirs, Permian Rotliegende, North Sea. Bull Am Ass Petrol Geol 72: 1074–1089.
Lysne P (1986) Determination of borehole shape by inversion of televiewer data. The Log Analyst 27: 64–71.
Maki V, Gianzero S, Strickland R, Kepple HN, Gianzero MV (1991) Dynamically focused transducer applied to the CAST imaging tool. Trans 32nd Symp Soc Prof Well Log Analysts: Paper HH.
McDougall JG, Howard, MG (1989) Advances in borehole imaging with second generation CBIL [circumferential borehole imaging log] borehole televiewer instrumentation. Ontario Petroleum Institute: 28th Annual Symposium Transactions, paper 12.
Mastin L (1988) Effect of borehole deviation on borehole breakout orientations. J Geophys Res 93: 9187–9195.
Menger S (1994) New aspects of the borehole televiewer decentralization correction. The Log Analyst 35 July/August: 14–20.
Paillet FL, Keys WS, Hess AE (1985) Effects of lithology on televiewer-log quality and fracture interpretation. 26th Symp Soc Prof Well Log Analysts: Paper JJJ.
Paillet FL, Kim W (1987) Character and distribution of borehole breakout and their relationship to in situ stresses in deep Columbia River basalts. J Geophys Res 92: 6223–6234.
Pasternak ES, Goodwill WR (1983) Application of digital borehole televiewer logging. Trans 25th Symp Soc Prof Well Log Analysts: Paper C.
Plumb RA, Hickman SH (1985) Stress-induced borehole elongation – a comparison between four-arm dipmeter and the borehole televiewer in the Auburn geothermal well. J Geophys Res 90: 5513–5521.

Plumb RA, Luthi SM (1989) Analysis of borehole images and their application to geologic modeling of an eolian reservoir. Soc Petr Eng Form Eval: 505–514.

Priest JF (1997) Computing borehole geometry and related parameters from acoustic caliper data. Trans 8th Symp Soc Prof Well Log Analysts: Paper G.

Rambow FHK (1984) The borehole televiewer: Some field examples. Trans 25th Symp Soc Prof Well Log Analysts: Paper C.

Seiler D, Edmiston C, Torres D, Goetz J (1990) Field performance of a new borehole televiewer tool and associated image processing techniques. Trans 31st Symp Soc Prof Well Log Analysts: Paper H.

Suter M (1987) Orientational data on the state of stress in northeastern Mexico as inferred from stress-induced borehole elongations. J Geophys Res 92: 2617–2885.

Taylor TJ (1991) A method for identifying fault related fracture systems using the borehole televiewer. Trans 32nd Symp Soc Prof Well Log Analysts: Paper JJ.

Verdur H, Stinco L, Naides C (1991) Sedimentological analysis utilising the circumferential borehole acoustic log. Trans 32nd Symp Soc Prof Well Log Analysts: Paper II.

Wiley R (1980) Borehole televiewer – revisited. Trans 21th Symp Soc Prof Well Log Analysts: Paper HH.

Wong SA, Startzman RA, Kuo TB (1989) Enhancing borehole image data on a high-resolution PC. Petroleum Computer Conference, 64th Ann Conf Soc Petr Eng: Paper 19124.

Yassir NA, Dusseault MB (1992) Stress trajectory determinations in South Ontario from borehole logs. In: Hurst A, Griffiths CM, Worthington PF (eds) Geological Applications of Wireline Logs II. Geol Soc Spec Publ 65, pp 169–177.

Ye SJ, Rabiller P, Keskes N (1997) Automatic high resolution sedimentary dip detection on borehole imagery. Trans 38th Symp Soc Prof Well Log Analysts: Paper O.

Zemanek J, Caldwell RL, Glenn EE, Holcomb SV, Norton LJ, Straus AJD (1969) The borehole televiewer – a new logging concept for fracture location and other types of borehole inspection. J Pet Tech 21: 762–774.

Zemanek J, Glenn EE, Norton LJ, Caldwell RL (1970) Formation evaluation by inspection with the borehole televiewer. Geophysics 35: 254–269.

Zoback, MD, Moss D, Mastin L, Anderson RN (1985) Wellbore breakouts and in-situ stress. J Geophys Res 90: 5523–5530.

2.4 Density Borehole Imaging

2.4.1 Measurement Principles

The density measurement is one of today's standard wireline logging services. Together with the neutron porosity measurement it is used to accurately determine the rock porosity and its matrix or "grain" density. The principle of the measurement is based on gamma ray interactions with the rock (Ellis, 1987). At relatively high energies, between about 100 keV and 10 MeV, the gamma rays interact primarily with electrons in a process called Compton scattering, which in essence attenuates the gamma rays in proportion to the bulk density of the material. It is this property that is used to determine the bulk density of rocks in the borehole. Gamma rays of a known energy are emitted from a source mounted on a pad that is pressed against the borehole wall. The gamma rays travel through the rock, but also the mudcake and the tool itself, interacting with these materials and losing energy in the process. Their attenuation is measured with two detectors at different spacings from the source in order to compensate for the influence of the mudcake. Shielding is used to eliminate the effects of the tool itself. This bulk density measurement is a function of the porosity of the material and the composition of the solid as well as the fluid volumes within the rock.

As the gamma rays reach lower energies (<100 keV), they start interacting with entire atoms, whereby the gamma rays get absorbed to produce higher-energy electrons in the atoms. This photoelectric absorption is strongly dependent on the atomic number of the material: It increases very rapidly with higher atomic numbers (aside from increasing rapidly with lower gamma ray energies). Measuring the absorption of these low-energy gamma rays, therefore, allows determination of the photoelectric absorption factor, which depends on the elementary composition of the material.

These two measurements are generally done with the same device. A "hard" window in the near detector responds solely to Compton-scattered gamma rays and therefore the density of the target material, while a "soft" window responds to both atomic number and density. The ratio of the two windows cancels effectively the density contribution out and yields a measurement of the photoelectric absorption index P_e. While the grain density of common rock minerals varies by considerably less than one order of magnitude, the photoelectric absorption index varies by several orders of magnitude: Light elements such as oxygen have a P_e of less than one, while the heavy element barium has a P_e of almost 500. The latter, unfortunately, is contained in barite, a commonly used mud additive which may therefore strongly affect the measurement. The photoelectric index of a mineral is obtained through computation from the P_e values of its individual components as well as their volume fractions and their densities. The P_e of quartz is 1.8, of dolomite 3.1 and of calcite 5.1.

Much of the gamma ray scattering takes place at a small distance from the measuring device, and the probed volume is thus generally quite small. These devices have therefore been used for a number of years in so-called "high-resolution" logging, whereby the sampling rate is increased from the standard 6 inches to 1.2 inches. This results in a better definition, particularly of thinly bedded layers.

Such high-resolution logging has also spurred interest in using these devices for borehole imaging. Recently, Schlumberger has developed a logging-while-drilling tool, called Azimuthal Density-Neutron tool (ADN*), which measures azimuthal density in a similar way as the electrical LWD imaging tool measures azimuthal resistivity. The density skid is mounted on an eccentered pad close to the borehole wall (figure 2.4.1), and as the bottom-hole assembly rotates, the sensor scans the borehole wall and records azimuthal changes in density and photoelectric absorption (Holenka et al, 1995; Evans et al., 1995; Carpenter et al., 1997). Additionally, an eccentered neutron porosity sensor is also placed on the tool and could theoretically be used to provide a neutron porosity image, but this mode is not implemented yet. Measurement of the orientation of the tool is provided by a two-coil magnetometer, which generates four quadrants around the borehole (figure 2.4.2). These are oriented with respect to gravity in non-vertical wells as the bottom, left, top, and right quadrant, and all measurement counts, i.e. the incoming gamma rays and neutrons, are grouped into these quadrants (figure 2.4.3). This coarse azimuthal sampling is necessary in order to ob-

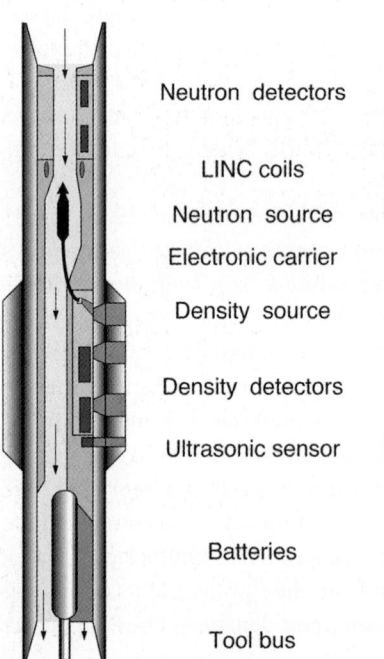

Figure 2.4.1. Sketch of the azimuthal density/neutron tool (ADN*) showing the density and neutron sources with their detectors. In rotating mode these sensors scan the borehole wall and record azimuthal changes in density, photoelectric absorption and neutron porosity. Configuration on left is with in stabilized mode with density sensors mounted on sleeve, while configuration on right is "slick" which allows higher build-up rates (from Evans et al., 1995).

* Mark of Schlumberger

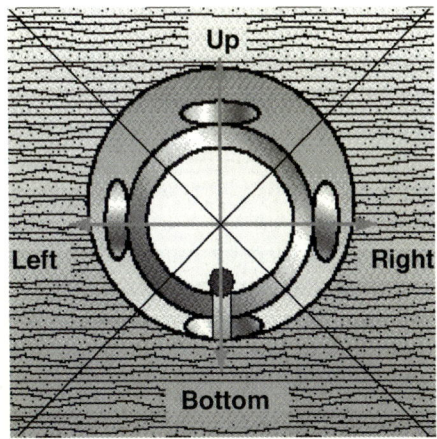

Figure 2.4.2. Division of the tool measurements into quadrants with respect to gravity tool face (GTF), a reference indicating the bottom of the well (from Evans et al., 1995).

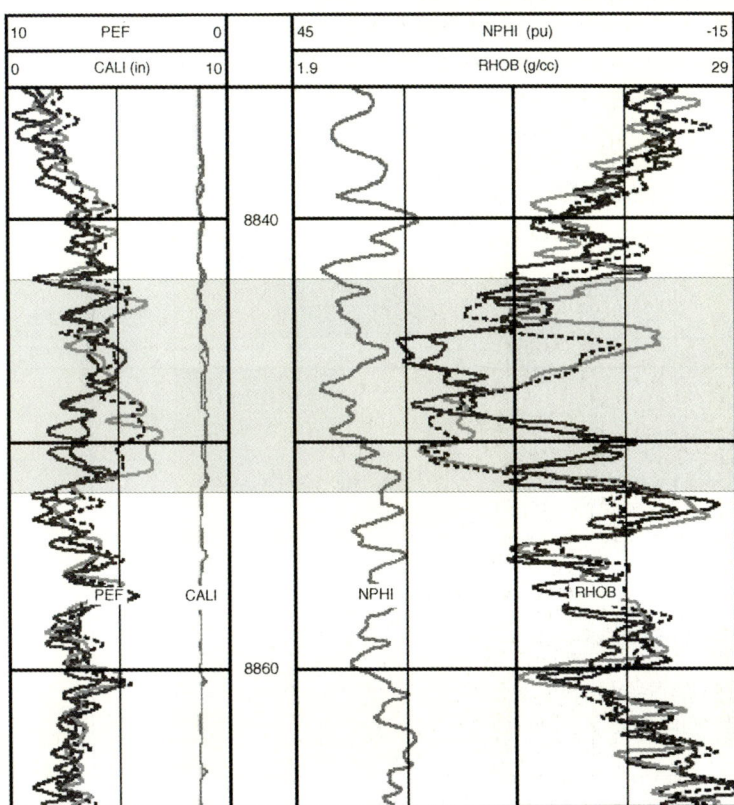

Figure 2.4.3. Example of density (RHOB) and photoelectric factor (PEF) curves for the four quadrants together with the averaged neutron porosity (NPHI) and the ultrasonic calipers (CALI) used to correct the neutron measurement for standoff. In these glauconitic and sideritic sandstones the four azimuthal curves generally track each other except in the interval indicated in grey, where high layer dips or formation heterogeneities cause them to diverge. For curve position refer to figure 2.4.4 (after Holenka et al., 1995). Depth in feet.

tain sufficient count statistics in the detectors, but the measurement resembles more a dipmeter than a borehole image. Ultrasonic stand-off measurement are done in each of the four quadrants using a reflection method similar to the one discussed in chapter 2.3. They are used to correct the neutron porosity measurement for the strong influence of the borehole fluid.

More recently, however, the density and photoelectric measurements are grouped into sixteen azimuthal bins during one rotation, i.e. each quadrant has been divided by four (Evans et al., 1995). This reduces the count rates, an important consideration in nuclear logging techniques, but the density and photoelectric measurement have often considerably better count rate statistics than the neutron porosity tool, for which such an increase in azimuthal sampling is not possible. The measurement thus becomes a borehole image (figure 2.4.4), with an azimuthal sampling slightly better than the wireline resistivity imaging (12 azimuthal samples) but below the LWD resistivity imaging (56 samples). The tool is mounted on a combination string of LWD tools developed for relatively small boreholes together with resistivity and gamma ray measurements. It represents a significant step for geologists and drilling engineers, because it allows better geometric and geological control in well drilling, particularly in deviated and horizontal wells. Unlike electrical imaging, this tool can be run in water- as well as oil-based muds.

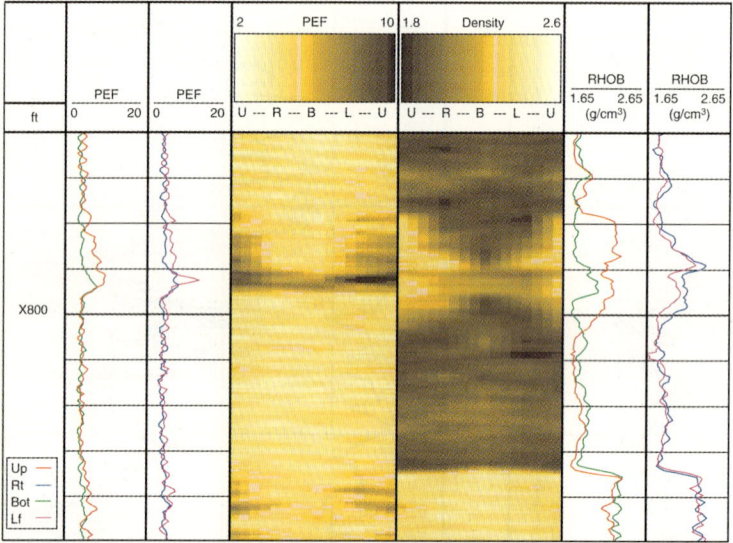

Figure 2.4.4. Density and photoelectric images, together with their average measurements for each quadrant (Up = U, Right = R, Bottom = B, Left = L). Notice how the two images differ from each other, indicating that porosity and mineralogy change independently from each other. This is best illustrated by the abrupt change seen on the density images close to the bottom but not seen on the P_e images. It may represent a lithological change without a corresponding change in the photoelectric absorption factor (from Carpenter et al., 1997).

As an operational precaution, the ADN tool contains a system that allows retrieval of the nuclear sources in case the drill string gets stuck. This avoids having to leave nuclear material in the well, an important safety consideration. A downhole memory of 10 Mbytes is capable of storing data over a long drill section, and the data can be downloaded directly using a wireline cable from the surface without "tripping", i.e. without pulling the entire drill string. In rotating mode, where the entire drill string turns, the slim version of these tools can support build-up rates of the well deviation of up to 16°/100 feet. In sliding mode, where a downhole motor turns the drill bit while the drill pipe slides, the build-up rate can be as high as 30°/100 feet. Accuracy of the measurements is similar to those of wireline tools, except at very high rates of penetration, where the count rates per unit interval decrease and the accuracy as well. Vertical resolutions are comparable to wireline tools, with the density measurement at around 6 inches (15 cm), the photoelectric measurement around 2 inches (5 cm), and the neutron porosity measurement around 12 inches (30 cm).

2.4.2. Applications

The ADN tool has many petrophysical applications that include the traditional uses of bulk density and photoelectric factor measurement in formation evaluation, (chapters 1.3 and 2.7). The azimuthal information provided by the images can be used to enhance the petrophysical analysis. There are several applications of geological interest that are similar although not identical to those of electrical and acoustic imaging (Bornemann et al., 1998). They include

- Structural definition of the reservoir
- Heterogeneity identification
- Geological well control in LWD

These three applications are naturally somewhat linked to each other. Structural dips can be determined from the layering on the images, or by correlating the four density or P_e, curves from the four quadrants in a manner similar to dipmeter curves. Since LWD measurements are often performed in deviated or horizontal wells, a conversion from apparent to true dip is necessary. The changes in bedding dips seen on figure 2.4.5 illustrate how geological structure can be defined with these measurements. Equally, figure 2.4.3 illustrates in the middle section of the interval how dipping layers or heterogeneities can result in unequal response of the four curves (Holenka et al., 1995; Wolcott et al., 1995). Such features can help the petrophysicist define an appropriate interpretation model, and the geologist can used them to identify lithofacies changes. Perhaps the most significant contribution of this new tool is its use for drilling decisions on the wellsite (Prilliman et al., 1995; Bornemann et al., 1998). The availability of borehole images at the surface – which may soon be available in real-time –

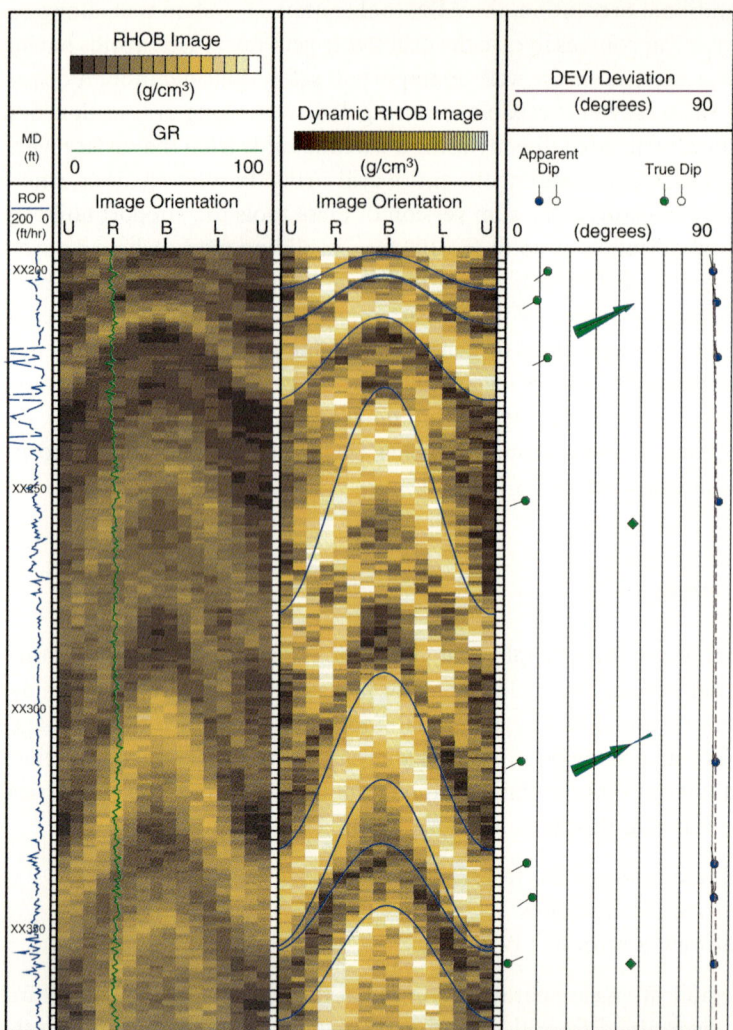

Figure 2.4.5. Photoelectric and density images from the ADN tool in a horizontal well with interbedded sand/shale layers dipping at about 10° (top of interval) to about 5° (bottom of interval) towards SW. Notice how the amplitudes of the sinusoids vary as a result of this dip change and a slight upward bend in the borehole deviation. MD = Measured depth along hole; ROP = Rate of penetration. Left image is raw data, right image is contrast-enhanced. Image orientation labeled as in figure 2.4.4.

allows drillers and geologists to make early drilling decisions and thus place the well in the desired target (chapter 3.5).

Fracture analysis is difficult with density imaging because of the small effect fractures have on the bulk density of the rock. However, exceptionally wide fractures may bee seen. Detailed sedimentological analyses are also not possible for lack of resolution of the images.

References

Bornemann E, Bourgeois T, Bramlett K, Hodenfield K, Magggs D (1998) The applications and accuracy of geological informaiton from a logging-while-drilling density tool. Trans 39th Symp Soc Prof Well Log Analysts: Paper L.

Carpenter WW, Best D, Evans M (1997) Applications and interpretation of azimuthally sensitive density measurements acquired while drilling. Trans 38th Symp Soc Prof Well Log Analysts: Paper EE.

Evans M, Best D, Holenka J, Kurkoski P, Sloan W (1995) Improved Formation Evaluation Using Azimuthal Porosity Data While Drilling. 70th Ann Conf Soc Petrol Eng: Paper 30546.

Holenka J, Best D, Evans M, Kurkoski P, Sloan W (1995) Azimuthal porosity while drilling. Trans 36th Symp Soc Prof Well Log Analysts: Paper BB.

Prilliman JD, Allen DF, Lehtonen LR (1995) Horizontal well placement and petrophysical evaluation using LWD. 70th Ann Conf Soc Petrol Eng: Paper 30549.

Wolcott DS, Shafer D, Vittachi A (1995) Characterization of heterogeneous formation using azimuthal logging-while-drilling (LWD) measurements. Trans 36th Symp Soc Prof Well Log Analysts: Paper Y.

2.5 Optical Borehole Imaging

2.5.1 Measurement Principles

Optical imaging is the oldest borehole imaging technique. The earliest devices were simply photographic cameras, which were lowered into the borehole on a cable. Photographic techniques have a much higher resolution than other borehole imaging techniques, but the early systems did not produce an image until after the device was retrieved to the surface and the film was developed. The advent of small video cameras in the 1960s allowed continuous recording of images downhole (Briggs, 1965) although transmission was limited and the temperatures and pressures in oil wells put serious constraints on the range of applicability. Additionally, the opaque nature of the muds in practically all oil wells posed a serious problem although Mullins (1966) describes a technique where a camera is mounted onto a packer system, and water is pumped in front of the camera lens to provide visibility. Despite this effort, however, optical borehole imaging stayed in the domain of environmental and mining applications where water is commonly used as drilling fluid. Cased-hole applications were also reported, notably in the form of casing inspection or to monitor fluid entries from perforations.

Improvements in lighting, electronics and transmission have significantly increased the capabilities of optical borehole imaging in the last two decades (Lau et al., 1987; Hawkins et al., 1989; Darilek, 1986; Cobb & Schultz, 1992; Maddox, 1996; Schultz, 1996, and Whittaker & Linville, 1996). Today, several companies offer commercial video camera services for borehole inspection in tubings, casings, and open holes in both oil and water wells.

One of the more widely available of these, the Optic Video System developed by DHV International, is rated at a borehole temperature of 125 °C (275°F) and a maximum pressure of 69 MPa (10,000 psi). Its outer diameter is 4.3 cm (1 11/16 inch) and is thus able to go through most tubings and casings. It is deployed on a fiberoptic cable with a maximum length of 4880 meters (16,000 feet) and records live video pictures in VHS format at a rate of 30 frames per second. The image size is 550×350 pixels, with a 55° viewing angle in water, and 73° in air. Illumination is provided by a 100 watt halogen lamp. At desired depths a still picture can be taken and transmitted to the surface. It is available in JPEG format and is 640×480 pixels in size.

The most important aspect of this system is the cable design, in which armor, insulators and buffers combine to give the fiberoptic cable a tensile strength of 2000 pounds per foot. This is much higher than in previous versions and increases the operational scope of the system considerably, although still being one of the limiting factors in the tool's range. Another interesting aspect of this system is provided by a special polymer surfactant applied to the optical view port, which repels oil and thus keeps the lens clear. The video camera is centered in the middle of the borehole, pointing downwards and providing continuous pictures while the sonde is moved in the borehole.

For greater depths, DHVI mounts the same camera on any type of logging cable and records videos at a rate of 1.7 frames per second with a slightly lower resolution than with the fiberoptic cable. Again, still pictures can be taken whenever needed. A high-temperature version is slightly bulkier (5.4 cm, or 2 1/8 inch) and can withstand temperatures up to 177 °C (350°F) up to four hours thanks to cooling with a Dewar flask. DHVI can also deploy the camera on a17,000 feet (5185 meters) long coiled tubing, which is of particular interest in horizontal wells.

Other companies providing downhole video services rated for pressure of at least 10,000 psi include Read Well Services, Sondex, and Venture Service. Some systems provide side-viewing cameras, which have the advantage of giving a relatively undistorted view of the borehole wall, but these are often bulkier and therefore do not fit through small tubulars.

2.5.2 Applications

The quality of optical borehole images is almost completely determined by the clarity of the fluid. The borehole muds used in the oil industry are usually opaque, which prevents a widespread use in open hole for geological applications. Even in wells drilled with water the fluid may be cloudy, and time-consuming well preparation methods have to be used (Whittaker & Linville, 1996).

There are, however, examples of optical images recorded in wells drilled with water showing geological features with astounding clarity and detail (Overbey et al., 1988; Palmer & Sparks, 1990). Figures 2.5.1 and 2.5.2 illustrate views of a fault

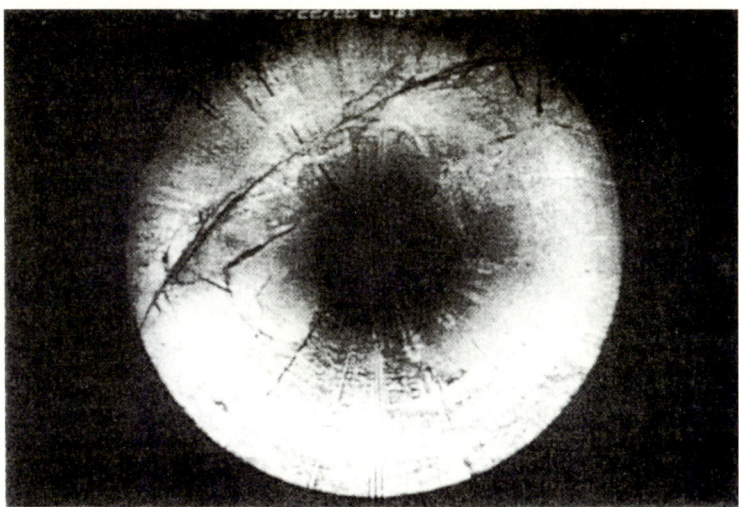

Figure 2.5.1. Fisheye video view in a borehole of a large displacement fault in volcanic rocks (from Lau et al., 1987)

Figure 2.5.2. Side view of the fault in figure 2.5.1 taken with an inspection lens (from Lau et al., 1987).

in volcanic sequence, taken with an older system featuring an underwater video camera with a fisheye lens and pan-and-zoom capabilities. The fisheye view (figure 2.5.1) allows identification of the fault, while the zoom picture (figure 2.5.2), taken with a special inspection lens looking sideways, shows the fault including its fault gauge in more detail. Notice that the resolution of these images is at least one order of magnitude better than other borehole imaging techniques. Figure 2.5.3 is recorded with DHVI's Optic Video System in the Clearfork formation, a naturally fractured dolomite that received additional fracture treatment. The image shows a fracture, most likely to be hydraulically induced, running down two sides of the borehole. Some layering may be present but is difficult to ascertain. In figure 2.5.4, recorded almost 30 meters higher up, significant borehole collapse is seen, with large dolomite blocks seemingly detached in the borehole. This may be a fault that has been additionally broken up by the hydraulic fracture treatment. A particularly interesting feature in this image is the stream of dark oil bubbles emanating from the fault and flowing uphole.

Most open-hole videos are run in geotechnical, environmental, and research wells. In oil wells, they are restricted to high water-cut wells. Downhole videos have been run in open holes in gas wells including coalbed methane wells, in air-drilled holes, and even in some research wells drilled through ice (Harper & Humphrey, 1994).

In summary, the main geological applications in open holes include the following:

- Identification of fractures and faults
- Identification of breakouts and other borehole damage

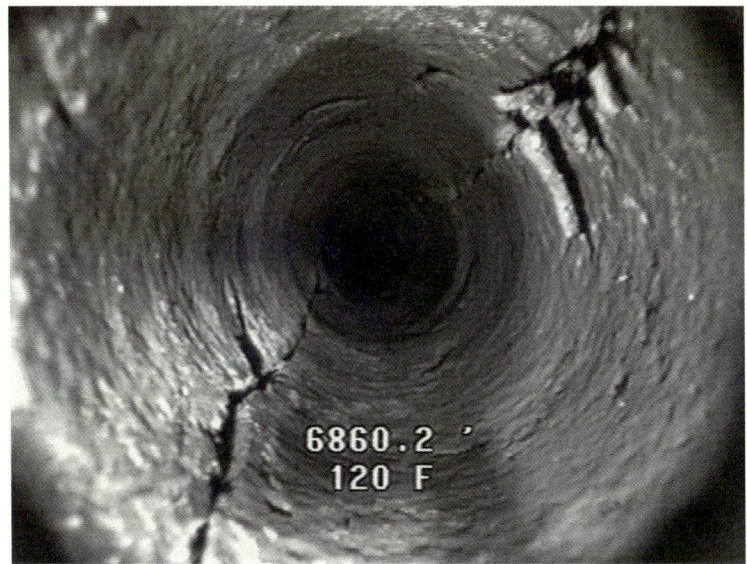

Figure 2.5.3. Downhole video still picture in the Clearfork dolomite (Texas) showing a hydraulic fracture with smaller joints branching off. Borehole depth and temperature as indicated on figure (courtesy DHV International)

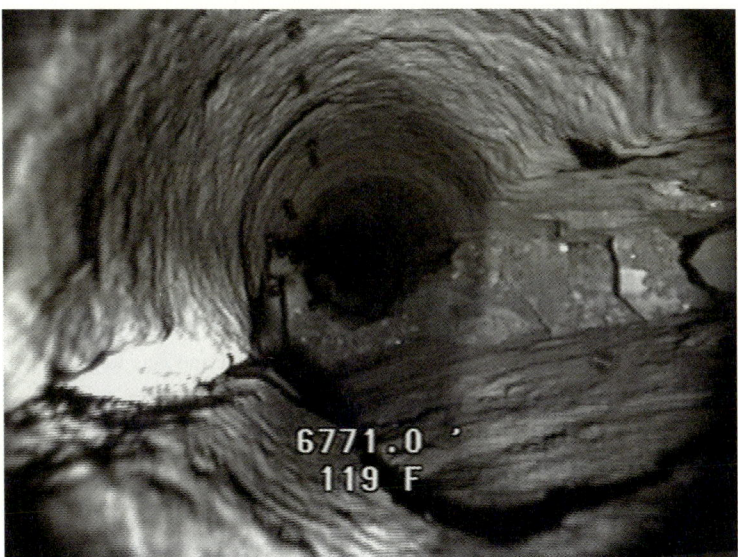

Figure 2.5.4. Downhole video still picture in the Clearfork dolomite (Texas) showing a partially collapsed borehole around a fracture or perhaps a fault zone. Oil bubbles are seen flowing out from the crack (center top). Borehole depth and temperature as indicated on figure (courtesy DHV International).

- Detection of productive zones in high water-cut wells
- Some information on bedding, grain size, and porosity types

In cased holes it is easier to obtain good visibility and, therefore, downhole videos are much more common than in open holes. Water can be introduced into shut-in well, or a flowing well may have predominantly water. The major applications of optical borehole imaging in cased holes include the following:

- Inspection of tubing and downhole equipment
- Identification and observation of perforations
- Detection of fluid entries
- Identification and orientation of stuck equipment ("fish")
- Detection of particulate entries

References

Briggs, RO (1964) The downhole TV camera. Trans 5th Symp Soc Prof Well Log Analysts: Paper N (abstract).
Cobb CC, Schultz PK (1992) A real-time fiber optic downhole video system. Proceedings 24th Offshore Technology Conference, Houston: Paper 7046.
Darilek GT (1986) A color borehole television system for developing gas production from Devonian shales. Symp on Unconventional Gas Technology, 61st Ann Conf Soc Petr Eng: Paper 15219.
Hawkins WL, Oliver RD, Lavelle MJ (1989) Borehole inspection system for large diameter boreholes. The Log Analyst 30, no. 1: 26–30.
Harper JT, Humphrey NF (1994) Borehole video observations within a temperate valley glacier – implications for englacial and subglacial processes (abstract). EOS Trans Amer Geophys Union 74 (supplement): 224.
Lau JSO, Auger LF, Bisson JG (1987) Subsurface fracture surveys using a borehole television camera and acoustic borehole televiewer. Can Geotech J 24: 499508.
Maddox S (1996) Visualizing production in flowing wells. Trans 37th Symp Soc Prof Well Log Analysts: Paper EEE.
Mullins JE (1966) Stereoscopic deep well photography in opaque fluids. Trans 7th Symp Soc Prof Well Log Analysts: Paper N.
Overbey WK Jr, Yost LE, Yost AB II (1988) Analysis of natural fractures observed by borehole video camera in a horizontal well. Soc Petrol Eng Gas Technology Proc: Paper 17760.
Palmer ID, Sparks DP (1990) Measurements of induced fractures by downhole television camera in coalbeds of the Black Warrior Basin. 65th Ann Conf Soc Petr Eng: Paper 20660.
Schultz PK (1995) DHV systems allow downhole vision. Amer Oil and Gas Reporter 38: 60–65.
Whittaker JL, Linville GD (1996) Well preparation – essential to successful video logging. Trans Soc Petr Eng 66th Western Region Meeting: Paper 35680, 297308.

2.6 Nuclear Magnetic Resonance Logging

2.6.1 Introduction

Nuclear magnetic resonance (NMR), or "nuclear induction" as it was initially called, was developed in the late 1940s (Bloch et al., 1946). Its principle is based on the response of nuclei to magnetic fields (for terminology, see table 2.6.1). Some nuclei, particularly protons – which form the nuclei of hydrogen atoms – have a magnetic moment which can be visualized as a spinning bar magnet. These magnetic moments can be detected with suitable measurement set-ups and an estimation of the location and amount of hydrogens in a sample can be obtained. As hydrogen is abundant in both oil and water, NMR logging was soon proposed for oilfield applications. Chevron developed and patented a NMR logging tool in the 1950s, and in 1960 ran the first NMR log (Brown & Gamson, 1960). One of their primary goals was to quantify the amount of tar in some of their Californian reservoirs. When compared to lighter oils and water, hydrogen atoms in heavy oil take longer to respond to a magnetic field and, conversely, lose an acquired precession faster when the magnetic field is turned off (they "relax" faster). To this day, this represents one of the unique and more interesting applications of NMR logging.

Although commercial NMR logging tools have been around since that time, they failed to become popular, mostly because they relied on the relatively weak magnetic field of the earth to make the protons precess. Additionally, the borehole signal from the hydrogens in the mud had to be suppressed by doping the mud with a material such as magnetite. This was generally quite inconvenient and therefore NMR logging was only rarely done. Laboratory research, however, continued and produced important insights into the use of NMR, notably for the estimation of permeability and the "free-fluid index" (Seevers, 1966; Timur, 1968;

Table 2.6.1. Terminology and synonyms used in NMR logging

Term	Synonyms
nucleus	spin
	magnetic moment
	hydrogen
	proton
T_1	longitudinal relaxation time
	spin-lattice relaxation time
	thermal relaxation time
T_2	transverse relaxation time
	spin-spin relaxation time
	Carr-Purcell-Meiboon-Gill (CPMG) decay time
T_2^*	free induction decay (FID) time
	dephasing time

Table 2.6.2. Main characteristics of the CMR and the MRIL NMR logging tools

	CMR[1]	MRIL[2]
Tool type	Pad tool	Mandrel tool
Frequency of measurement	2.2 MHz	550–750 kHz
Shape of sensitive volume	Saddle zone in front of pad	Tubes coaxial with tool axis
Vertical extent of sensitive volume	15 cm	60 cm
Depth of investigation	2.8 cm from pad	20 cm from tool axis
Wait time (range)	1.3–6.0 s	1–12 s
Echo spacing T_E	0.2 ms	0.6/1.2/2.4/3.6/4.8 ms
Number of echo spacings per measurement	600	400–900

[1] CMR-D tool
2 MRIL-Prime tool

Loren & Robinson, 1969; Loren, 1972). These efforts were crucial in further developing the tool. In 1984, Jackson presented a new NMR logging tool design for which he had previously obtained a patent (Jackson, 1984). His device uses a permanent magnet many times stronger than the earth's magnetic field to align the protons, and a pulse-echo measurement cycle with a radiofrequency antenna similar to what was used in the laboratory. Today, two commercial tools – both inspired by Jackson's ideas – are available on the market, Schlumberger's Combinable Magnetic Resonance Tool CMR* (Kleinberg et al., 1992; Morriss et al., 1993; Freedman et al., 1997), and NUMAR's Magnetic Resonance Imager Log MRIL** (Miller et al., 1990; Prammer et al., 1996; Prammer et al., 1999). Although different in geometry and in several other design aspects (table 2.6.2) both tools probe a very well defined volume of the formation, thereby eliminating – or at least largely reducing – the borehole signal and the need to dope the mud. This is achieved by accurately tuning the measurement frequency to the strength of the magnetic field in the rock volume of interest. These tools have made NMR logging very popular to the extent that it has become part of the standard measurement suite in many areas[1].

Although nuclear magnetic resonance (NMR) logging is mostly used for petrophysical purposes, it can provide very useful geological information. While nuclear porosity logs respond to bulk porosity, the NMR measurement is sensitive to the *pore sizes* and is, therefore, the only log responding to microscopic,

* Mark of Schlumberger

** Mark of Numar, a wholly-owned subsidiary of Halliburton

[1] Notice that from conception of the idea to availability of a reliable commercial tool about 30 years have elapsed. Many service companies strive for "product cycles" of 2 to 5 years, but reality often looks different.

textural properties of the rock. Its primary application, however, is to determine the total porosity, the amount of bound water (which is contained in pores below a certain size), and to estimate the rock permeability.

NMR logging applications have been the subject of a large number – if not the majority – of recent publications in the logging literature. The field, however, is far from being fully explored, and this chapter is therefore merely a snapshot in time. Its purpose is to discuss briefly the measurement principles and to summarize the various applications of NMR logging, including those of more geological interest.

2.6.2 Measurement Principle

NMR logging consists of a rapid sequence of manipulations of the hydrogen nuclei contained in the pores of the rock. In the following, we discuss the various steps in this measurement cycle. For a list of the terminology and synonyms used in NMR refer to table 2.6.1.

Proton Alignment. The current NMR logging tools contain a permanent magnet which, as the tool approaches a given rock volume, aligns the proton spins in the direction of this strong static field B_0. In the CMR, the magnetic field is about 550 gauss, or 0.055 tesla, which is about 1,000 times larger than the earth's magnetic field. The direction of B_0 is perpendicular to the borehole axis, and the protons precess around an axis parallel to this direction (figure 2.6.1). The net magnetization is determined by the number of protons aligned in this direction, a process which occurs exponentially with time following

$$M = \Phi\left[1 - e^{-WT/T_1}\right] \qquad (2.6.1)$$

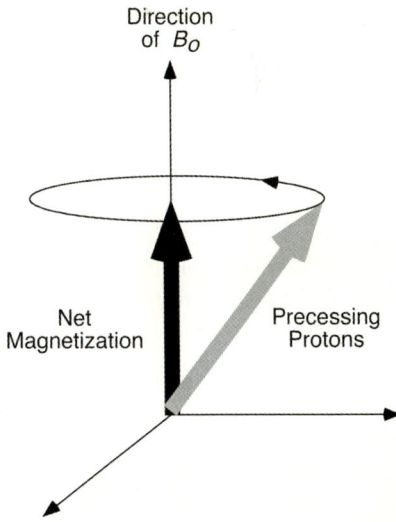

Figure 2.6.1. The magnetic moments of the precessing protons combine to produce a net detected magnetization transverse to the external magnetization B_0.

where M is the NMR magnetization signal, Φ the porosity, WT the wait time, or exposure time of the protons to the magnetic field, and T_1 the time constant at which the exponential alignment of protons occurs. Thus, after wait time $WT = T_1$, 63.2 % of the protons are aligned, after $2T_1$ 87.7 % and after $3T_1$ 96.1 %. The time constant T_1 is called *longitudinal relaxation time* but has other synonyms too (table 2.6.1). This proton alignment is reversible, i.e. when the magnetic field is switched off, the protons relax or dephase with the same time constant and the NMR signal decays accordingly.

If no rock is present, T_1 is a property of the fluid and a function of its temperature, and viscosity and is called *bulk relaxation*. Typically, T_1 increases with increasing temperature and with decreasing viscosity. For water at room temperature, T_1 is about 3 seconds. Differences in T_1 are used for discriminating hydrocarbons from water with a method called "direct hydrocarbon typing technique" that will be described below.

In a porous rock filled with water, T_1 is dependent on the pore size, or, more accurately, on the ratio of volume over surface of the pores. This ratio has the dimensions of length and, for a given pore shape, is directly proportional to pore size. In addition, T_1 is inversely proportional to the *surface relaxivity* ρ of the rock according to

$$\frac{1}{T_{1s}} = \rho_1 \left(\frac{S}{V}\right)_{pore} \qquad (2.6.2)$$

ρ is the rate at which the surface of the minerals constituting the rock influences the magnetic moment of the protons in the fluid, and it is usually correlated with the iron contents in the framework and accessory minerals. Sandstones are found to have higher values of ρ than carbonates, meaning that for a given pore structure the protons align and relax faster in sandstones than in carbonates. Typical values are $\rho \approx 5\ \mu m/s$ in sandstones and $\rho \approx 1.7\ \mu m/s$ in carbonates (Kleinberg, 1999). This difference compares well with the average iron contents for the two rock groups. Pettijohn et al. (1973, p. 60) report an average iron oxide content of 1.38 % in 253 sandstone samples, while Pettijohn (1975, p. 327) gives an average iron oxide content of 0.54 % in 345 limestones, or almost three times less than in sandstones.

Figure 2.6.2 shows the increase in the NMR signal due to the proton alignment for two different relaxation times, a short one (0.2 seconds) and a long one (2 seconds). In order to has a good signal in the subsequent measurements, a significant number of protons have to be aligned with the magnetic field. The exposure time of the protons to the magnetic field, therefore, has to be adapted to the expected values of T_1. This delay, called *wait time*, controls the logging speed in a single-frequency tool such as the CMR, and depends on the purpose of the measurement. For example, in "bound fluid logging", a measurement where only the capillary-and clay-bound pore fluids are determined, the expected relaxation

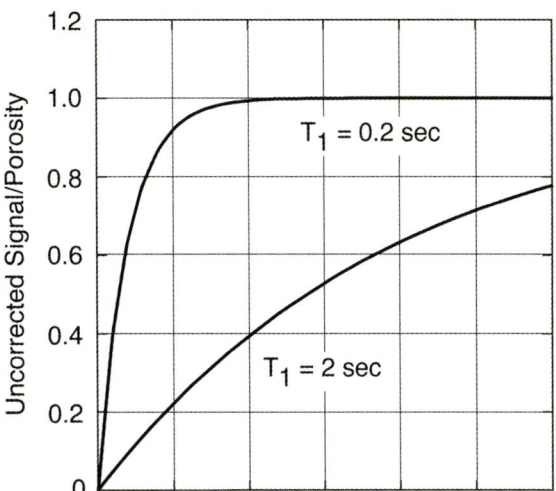

Figure 2.6.2. Example of the proton alignment for two longitudinal relaxation times T_1.

times are very short and the logging can therefore proceed at relatively high speeds (Singer et al, 1997; Flaum et al, 1998).

Sensitive Volume. The permanent magnets used in NMR logging do not produce a field of constant strength, but contain gradients that depend on the geometry and type of the magnets (figure 2.6.3). In the case of the MRIL, which is a centralized "mandrel" tool, the field strength decreases radially away from the tool axis in a more or less monotonic way (Miller et al., 1990; Prammer et al, 1999). The CMR, by contrast, is a pad-type tool and its magnets produce a complicated magnetic field, which, in front of the pad, has a "saddle", or a zone of relatively constant field strength. The precession rate of the protons is given by

$$f_L = (\gamma/2\pi) \cdot B_0 \qquad (2.6.3)$$

where the precession rate f_L is called "Larmor frequency", B_o is the external magnetic field, and γ the "gyromagnetic ratio" which is a property of the nucleus. For a proton, $\gamma = 2\pi \cdot 4258 \text{ s}^{-1} \cdot \text{gauss}^{-1}$ and the Larmor frequency in a field corresponding to the saddle zone of the CMR is around $f_L = 2.3$ MHz. For the earth's magnetic field, this frequency is typically about one thousand times smaller. This dependence of the precession frequency on the magnetic field allows an accurate control of the volume that is measured by the NMR logs. Thus, the MRIL measurement takes place at several frequencies, which correspond to several thin tubes within the formation and centered with the tool axis. The CMR measurement, on the other hand, takes place at the frequency corresponding to the saddle zone, and the sensitive volume in three dimensions looks somewhat like a cigar.

CPMG Measurement Cycle. Aside from using strong permanent magnets, the CPMG measurement cycle is the second important innovation that sets the new

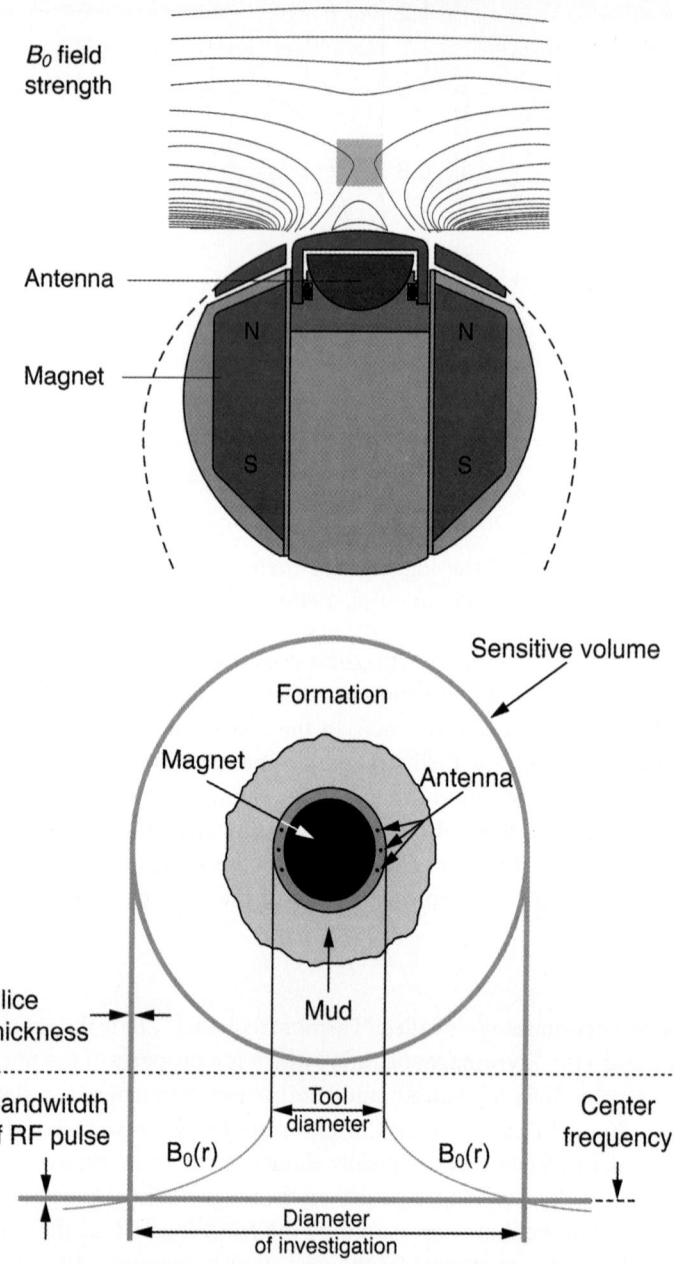

Figure 2.6.3. Top: Sketch of Schlumberger's CMR tool with the magnetic field strengths showing a saddle zone in front of the pad. The measurement takes place in this zone over a vertical distance of about 15 cm (Courtesy Schlumberger). Bottom: Sketch of NUMAR's MRIL tool with the magnetic field lines arranged concentric with the borehole. The measurement takes place in tube-shaped volumes shown in grey (reproduced by permission of Halliburton Energy Services, Inc.).

NMR logging tools apart from the older ones. The acronym stands for its inventors, Carr, Purcell, Meiboom, and Gill. The CPMG measurement sequence consists first of a tilt of the proton spin alignment by 90° into the transverse plane (the horizontal plane in figure 2.6.1), followed by a series of 180° rotation pulses. These pulses are made with a radio-frequency antenna whose position for the CMR is shown in figure 2.6.3 at a frequency corresponding to the field strength in the sensitive volume discussed above. The 180° pulses last about twice as long as the initial tilt by 90°, or a few tens of microseconds. The intervals at which they occur – the *echo spacing* – is 200–320 μs for the CMR and 600–1200 μs for the MRIL (see table 2.6.2). Typically, several hundred pulses are made in one measurement cycle.

The effect of the CPMG pulses is first, to rotate the protons into the plane perpendicular to the B_o field. In fact, the protons then rotate like tilted spinning tops in a gravity field. Initially, as they all have been tilted at the same time, they precess in a synchronous manner, thereby generating a small magnetic field at the Larmor frequency that can be detected by the radiofrequency antenna. However, due to imperfections such as gradients in the magnetic field, the protons will increasingly get out of synchronization with each other. This *dephasing* causes the signal in the antenna to decay at a time constant called the free induction decay (FID) or T_2^* (table 2.6.1). By flipping the spins of the protons by 180°, this dephasing is reversed. Ideally, the protons will completely regroup after a time equal to the one elapsed during dephasing.

While the dephasing caused by irregularities in the magnetizing field are completely reversible, there are dephasing effects that are not. These are caused by interactions among molecules, and between molecules and pore surfaces. This dephasing is measured as a decay of the signal amplitudes after each rotating pulse, the *spin echoes*, and the time constant T_2 of this decay is called *transverse*

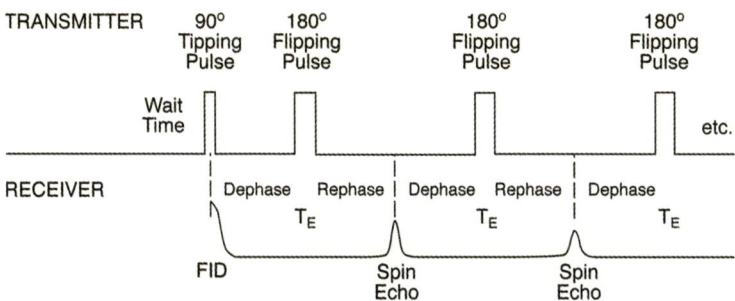

Figure 2.6.4. The CMPG measurement cycle. After the initial tipping pulse the free induction decay (FID) falls into the dead time of the electronics and is not recorded. The flipping pulse is usually repeated several hundred times, and an equal number of spin echo amplitudes is collected for each measurement. Typically, the wait time is in the order of a few seconds (which determines the logging speed), the pulse duration several tens of microseconds, and the echo spacing T_E between 0.2 (CMR) and 0.6 milliseconds (MRIL).

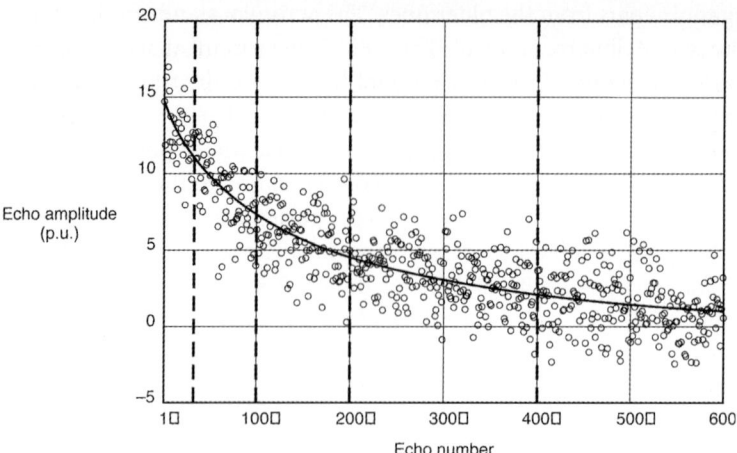

Figure 2.6.5. An example of the decay of echo amplitudes with time (or echo number). 600 echo amplitudes have been collected, resulting in a total measurement time of about 0.2 seconds. A smoothed average is shown by the solid line, and its average value is calculated for the windows delimited by vertical dashed lines. These values are used in the multiexponential fits to obtain T_2 distributions.

relaxation. The entire CPMG measurement cycle is illustrated in figure 2.6.4 and an example of the decay of the spin echo amplitudes is shown in figure 2.6.5.

Transverse Relaxation. The several hundred spin echoes collected during each CMPG measurement cycle are first smoothed and for a small number of decay time steps the average echo amplitude values are obtained (see figure 2.6.5). This smoothed amplitude decay curve is then decomposed into its time decay components with a multiexponential fit so that a distribution of the transverse relaxation times T_2 is obtained.

T_2 is the quantity of most interest in modern NMR logging as it contains useful petrophysical and geological information, and is quite easily measurable. In the presence of a fluid only, similar to longitudinal relaxation, it is a function of the type of fluid, its temperature, and viscosity, and it is called *transverse bulk relaxation* T_{2b}. In viscous oils, T_{2b} is equal to the longitudinal bulk relaxation time T_{1b}, but in lighter oils and water it is smaller due to diffusion effects which will be discussed below. Otherwise, the dependencies on viscosity and temperature are the same as for the longitudinal bulk relaxation.

In a water-filled rock, T_2 depends, like for longitudinal relaxation, on the surface relaxivity of the rock and the ratio of pore volume over pore surface:

$$\frac{1}{T_{2s}} = \rho_2 \left(\frac{S}{V}\right)_{pore} \tag{2.6.4}$$

T_{2s} is called the *transverse surface relaxation*. Since in most rocks there is not just one pore size present, but an entire range of pore sizes, the T_2 distribution ideally reflects the pore size distribution. In order to illustrate how the T_2 distributions are related to pore size, consider the schematic depicted in figure 2.6.6. Pores are shown in three different sizes, each corresponding to a characteristic relaxation

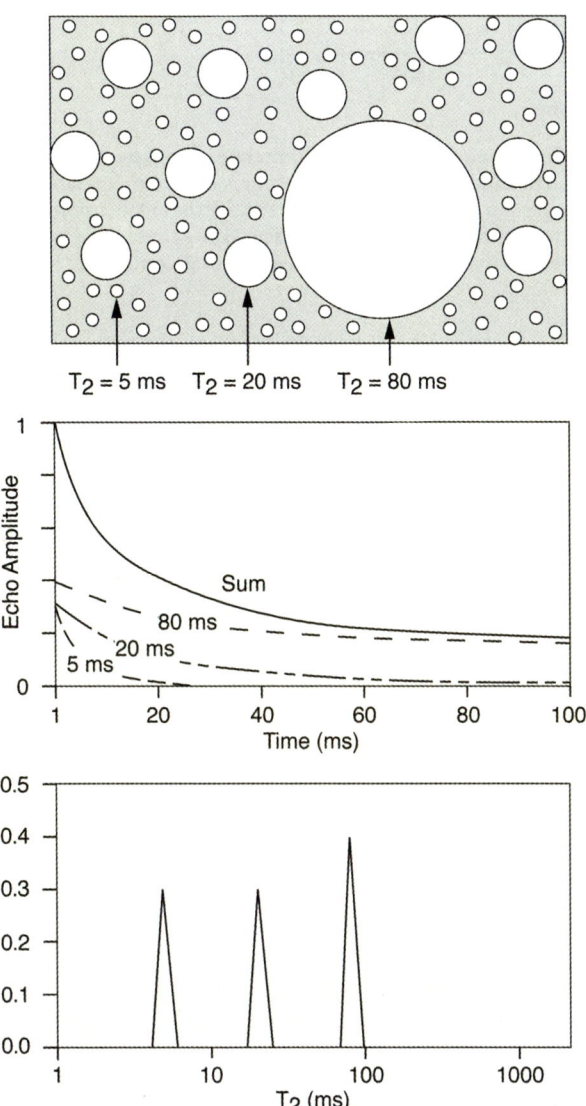

Figure 2.6.6. A water-filled rock containing one pore type in three different sizes (top) shows a total transverse relaxation of the spin echoes equal to the sum of the individual decays (middle). After decomposition into its components, a T_2 distribution is obtained which reflects the pore size distribution (bottom).

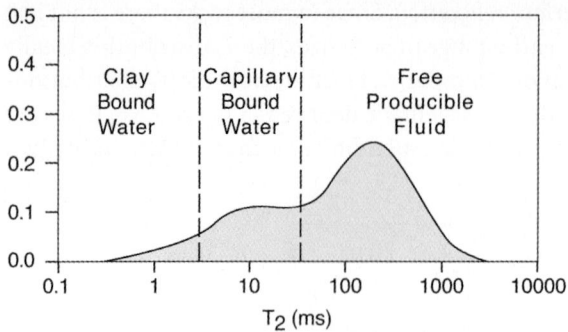

Figure 2.6.7. Example of a T_2 distribution. Shorter relaxation times correspond to smaller pores following equation 2.6.4. The fluids contained in these pores are often separated into free, producible fluid for the larger pores, capillary-bound water for the smaller pores, and clay-bound water for shortest relaxation times. Thresholds separating these three types of fluids are indicated on the figure but are generally adjusted to local conditions.

time T_2 (figure 2.6.6 top). The proton relaxation is uniform within each pore but differs from pore to pore (Kleinberg, 1994). The protons in the small pores relax very quickly, while in the largest pores they take longer and the decay curve is correspondingly flatter. The sum of the individual decay curves, weighted by their relative abundance, gives the total decay curve (figure 2.6.6 middle). The T_2 distribution is obtained by extracting the decay components in this curve through a multiexponential fit. In this example, the result is three discrete delta-functions whose peaks are at the T_2 values corresponding to the three pore sizes, and whose heights correspond to the volumetric abundance of each pore type.

A more realistic example of a T_2 distribution in a water-filled rock is depicted in figure 2.6.7. Thresholds are used to delimit various pore fluids contained in pores of different sizes. These are, from large to small pore sizes, free fluid, capillary bound fluid, and clay-bound fluid. The cutoffs depend on lithology and other factors, and often have to be fine-tuned locally. The cutoff between capillary-bound water and free fluids, for example, is usually taken at 33 ms in sandstones and somewhere from 100 to 200 ms in carbonates.

Complicating Factors. If rocks consisted of simple minerals and a well-defined pore system containing only water, such as the situation depicted on the left in figure 2.6.8, NMR log interpretation would be quite straightforward. The principal relaxation mechanism would be surface relaxation whose relationship to pore size is quite well understood. In reality, however, several complicating factors interfere. For example, two or even three different fluids may be present in the pore space resulting in partial saturations. A simple but common case is a water-wet rock containing oil, shown on the right in figure 2.6.8. The protons in the oil phase cannot reach the sites on the pore surface, and the oil-water interface does not act as a relaxation surface. Therefore, these protons relax at the bulk

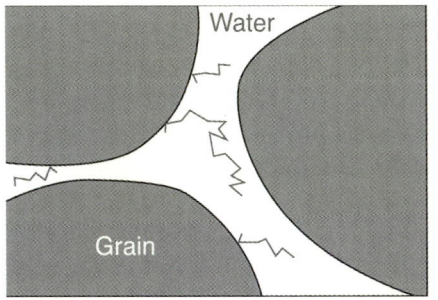 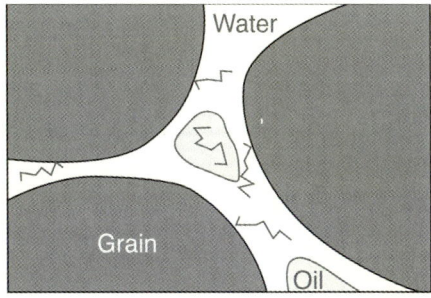

Figure 2.6.8. Water-wet rock (left) and partially filled with oil (right). The molecular diffusion is schematically indicated. Surface relaxation takes place when molecules collide with the grain surface. In smaller pores, this occurs more often than in larger pores. In the oil blobs, no collision with a grain surface is possible, and bulk fluid relaxation dominates.

relaxation rate, which is usually different from the surface relaxation rate. The protons in the water phase, on the other hand, have a much more confined pore space available, and their surface relaxation proceeds at a faster rate than in the water-filled scenario. Straley et al. (1991) have studied the NMR response of partially saturated rocks in the laboratory. They filled the samples first with water and then injected oil (actually kerosene) in stepwise increments, each followed by a T_2 measurement. A result is shown in figure 2.6.9. In the fully water-saturated case, surface relaxation dominates. As soon as kerosene is introduced, a bulk relaxation peak appears at about 1 second. It keeps growing in amplitude as more kerosene is injected while the surface relaxation signal of the water becomes smaller and shifts towards shorter T_2 times because of the smaller pore space available with decreasing water saturations. When there is no water left, the kerosene fills the entire pore space and its protons now relax at the surface relaxation rate, which, as is shown in the uppermost curve in figure 2.6.9, is faster than the bulk relaxation.

Diffusion is the second complicating factor. It affects only the T_2, not the T_1 measurement, and it is caused by small gradients in the magnetic field (Kleinberg & Vinegar, 1996). The protons, which diffuse with their molecules in the fluid, are exposed to changes in the magnetic field and relax faster than they would in a homogenous field. The effect is stronger in oils with low viscosities and in gas. Additionally, longer echo spacings increase the effect too because the molecules have more time to diffuse. Special acquisition and processing techniques can be used to take advantage of this effect (Akkurt et al., 1995; Prammer et al., 1995; Flaum et al., 1996)). Notably, the *shifted spectrum* method consists of making two logging passes, both with relatively long wait times, but one with a short and the other with a long echo spacing. Diffusion will affect the results with the long echo spacing more, resulting in a shortened relaxation of the peak corresponding to the gas or light oil. By calculating the difference between the two passes, a mea-

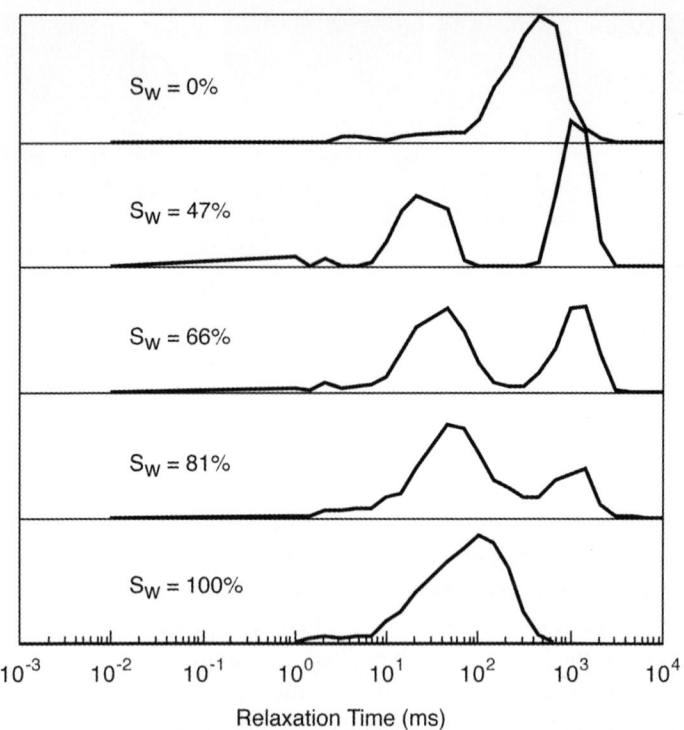

Figure 2.6.9. NMR laboratory measurements on a rock whose initial water saturation was stepwise replaced by oil (kerosene). The resulting T_2 distributions show surface relaxation only for the fully water-saturated case. As soon as oil is introduced, a bulk relaxation peak appears at around one second and grows in size as the amount of oil increases. Simultaneously, the surface-relaxation peak of water gets smaller (because of the smaller volume water occupies) and shifts towards shorter T_2 values because of the smaller water-filled pore space (from Straley et al., 1991). Refer also to figure 2.6.8.

sure of the amount of this fluid is obtained. Another method, the *differential spectrum* technique, consists of two passes with equal echo spacings but different wait times. Gas and light oil have long T_1 relaxation times, while water is comparatively short, so the additional polarization in the pass with the long wait time can be taken as a measure of the volume of light oil or gas. Both methods are merely qualitative and need local fine-tuning. They work better with strong tool gradients since diffusion relaxation times are inversely proportional to the square of the magnetic field gradient.

Ramakrishnan et al. (1998) have pointed out another complication in the interpretation of NMR signals which arises from the diffusive coupling between pores of variable sizes, for example in carbonates with intergranular pores and micropores. The result is a blurring of the T_2 spectrum, often associated with a shift of the peaks. However, they found that the relative volumes of these two pore types can still be determined.

2.6.3 Applications

Bound and Free Fluids

The quantification of these two types of fluids is one of the main applications of NMR logging. It is based on the principle that fluids – generally water – in the smaller pores relax faster and thus have shorter T_2 times (Timur, 1969). The threshold below which fluids are bound by capillary forces and therefore cannot be produced has been determined with laboratory and field experiments. It is found to be typically around 33 ms for sandstones and around 100 to 200 ms for carbonates (see figure 2.6.7). Water contained in the interlayers of clay minerals is often designated as clay-bound and its hydrogens have even shorter relaxation times, typically below 3 ms (Straley et al., 1997). Hydrogens atoms in the minerals themselves (such as in the hydroxyls of the clay lattices) have such short relaxation times that they are usually well below the detection limit of current NMR logging tools, which is about $T_2 \approx 0.2 - 0.5$ ms.

It is quite common in NMR logging today to represent the various fluid volumes obtained from thresholding the T_2 distributions with different curves. Figure 2.6.10 shows an example where the pore fluids have been grouped into free fluid (above 30 ms), capillary-bound fluid (3 to 30 ms), small pore-bound fluid (0.28 – 3 ms), and very small pore-bound fluid (0.2 – 0.28 ms). The latter had been obtained with an additional logging pass in which the echo spacing was reduced to 0.2 ms. The two small pore-bound fluids are water molecules trapped in the interstices of clay platelets and correspond to the "clay-bound fluid" mentioned earlier. This example is from a predominantly shaly section so that the differences in the bound fluids are more obvious. In fact, none of the small pore-bound fluids could be measured, with the older NMR logging tools, and the total NMR porosity was different from those obtained from nuclear and other porosity logs. Two sand-rich, thin layers can be seen in figure 2.6.10, both of which have pronounced peaks in the higher T_2 range well above the bound-fluid cutoff.

Since bound fluids have short polarization and decay times, their volumes can be determined with a reduced measurement cycle, notably by shortening the wait times and by reducing the number of echoes (Singer et al, 1997). This "bound fluid" logging procedure is therefore much faster than full NMR logging, often allowing logging speeds of up to 1100 m/hr. The purpose of the measurement is to obtain only those fluids decaying at 33 ms or less, i.e. the small pore-bound and capillary-bound fluids. It can be combined with a total porosity measurement in order to obtain the free fluid volumes.

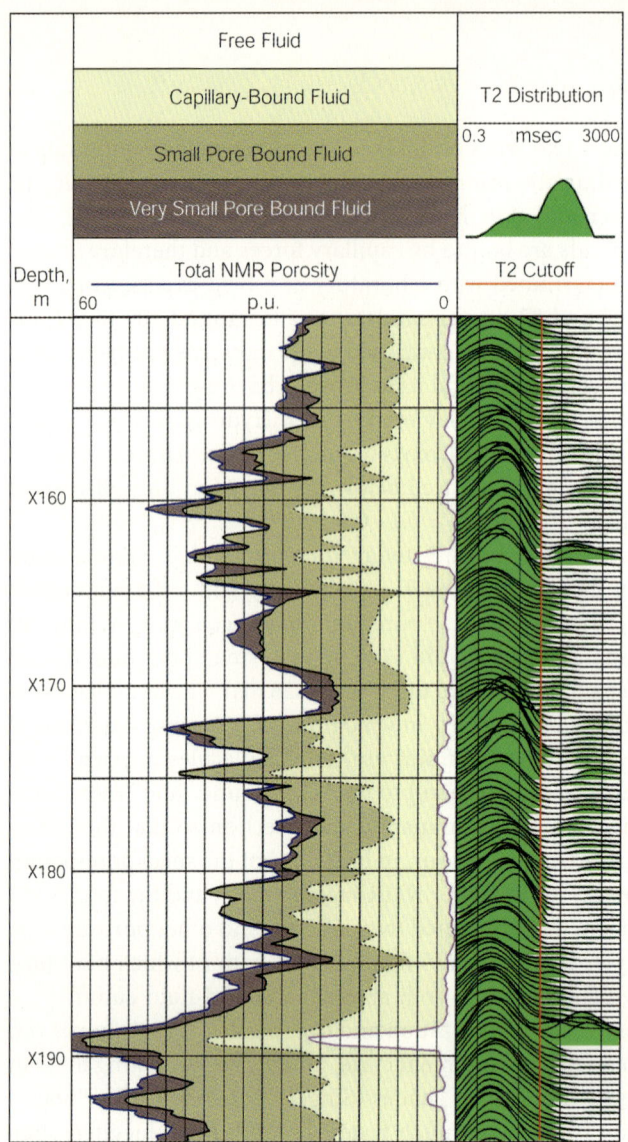

Figure 2.6.10. A NMR log example where the pore fluids have been grouped into free fluid (above 30 ms), capillary-bound fluid (3 to 30 ms), small pore-bound fluid (0.28 – 3 ms) and very small pore-bound fluid (0.2-0.28 ms). The latter had been obtained with an additional logging pass in which the echo spacing was reduced to 0.2 ms. The two small pore-bound fluids correspond to water molecules contained in the interstices of clay platelets. This example is from a predominantly shaly section here the differences in the bound fluids are more obvious.

Heavy Hydrocarbons

Very heavy hydrocarbons such as tar and bitumen are almost solids and the hydrogens they contain relax very fast because of their strong bonding forces. At higher temperatures, however, their viscosities are reduced and they contribute to the low end of the T_2 distribution. Figure 2.6.11 shows NMR laboratory measurement on a bitumen sample at three different temperatures. It can be seen that at room temperature much of the signal lies between 0.1 and 1 ms and is thus potentially measurable with a NMR log. With increasing temperature the distribution shifts towards longer relaxation times. At the highest temperature, corresponding to reservoir conditions, almost all of the signal is above 0.3 milliseconds and about half is above 3 milliseconds. This is well within the measurement range of NMR tools, but in practice the effect is small. Although the chemically determined hydrogen index (the concentration of hydrogen in the hydrocarbon relative to the one in water) is 0.3, the hydrogen index visible by NMR is about an order of magnitude lower. In this example, it decreases from about 0.05 to 0.02 with the increase in temperature. Some tars and bitumen, however, have larger hydrogen indices and produce a distinct bulk relaxation signal at short T_2 times. This signal is independent of whether the rock is water- or oil-wet.

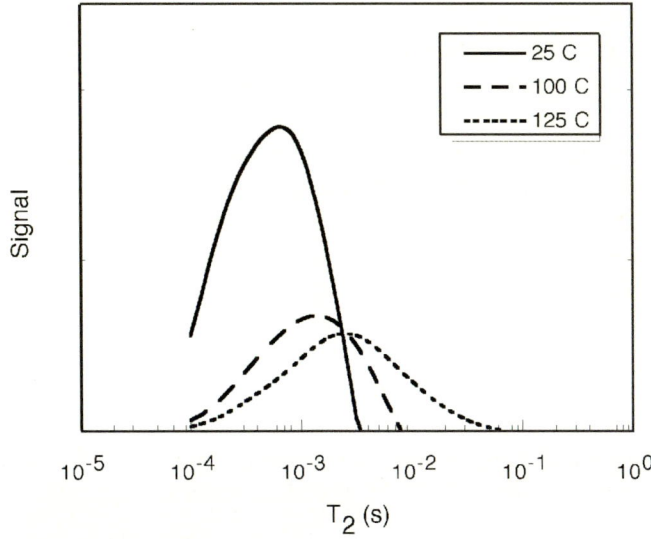

Figure 2.6.11. NMR measurements on a bitumen sample from a low porosity reservoir in Central Asia, where bitumen quantification is important to assess reservoir quality. The T_2 relaxation times were obtained in the laboratory and become longer with increasing temperatures. However, the low hydrogen index, the low porosity and very short NMR relaxation time make the bitumen virtually undetectable by NMR techniques – but not with nuclear logs. The difference can be used to estimate the amount of bitumen (courtesy S. Farooqui and R. Kleinberg).

In order to detect and quantify heavy hydrocarbons, therefore, a bulk relaxation signal has first to be looked for in the very short T_2 range. In fact, the bound fluid logging method described above can be used for this. If none is found, a different approach can be taken: First, a total porosity is obtained in the standard manner from density or a combination of the neutron/density measurements. Then the NMR measurement, which is unlikely to contain a heavy hydrocarbon signal, is compared to the total porosity, and the deficit is used to estimate the volume of heavy hydrocarbons. Appropriate corrections have to be made for the different hydrogen indices in each of these fluids.

The use of NMR logging for source rock quantification remains to be investigated.

Pore Size Distribution

The relationship of both longitudinal and transverse relaxation times with pore size has been given in equations 2.6.2 and 2.6.4 and was briefly discussed. It is important, however, to realize that the ratio of pore volume over pore surface is not the same as pore size or pore diameter. It merely reflects a metric for the general size of the pore, due to its dimension of length. In addition to this, it also depends strongly on the *pore shape*. The highest value is found in spheres (bubbles minimize their surface per volume), and with increasingly complex pore geometries this ratio becomes smaller. Much lower values are likely to be found in intergranular pores, which are by far the most common pores in reservoir rocks, both at the micro- as well as the mesopore scale. Their branched-out nature adds considerable surface to the pore, which will contribute to the surface relaxation of the hydrogen nuclei. Since the NMR T_2 distributions are essentially plots of volume (on the vertical) versus V/S (on the abscissa), the response of two pores of equal volume but different shapes may therefore fall into different places. Thus, pore shape modulates the relationship between relaxation times and pore size.

If all pores are geometrically similar such as, for example, in a poorly-sorted sandstone with predominantly intergranular porosity, the T_2 distribution can indeed be thought of as reflecting the pore size distribution. The exact conversion not only depends on the pore shape but also on the surface relaxivity of the rock minerals, a parameter that can be difficult to estimate (Basan et al., 1997).

Figure 2.6.12 shows a clastic sequence with shales overlying a sandstone in a water zone. The volumetric calculations in the left track indicate an upward-decreasing amount of clay in the sandstone interval, and conventional geological log analysis suggests that this be interpreted as a coarsening-upward sequence. Such deposits can be attributed to barrier bars and similar shallowing-upward settings. While this is all based on inference, the T_2 distribution shows that the relaxation times increase gradually upwards, a trend which can only be explained

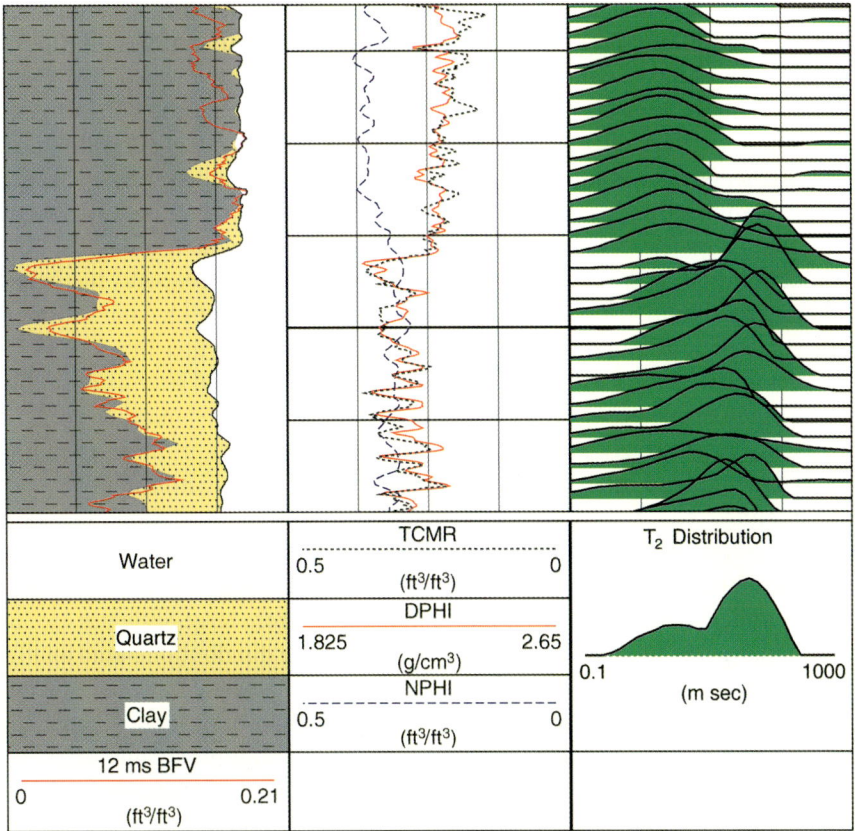

Figure 2.6.12. NMR T_2 results over a clastic sequence with a sandstone overlain by a shale. Since the entire section is water-filled, the increase in T_2 times towards the top of the sandstone directly indicates larger pore sizes and, therefore, larger grain sizes. Such coarsening-upward trends can only be inferred from other logs.

by an increase in the pore size and, therefore, an increase in grain size. Thus, while all other logs indirectly imply the coarsening-upward trend, the NMR log measures it quite directly and reinforces the geological interpretation of the nature of the deposit.

Carbonates, on the other hand, often feature a variety of pore types such as moldic, intercrystalline and interparticle pores. The relationship between relaxation times and pore size is therefore complex and certain assumptions regarding the pore type distributions need to be made. It is conceivable that a relatively broad T_2 distribution can be caused entirely by different pore shapes, but there are no published examples so far. A simple way of characterizing pore size is through the relative amount of microporosity. Because of its small size, microporosity is often water-filled and can strongly reduce the electrical resistivities such that the rock appears water-filled. Kenyon et al. (1995) have demonstrated the use of NMR data in quantifying the amount of microporosity in carbonates, using a thresholding procedure similar to the determination of bound fluids.

Figure 2.6.13 illustrates data from three carbonate rocks for which digital petrographic image analysis (PIA) data have been collected and the pore size distribution calculated, albeit in two dimensions only (Anselmetti et al., 1998). NMR measurements were made on the samples under fully water-saturated conditions, allowing a direct comparison with the PIA pore size distribution. Sample BA-1016, a sucrosic dolomite, has intercrystalline pores with a unimodal pore size distribution in the meso-pore range, i.e. larger than micropores but below the eye's resolution. The NMR T_2 distribution matches the PIA data well for a shift corresponding to $d = 90T_2$. Sample BA-2184, on the other hand, is a foraminiferal wackestone with abundant microporosity, defined here as pores below 20 μm in diameter. Again the NMR distribution can be well matched with the same shift. Interestingly, even the small peak just below 100 μm, corresponding to fossil moulds and interparticle pores, shows up on the T_2 distribution at about one second.

For textures with variable pore types and sizes, however, the picture becomes considerably more complex: Sample BA-526 in figure 2.6.13 contains an exceptionally wide range of pore types and sizes, ranging from large-scale vugs to intergranular pores, molds, and micropores. Its pore size distribution based on PIA spans three orders of magnitude, from almost a millimeter to less than one micron. The T_2 distribution shows only very vague similarities, even though the same shift had been applied as before. The peak distribution is quite different, and the bimodal nature of the pore size distribution is not obvious. Several reasons may account for these discrepancies:

- Different pore shapes, resulting in different values for V/S_{pore},
- Different surface relaxivities, caused by different diagenetic mechanism acting on the pores
- Coupling effects mentioned above, which occur when small and large pores are in hydraulic contact and the magnetic decays in the two pore types influence each other (Ramakrishnan et al., 1998).

These examples suggest that T_2 distributions can be used for the textural characterization of rocks and for facies identification. The NMR logs provide at each depth a spectrum, or a waveform, reflecting in some way the microscopic properties of the rock. Kenyon et al. (1989) have made promising steps in this direction, but more case studies are needed. Another possible use of the textural information contained in NMR log data is estimating the m and n coefficients used for resistivity interpretation (chapter 1.3). Allen et al. (1998) have started working in this direction, but again more studies are needed.

An impressive aspect of NMR-derived pore size distributions is the lower detection limit. For a minimum value of T_2 of 0.2 milliseconds and the relationship obtained from figure 2.6.13 we obtain a minimum detectable pore size of

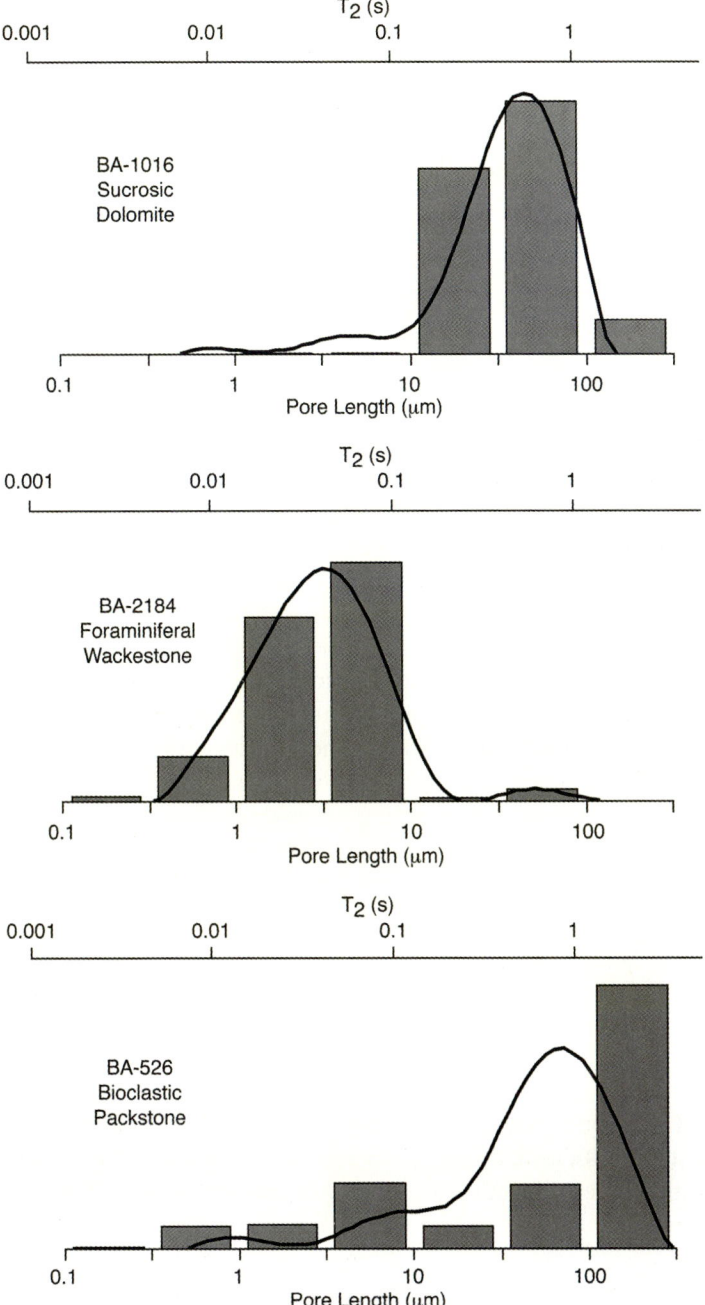

Figure 2.6.13. Three carbonate rock samples whose pore size distribution was quantified with digital petrographic image analysis (bar plots) and NMR T_2 measurements (line plots). Notice the good match when unimodal pore size distributions are present (upper two samples), while the polymodal sample presents a more difficult comparison (bottom).

$d = 0.02\,\mu$m. This is about one order of magnitude better than the resolution of a conventional scanning electron microscope (SEM) and two to three orders of magnitude better than thin section analysis with optical microscopes. On the other extreme, however, we find for a maximum measurable T_2 of 3 seconds the corresponding pore size to be 270 μm. All pores bigger than this will look the same, and in fact will show polarization decays equal to the bulk relaxation time of the fluid it contains[2].

Permeability

One of the earliest suggested uses of NMR rock measurements was the estimation of permeability (Seevers, 1966; Timur, 1968; Loren & Robinson, 1969; Loren, 1972). Relationships between porosity and permeability had been investigated long before that, but they were found to have a large scatter and to be rock-texture dependent. Wylie & Rose (1950) proposed using the irreducible water saturation S_{wirr}, together with the porosity, in order to estimate permeability in a general equation of the form

$$k = C \cdot \Phi^x \cdot S_{wirr}^{-y} \tag{2.6.5}$$

Subsequently, several workers proposed specific forms of this equation, usually based on laboratory data from rock samples (Timur, 1968; Coates & Dumanoir, 1974). With the advent of NMR logging came a more reliable way of estimating the irreducible water saturation, and more work was done to refine the methodology. Currently, two methods of estimating permeability from NMR logs are in use. One is based on equation 2.6.5 and is called the Timur/Coates equation. Instead of the irreducible water saturation it uses the bound fluid index BVI, with the two related to each other through $BVI = \phi \cdot S_{wirr}$, as well as the free fluid index $FFI = \phi \cdot (1 - S_{wirr})$. The equation reads

$$k = a \cdot 10^4 (\Phi_{NMR})^4 \left(\frac{FFI}{BVI}\right)^2 \tag{2.6.6}$$

with $a = 1$ mD. The second equation was developed by researchers at Schlumberger-Doll Research in Ridgefield (USA) and uses the logarithm of the average transverse relaxation time T_2, together with the NMR-derived porosity. The SDR equation has the form

$$k = b \cdot (\Phi_{NMR})^4 (T_{2\,\log})^2 \tag{2.6.7}$$

---

[2]Such pores are often referred to as "vugs" in NMR literature, while geologists prefer to use the term vug for "small cavities...from the size of a small pea upwards" (Bates & Jackson, 1980).

where $b = 4$ mD/(ms)^2. The two relationships are somewhat similar and strongly dependent on the total porosity. The Timur/Coates equation has the advantage of being applicable even if a strong bulk relaxation signal in the high T_2 range is present which shifts the mean relaxation time towards higher values. The SDR equation, on the other hand, avoids having to determine a free-fluid cutoff. In some cases both permeability estimates are made and shown on the same plot. Figure 2.6.14 contains an example of NMR-derived permeability using equation 2.6.6, together with volumetric estimates of fluids and minerals based on all open-hole logs. The well is from a clastic sequence in the Gulf of Mexico and

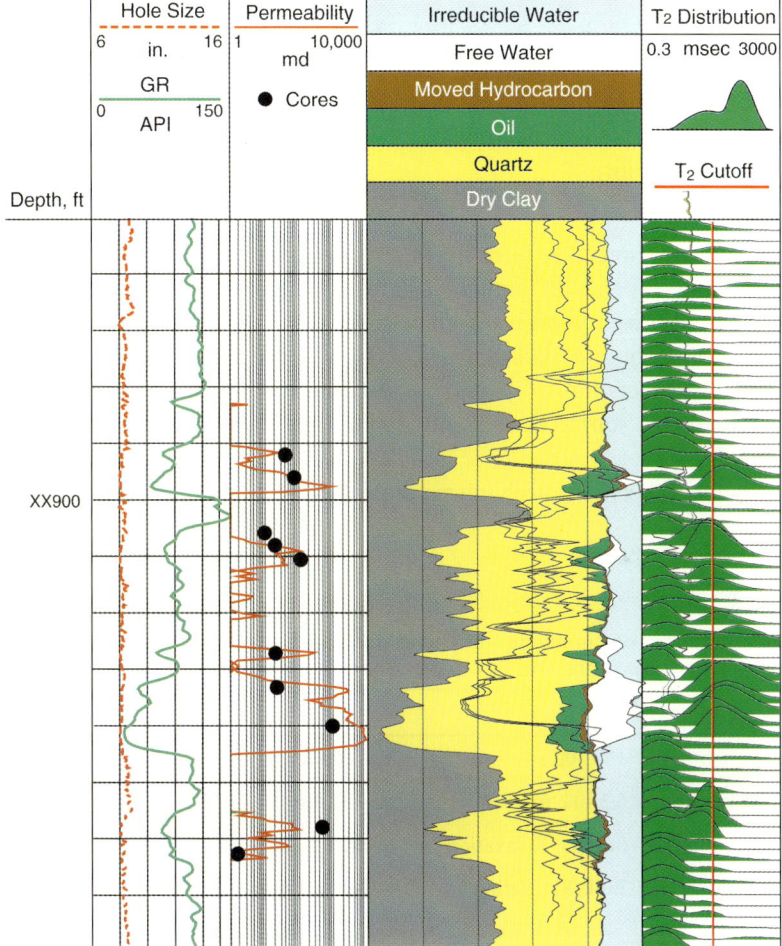

Figure 2.6.14. Example of permeabilities calculated from NMR logs using equation 2.6.6. Comparison with core-measured permeabilities (black circles) is good. Among the permeable zones the lowermost one is mostly water-filled, while the upper ones are oil-filled but thinner and less permeable.

lies in a mostly depleted field. Among the zones with higher T_2 times the lowermost interval is mostly water-filled, perhaps caused by the waterflood, while the upper, thinner zones seem to contain producible oil although their NMR-derived permeabilities are somewhat lower. Subsequent testing confirmed this interpretation. The figure also shows a good match between NMR-derived and core-measured permeabilities and water saturations.

2.6.4 Summary

NMR logging has made significant progress in recent times. Today, reliable acquisition, processing, and interpretation techniques are available which allow quantifying or estimating the following properties

- Total porosity
- Bound and free pore fluids
- Presence and type of hydrocarbons
- Intrinsic permeability
- Pore size distribution

The interactions between the pore fluids and the external magnetic forces are complicated and depend on numerous factors. Therefore, the interpretation of NMR logs requires a good basic knowledge of the measurement principle, as well as experience and skill. Some aspects of more geological interest are the textural interpretation of pore size distributions, or the application of NMR techniques to source rock identification, but more research is needed in these fields.

The principal relaxation mechanisms of protons in pore fluids are summarized in table 2.6.3.

Table 2.6.3 Summary of dominant NMR relaxation mechanisms in rocks. Notice that bulk and diffusion relaxations are only dependent on the fluid, not the rock.

Hydrogen in	T_1	T_2	T_1 / T_2
Minerals	10–100 s	10–100 μs	$\sim 10^6$
Clay-Bound Water	0.5–12 ms	0.5–12 ms	~ 1
Pore Water	Surface Relaxation	Surface Relaxation	~ 1.5
Water in "Vugs"	Bulk Relaxation	Bulk/Diffusion Relaxation	$T_1 = T_2$
Medium/Heavy Oil	Bulk Relaxation	Bulk Relaxation	$T_1 = T_2$
Light Oil	Bulk Relaxation	Bulk/Diffusion Relaxation	$T_1 = T_2$
Gas	Bulk Relaxation	Diffusion Relaxation	$T_1 \gg T_2$

References:

Akkurt R, Vinegar HJ, Tutunjian PN, Guillory AJ (1995) NMR logging of natural gas reservoirs. Trans 36th Symp Soc Prof Well Log Analysts: Paper N.

Allen DF, Basan PB, Herron MM, Kenyon WE, Matteson A (1998) Petrophysical significance of the fast components of the NMR T2 spectrum in shales sands and shales. 73rd Ann Conf Soc Petr Eng: Paper 49307.

Basan PB, Lowden BD, Whattler PR, Attard JJ (1997) Pore-size data in petrophsyics: a perspective on the measurement of pore geometry. In: Lovell MA, Harvey PK (eds) Developments in Petrophysics. Geol Soc Spec Publ 122, pp 47–67.

Bates RL, Jackson JA (1980) Glossary of Geology. American Geological Institute, Falls Church VA.

Bloch F, Hansen W, Packard ME (1946) The nuclear induction experiment. Physical Review 70, 474–485.

Brown RJS, Gamson BW (1960) Nuclear magnetism logging. Jour Petrol Technol 12, 199–207.

Coates GR, Dumanoir JL (1974) A new approach to improved log derived permeability. The Log Analyst, 15, 1.

Coates GR, Xiao LZ, Prammer MG (1999) NMR logging Principles and Applications. Halliburton.

Flaum C, Kleinberg RL, Hurlimann MD (1996) Identification of gas with the combinable magnetic resonance tool (CMR). Trans 37th Symp Soc Prof Well Log Analysts: Paper L.

Flaum C, Kleinberg RL, Bedford J (1998) Bound fluid volume, permeability and residual oil saturation from incomplete magnetic resonance logging data. Trans 39th Symp Soc Prof Well Log Analysts: Paper UU.

Freedman R, Boyd A, Gubelin G, Morriss C, Flaum C (1997) Measurement of total NMR porosity adds new value to NMR logging. Trans 38th Symp Soc Prof Well Log Analysts: Paper OO.

Kenyon WE, Howard JJ, Sezginer A, Straley C, Matteson A, Horkowitz K, Ehrlich R (1989) Pore size distribution and NMR in microporous cherty sandstones. Trans 30th Symp Soc Prof Well Log Analysts" Paper LL.

Kenyon WE, Takezaki H, Straley C, Sen PN, Herron M, Matteson A, Petricola MJ (1995) A laboratory study of NMR relaxation and its relation to depositional texture ad petrophysical properties – carbonate Thamama Group, Mubarraz field, Abu Dhabi. Trans Soc Petr Eng Middle East Oil Show: Paper 29886, 477–502.

Kleinberg RL (1994) Mechanism of NMR relaxation of fluids in rock. J of Magnetic Resonance, Series A, 108, 206–214.

Kleinberg RL (1999) Nuclear magnetic resonance. Experimental Methods in the Physical Sciences 35, chapter 9, Academic Press.

Kleinberg RL, Sezginer A, Griffin DD, Fukuhara M (1992) Novel NMR apparatus for investigating an external sample. J of Magnetic Resonance 97: 466–485.

Kleinberg RL, Vinegar HJ (1996) NMR properties of reservoir fluids. The Log Analyst 37, 6, 20–32.

Loren JD, Robinson JD (1969) Relations between pore size, fluid and matrix properties, and NML measurements. 44th Ann Conf Soc Petr Eng: Paper 2529.

Loren JD (1972) Permeability estimates from NML measurements. J Petrol Technol 24, 923–928.

Miller MN, Paltiel Z, Gillen ME, Granot J and Bouton JC (1990) Spin echo magnetic resonance logging: Porosity and free fluid index determination. 65th Ann Conf Soc Petr Eng: Paper 20561.

Morriss CE, MacInnis J, Freedman R, Smaardyk J, Straley C, Kenyon WE, Vinegar HJ, Tutunjian PN (1993) Field test of an experimental pulsed nuclear magnetism tool. Trans 34th Symp Soc Prof Well Log Analysts: Paper GGG.

Pettijohn FJ (1975) Sedimentary Rocks. Harper & Row, New York.

Pettijohn FJ, Potter PE, Siever R (1973) Sand and Sandstones. Springer-Verlag, New York.

Prammer MG, Mardon M, Coates GR, Miller MN (1995) Lithology-independent gas detection by gradient-NMR logging. 68th Ann Conf Soc Petr Eng: Paper 30562.

Prammer MG, Drack ED, Bouton JC, Gardner JS, Coates GR, Chandler RN, Miller MN (1996) Measurement of clay-bound water and total porosity by magnetic resonance logging. 71st Ann Conf Soc Petr Eng: Paper 36522.

Ramakrishnan TS, Schwartz LM, Fordham EJ, Kenyon WE, Wilkinson DJ (1998) Forward models for nuclear magnetic resonance in carbonate rocks. Trans 39th Symp Soc Prof Well Log Analysts: Paper SS.

Seevers DD (1966) A nuclear magnetic method for determining the permeability of sandstones. Trans 7th Symp Soc Prof Well Log Analysts.

Singer JM, Jonhson L, Kleinberg RL, Flaum C (1997) Fast NMR logging for bound-fluid and permeability. Trans 38th Symp Soc Prof Well Log Analysts: Paper YY.

Straley C, Morriss CE, Kenyon WE, Howard JJ (1991) NMR in partially saturated sandstone: Laboratory insights into free fluid index and comparison with borehole logs. Trans 32nd Symp Soc Prof Well Log Analysts: Paper CC.

Straley C, Rossini D, Vinegar HJ, Tutunjian PN, Morriss C (1997) Core analysis by low field NMR. The Log Analyst 38 (2): 84–94.

Timur A (1968) Effective porosity and permeability of sandstones investigated through nuclear magnetic resonance principles. Trans 9th Symp Soc Prof Well Log Analysts: Paper K.

Timur A (1969) Pulsed nuclear magnetic resonance studies of porosity, movable fluids, and permeability in sandstones. J Petrol Technol 21: 775–786.

2.7 Nuclear Spectroscopy Logging

2.7.1 History of Lithology Computations

Lithology is one of the most important geological attributes of rocks, yet well logging with traditional electrical, acoustic and nuclear tools is mostly sensitive to porosity and fluid contents. Although the "rock signal" from the solid components is usually weak, it is not negligible. Thus, for example, the presence of clays may lower the resistivity, or a higher mineral density may increase the bulk density or the propagation speed of acoustic waves. The only "conventional" log that seems to carry a strong lithological signal is the gamma ray, whose response often correlates well with the volume of clay. This and its widespread availability are the major reasons why the gamma ray log is the tool of choice in well correlation.

Early Lithology Estimations. It was a challenging task to estimate the lithological composition using conventional well logs, but in the 1960s and 1970s several methods were developed, published and put to commercial use. Already a simple crossplot of neutron porosity and bulk density can reveal differences in the grain density and thus the lithology. This is still today one of the common quick-look methods used by log analysts. The M-N plot (Burke et al., 1969) uses a combination of the three porosity measurements from the sonic, the density and the neutron log to extract two parameters (M and N) sensitive to lithology. They are defined as

$$M = 0.01(t_f - t)/(\rho_b - \rho_f) \qquad (2.7.1)$$

$$N = (\phi_{Nf} - \phi_N)/(\rho_b - \rho_f) \qquad (2.7.2)$$

Here, t, ρ_b and ϕ_N are the sonic travel time (in μsec/ft), the bulk density and the neutron porosity as measured by the three porosity logs, while t_f, ρ_f and ϕ_{Nf} are the corresponding values for the pore fluid. Porosity variations affect both the numerators as well as the denominators of M and N, making them almost independent of porosity. Most minerals have well-defined values of M and N, some of which are listed in table 2.7.1[1]. They are plotted on a reference chart which can be found for example in Dewan (1983, figure 6–5). If a lithology consists of two minerals, a linear interpolation between the two mineral points is made to calculate their relative proportions; with three minerals, a triangular interpolation is possible. The M-N method, however, is not accurate because the points are quite closely spaced as seen in table 2.7.1. Additionally, gas and clays can strongly shift the points, leading to erroneous mineral assemblages.

---

[1] The minerals listed in tables 2.7.1, 2.7.2 and 2.7.4 are thought to represent over 90 % of the minerals occurring in all hydrocarbon reservoirs

Table 2.7.1. M and N values for some common minerals in freshwater mud ($\rho_f = 1.0$ g/cc) (modified after Dewan (1983) and Serra (1984))

	Mineral	Composition	M	N
	Quartz	SiO_2	0.81	0.64
Feldspars	Orthoclase	$KAlSi_3O_8$	0.79	0.68
Feldspars	Albite	$NaAlSi_3O_8$	0.88	0.64
Feldspars	Anorthite	$CaAl_2Si_2O_8$	0.83	0.59
Clays	Illite	$K_{1-1.5}Al_4(Al_{1-1.5}Si_{7-6.5})O_{20}(OH)_4$	~ 0.6	0.49
Clays	Kaolinite	$Al_2Si_2O_5(OH)_4$	~ 0.6	0.45
Clays	Smectite	$R^b_{0.33}Al_2Si_4O_{10}(OH)_2 \cdot nH_2O$	~ 0.6	0.50
Carbonates/Sulfates	Calcite	$CaCO_3$	0.83	0.59
Carbonates/Sulfates	Dolomite	$CaMg(CO_3)_2$	0.78	0.49
Carbonates/Sulfates	Anhydrite	$CaSO_4$	0.70	0.50
Carbonates/Sulfates	Gypsum	$CaSO_4 \cdot 2H_2O$	1.01	0.30
Carbonates/Sulfates	Halite	$NaCl$	1.27[a]	1.09[a]

[a] In salt mud
[b] R can include Na, K, Mg and Ca

Clavier & Rust (1976) utilize the same logs in the MID (Matrix Identification) plot, but their method does not imply any calculations and has a closer link to actual physical properties. They obtain the apparent matrix density, and the apparent matrix travel time from a density-neutron and a sonic-neutron cross-plot respectively. For this, published reference charts are available (e.g. Schlumberger, 1989) and interpolation methods are used to calculate mixed lithologies. The presence of clays and gas again complicates the procedure. Poupon et al. (1971) propose a two-step procedure for complex lithologies, whereby first the amount of clay is obtained from a number of clay indicators, most notably the gamma ray log. In a second step, the porosity and electrical logs are used to solve for two minerals and the fluid saturations. This was the first computerized log interpretation (CPI, or *computer-processed interpretation*) and was commercialized under the name CORIBAND* (Schmidt et al., 1971).

New Lithological Indicators. In the 1980s two logging measurements became available which were strongly sensitive to the lithological composition of the rock. The first was the *natural gamma ray spectroscopy* log (Serra et al., 1980), a measurement which quantifies the three elements with unstable isotopes that emit natural radioactivity: thorium, uranium, and potassium. This measurement will be described in more detail below.

---

* Mark of Schlumberger

Table 2.7.2. P_e and ρ_b values for some common minerals (Serra, 1984)

	Mineral	P_e (barns/el)	ρ_b (g/cc)
	Quartz	1.81	2.65
Feldspars	Orthoclase	2.86	2.55
	Albite	1.68	2.62
	Anorthite	3.13	2.76
Clays	Illite	3.45	2.53
	Kaolinite	1.83	2.42
	Smectite	2.04	2.12
Carbonates/Sulfates	Calcite	5.08	2.71
	Dolomite	3.14	2.87
	Anhydrite	5.05	2.96
	Gypsum	3.99	2.32
	Halite	4.65	2.17

The second was the development of a *photoelectric absorption factor* measurement on the density tool (Gardner & Dumanoir, 1980; Ellis et al., 1983). Density logging is made in a relatively high-energy range where gamma rays get scattered through interaction with electrons (so-called Compton scattering), and where the main information is related to the bulk density of the rock (Tittman & Wahl, 1965; see also chapter 2.4.1). Photoelectric absorption, on the other hand, takes place at lower energies where the gamma rays get absorbed rather than scattered. Their probability of absorption increases strongly with the atomic number, a relationship often expressed in a visually useful way as a *capture cross-section*. The higher the atomic number, the larger the cross-section becomes. The density tool is designed to make two measurements, one in a high-energy range where scattering prevails, providing bulk density information, and another in a low-energy range where both scattering and absorption take place and where information on both density and the atomic number of the material is gained. The ratio of the two is then used to calculate the photoelectric factor (P_e) whose dimension is expressed in barns/electron (1 barn = 10^{-24} cm^2). Values of P_e for the common minerals already listed previously are shown in table 2.7.2 Notice that, unlike for the values of M and N, the photoelectric factor shows a good spread, making it a useful lithological indicator if logging conditions are good. For comparison, table 2.7.2 also lists the bulk densities of the minerals[2]. Two observations

---

[2] Density tools actually measure the electron density $\rho_e = \rho_b \cdot 2Z/A$ (Z = atomic number, A = atomic mass) which is generally very close to the bulk density.

can be made when comparing the two lists: First, the photoelectric factor varies much more than the mineral density. In fact, the ratio from highest to lowest value is almost 3 for the P_e and less than 1.5 for ρ_b. Second, the orders are not the same. This is best illustrated by dolomite, which has the second-highest density but the third lowest photoelectric factor of the minerals listed in table 2.7.2. Therefore, a combination of density and photoelectric factor measurement is often a useful quick-look indicator of lithology. These two measurements are not only available as wireline logs, but also as borehole images in logging-while-drilling mode (see chapter 2.4).

For multiple mineralogies the photoelectric factor cannot be obtained from the relative volumes of each mineral, but rather from a combination of the individual photoelectric factors weighted by their respective electron densities ρ_e and volumes V (Ellis, 1987). For this, a parameter U has been defined as

$$U = P_e \cdot \rho_e \qquad (2.7.3)$$

U is called the volumetric cross-section and has the dimensions of barns/cm³. For a lithology with n minerals, the average photoelectric factor is obtained as

$$\overline{P}_e = \frac{\sum_n U_n V_n}{\sum_n \rho_{en} V_n} = \frac{U_{total}}{\overline{\rho}_e} \qquad (2.7.4)$$

These two lithology indicators have been used to enhance computer-processed log interpretation products. Mayer & Sibbit (1980) for example propose a method, called GLOBAL*, wherein the set of log measurements is related to the set of desired petrophysical properties through mixing laws and tool equations. Examples are equation 2.7.4 shown above, or equation 1.3.19, which relates bulk density and neutron porosity to the fluid and solid volumes, or the resistivity relationships discussed in chapter 1.3. GLOBAL also uses rules to constrain the likely answer at a given depth. For example, total water saturation cannot exceed 1, or, more geologically relevant, the bulk density cannot exceed a certain value, such as 2.87 if only carbonates are present. The program is constructed to have more equations than unknowns so that the solution is overdetermined, a condition which the log analyst can control, for example by appropriately selecting the number of minerals in the model. The program then uses an error model called "incoherence function" to search for the most probable solution. This is achieved by comparing the actual log measurements at a given depth with the theoretical values one would obtain from the formation properties calculated in a first iteration, the tool response equations, and the constraints. The incoherence function is simply a tribute to the fact that tool responses are merely approximations and mineral models are often highly simplified representations of the actual complexities found in real rocks. The more recent ELAN* program is similar in nature to GLOBAL, but uses a linear as well as a non-linear solver and can include all the modern logs.

The basic workflow of the log analyst in these programs consists of the following steps:

- Make quality control and choose suitable logs
- Select appropriate mineral model for each geological zone
- Select fluid model (types and properties of formation water, hydrocarbons)
- Select appropriate computational models (e.g. resistivity equations, sonic response equations etc.)
- Calculate first pass results and analyze quality
- Change models and response equations where needed, repeat calculation until satisfied with results

The Advent of Nuclear Spectroscopy. Lithological analysis from well logs made a giant leap forward with the development of neutron-induced gamma ray spectroscopy (Ellis, 1987; 1990). The technique is based on the prompt emission of characteristic gamma rays when neutrons emitted from a source on the tool are scattered or captured by formation nuclei after they have slowed down to thermal energies. Spectroscopic detection of these gamma rays allows the identification of the nuclei and the quantification of their abundance. The technique is similar to neutron activation analysis done in laboratories for determining elemental concentrations, but it has to cope with the inadequacies of the borehole environment. Nevertheless, accurate elemental concentrations are obtained with logging systems that varied considerably through time.

The first tools, the GST* (gamma ray spectroscopy tool) and Carbon/Oxygen Tool**, emit neutrons in pulsed bursts, and two measurements can be made (Hertzog, 1978; Westaway et al., 1980):

1. Shortly after the burst inelastic scattering of the neutrons with formation nuclei, notably carbon and oxygen produces characteristic high-energy gamma rays. A spectral measurement is made from which the C/O ratio is obtained, a parameter sensitive to the volume of hydrocarbons and thus to water saturations. This saturation monitoring technique is applicable in cased holes where electrical logs cannot be run.
2. At a later time after the burst, the neutrons have slowed down to lower ("thermal") energies and produce gamma rays when captured by formation nuclei. A spectral measurement is made and the elements H, Ca, Si, S, Fe and Cl can be quantitatively obtained. Such "geochemical logging" is used to better evaluate the formation lithologies.

---

* Mark of Schlumberger
** Mark of Baker-Hughes

Oftentimes, very subtle peaks have to be detected in the spectra of these measurements, necessitating good count statistics and therefore slow logging speeds. But the relatively direct link between measured elements and minerals in the formation suggested that these tools have the potential of greatly improving lithological characterization (Flaum & Pirie, 1981; Gilchrist et al., 1982; Herron, 1986). Which elements are measured is unfortunately not determined by a geologist, but by what is physically and technically feasible. Thus, elements such as sodium, magnesium or aluminum, which are all important in the earth's crust and abundant in reservoir rocks, do not interact with neutrons in a way that produces strong characteristic gamma rays. Gadolinium (Gd), on the other hand, is readily detectable because of its large capture cross section although its abundance in most rocks is only measured in parts per million.

Subsequent tool development focussed on improving the list of elements to be measured. In the GLT* (Geochemical Logging Tool), a californium-252 low-energy neutron source was added to the tool, and a separate detector (identical to the natural gamma ray detector) was used to measure the weak but characteristic capture gamma rays of aluminum (Hertzog et al, 1987). This element is an excellent indicator of alumosilicates such as clays and feldspars. The tool also has a natural gamma ray spectroscopy sensor and thus consists of three separate measurement parts. However, it was too complicated for field operations, and in the late 1990s it was replaced by the ECS* (Elemental Capture Spectroscopy) sonde (Horkowitz & Cannon, 1997). This tool provides greatly improved spectroscopy data and can be logged at high speeds. It measures the elements Si, Ca, S, Fe, Ti and Gd.

For cased holes, the RST* (Reservoir Saturation Tool) and the RMT** (Reservoir Monitoring Tool) have been developed. They are designed to measure the C/O ratio from inelastic scatter spectra, but can also record elemental concentrations like the ECS, although at a much reduced logging speed (Horkowitz & Cannon, 1997).

In the following we discuss the principles and applications of natural and induced gamma ray spectroscopy measurement, whereby emphasis is placed on the latest generation tools.

2.7.2. Origin and Nature of Natural Radioactivity

Natural Radiation. There are three types of natural nuclear radiation: *Alpha-radiation*, which consists of helium atoms stripped of their electrons, *beta-radiation*, consisting of electrons, and *gamma-radiation*, consisting of photons. The first two types have little significance in well logging because their interac-

---

* Mark of Schlumberger
** Mark of Halliburton

tion with matter is dominated by Coulomb forces – due to their charged nature – and their range of penetration into rocks is consequently very small. Gamma rays, however, loose relatively little energy as they pass through rocks, and occur naturally. The measurement of natural gamma radiation in boreholes, developed in the 1930s, was the first non-electrical well log.

Gamma rays may be considered as either electromagnetic waves, similar to light and x-rays, or particles (photons). Their frequency is by definition between 10^{19} and 10^{21} sec^{-1}, while their wavelength ranges from 10^{-9} to 10^{-11} cm. Their energy is expressed in eV, with 1 eV corresponding to the energy required to impart a potential of one Volt to one electron. Typically, gamma ray energies range from keV to MeV. They are the most energetic types of electromagnetic radiation.

Radioactivity in rocks is gamma radiation caused by the decay of naturally occurring isotopes. The law of radioactive decay states that the rate of production of daughter isotopes is proportional to the amount of parent isotopes. The rate of decay is usually expressed by the half-life, or the time at which half of the initial radioactive isotopes have decayed. In rocks, only isotopes whose half-lives are of the same order of magnitude or larger than the earth's age (4.5 billion years) contribute to natural radioactivity. Isotopes with shorter half-lives already have decayed to a large degree into more stable daughter products.

Only isotopes of thorium, uranium, and potassium are found to contribute to naturally occurring radioactivity:

1. The ^{232}Th isotope decays to ^{208}Pb through a decay series involving numerous intermediate products. The half-life of the isotope is 14 billion years. A major gamma ray peak at 2.36 MeV is caused by ^{208}Tl. Many more peaks of this series occur at energies of less than 1 MeV.
2. The ^{238}U isotope decays in an even more complex radioactive series to ^{206}Pb, with a half-life of 4.4 billion years. The ^{238}U isotope constitutes more than 99 % of all naturally occurring uranium, while the equally radioactive ^{235}U isotope (used in nuclear reactors, for example, and leading to the "actinide" decay series) constitutes less than one percent and is generally negligible in well logging. The uranium series has a distinct peak at 1.76 MeV caused by the decay of ^{214}Bi.
3. The ^{40}K isotope constitutes about 0.012 % of all potassium present in the earth's crust, the remainder being ^{39}K. It decays to ^{40}Ar by emitting beta- and gamma rays with a single spectral peak at 1.46 MeV. This explains the relatively high concentration of argon in the earth's atmosphere. The half-life of this reaction is 1.3 billion years. ^{40}K also decays to ^{40}Ca but without any gamma ray emission.

A radioactive decay series is said to be in secular equilibrium if all its intermediate products are produced at the same rate as they decay. This is the case for most rocks older than one million years. The relative amounts of parent and daughter products are used to determine the age of rocks, whereby the decay rate (the half-life) of the series has to be known. For the case of very long half-lives compared to the probable age of the rock, most of the parent product is still around. From a natural radioactivity measurement one can therefore determine the abundance of this isotope and, knowing its relative abundance compared to other isotopes of the element, the concentration of the element in the rock can be determined.

The amount of decaying nuclei is expressed in Curies, with one Curie defined as the number of disintegrations per second occurring in one gram of ^{226}Ra, or $3.7 \cdot 10^{10}$ disintegrations per second. This is a large number, and most rocks have radioactivities that can be expressed in milliCuries or microCuries. Since a Curie is defined in terms of disintegrations, not in number of emissions, an intrinsic specific activity is used to relate it to the radioactive strength of the radioactive isotope. For the total element (i.e. the radioactive and the stable isotopes combined) relative activities are used. These are 1300, 3600, and 1 for Th, U, and K respectively (Adams & Weaver, 1958). The number of potassium is low because the radioactive ^{40}K isotope is rare compared to the stable ^{39}K isotope. Gamma radioactivity in well logging is expressed in μg Radium-equivalent per metric ton, or, more commonly, in API units. The latter is a unit established by the American Petroleum Institute in a test pit at the University of Houston facility which contains an artificial formation with approximately 24 ppm Th, 12 ppm U, and 4 % K (Belknap et al., 1959). Its radioactivity difference to the surrounding material is defined as 200 API units and is supposed to correspond to about twice the level of radioactivity found in an average shale.

Geochemistry of Naturally Radioactive Isotopes. The origins and geochemical occurrences of the naturally radioactive minerals are fairly well known:

- *Thorium* is linked to accessory minerals, mostly in intrusive rocks such as granite, granodiorite, and syenite. Monazite is the major thorium-bearing mineral, but zircon also contains it as a substitute of zirconium. It is almost insoluble and is found mostly in heavy minerals of the silt- and clay fraction. It can adsorb to some degree onto the surface of clay minerals.
- *Uranium* has the most complex geochemistry of the three radioactive elements. Its source is also mostly from acid igneous rocks, but unlike thorium it is soluble under certain conditions of pH and eH. Precipitation occurs in reducing environments, often of acidic nature and typically in organic-rich sediments like marine shales and carbonates. It also adsorbs readily onto clays and organic phosphates. Some uranium is found in silt- and clay-sized minerals, notably monazite and zircon, where it often substitutes into sites of thorium.
- *Potassium*-bearing minerals include principally clays (illite, glauconite), micas (muscovite, biotite) and feldspars (orthoclase). The alteration of K-feldspars to kaolinites is a major source of potassium in pore waters. In evaporitic sequences, sylvite as well as certain sulfates contain potassium. Potassium is readily transported in diagenetic processes.

Table 2.7.3. The geochemistry of naturally occurring radioactive elements

Element	Common Occurrence	Accessory/Rare Occurrence[a]
Thorium Th	Monazite $(Ce,La,Nd,Th)(PO_4)$ Adsorbed onto clays	Zircon $(Zr,Th)SiO_4$ Xenotime $(Y,Th)PO_4$ Thorianite ThO_2 Thorite/Huttonite $ThSiO_4$
Uranium U	Adsorbed onto clays Organic phosphates Monazite $(Ce,La,Nd,Th,U)(PO_4)$	Zircon $(Zr,U)SiO_4$ Xenotime $(Y,U)PO_4$ Thorianite $(Th,U)O_2$ Thorite $(Th,U)SiO_4$
Potassium K	Orthoclase $KAlSi_3O_8$ Muscovite $KAl_2Si_3O_{10}(OH,F)_2$ Biotite $K(Mg,Fe)_3(AlSi_3O_{10})(OH,F)_2$ Illite $K_{1-1.5}Al_4(Al_{1-1.5},Si_{7-6.5})O_{20}(OH)_4$	*Evaporites* (Sylvite, Kainite, Carnallite, Polyhalite) Feldspathoids (Leucite, Nepheline, Kaliophylite)

[a] The mineral is rare, or the element only partially substitutes another element in the crystal lattice, or both.

Table 2.7.3 summarizes the geochemical occurrences of the naturally radioactive isotopes. More detailed lists can be found in Serra (1984, tables 6–5 to 6-11), where also comprehensive listings of all minerals containing any of the three elements are given. The major observation to be made in table 2.7.3 is that only potassium occurs in major rock-forming minerals; thorium and uranium either occur only in accessory minerals, or substitute sites of other elements in so-called solid solutions. But because of their higher relative activities, thorium and uranium often contribute more than potassium to the total natural radioactivity in rocks.

In summary, significant levels of natural radioactivity can principally be found in the following rocks:

- Shales
- Arkoses and graywackes
- Argillaceous sandstones
- Organic-rich sedimentary rocks
- Igneous rocks, particularly acidic intrusives
- Potassium salts and phosphates

2.7.3 Total Gamma Ray Logging

Gamma Ray Detectors. The gamma ray log was the first non-electrical logging measurement. Like all nuclear tools it requires the use of a detector, which in this case has to be sensitive to the radiation caused by the decay of naturally occurring unstable isotopes. The earliest detectors were *gas-discharge counters*, such as the Geiger-Mueller counter, which take advantage of the photoelectric effect oc-

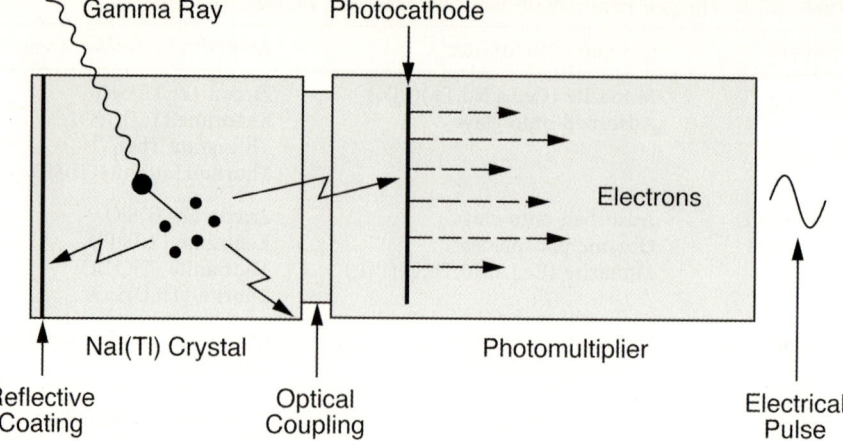

Figure 2.7.1 Sketch of a gamma ray scintillation detector. Photoelectric absorption or Compton scattering of an incoming gamma ray emits light flashes, which cause a measurable electrical pulse in the photomultiplier (after Ellis, 1987).

curring at low energies of the incident gamma rays. Although rugged and adapted to the borehole environment, these detectors suffer from poor statistics because of their intrinsic inefficiency of detection and a long dead time after each incoming and recorded gamma ray. *Scintillation detectors* are widely used today and consist of a single crystal, for example of sodium iodide (NaI) which is doped with impurities such as thallium (Tl). Incident gamma rays produce light flashes when they transfer their energy to electrons in the crystal. A photomultiplier which is optically coupled to the detector crystal transforms the flashes into electric pulses (figure 2.7.1). Ideally, the energy of these pulses is proportional to the energy of the gamma ray that produced it, but in reality other processes occur and the energy is generally smaller. The resulting energy spectrum is therefore quite blurred but can show distinct peaks. However, scintillation detectors are more efficient than gas discharge counters and are relatively small in size (typically 10–20 cm long) but have the drawback of being brittle. Bismuth germanate (BGO) has recently been used as detector crystal because of its higher detection efficiency and better spectral resolution. *Semiconducting detectors* such as germanium have an even better resolution and can produce excellent spectral details, but they have to be kept at very low temperatures and are actually less efficient because of their smaller size. They are not being used in commercial well logging.

Gamma Ray Tools. Total gamma ray measurements, sometimes referred to as gross-count gamma ray systems, measure all natural gamma rays without discriminating between different energies. In oilfield applications, where count rates are generally low, a 20 cm long NaI(Tl) scintillation detector is used which records pulses above a few hundred keV. In mineral exploration, on the other hand, count rates are often very high and the crystals may be as small as 2 cm (Hallenburg, 1984). The electrical pulses are summed up over a sufficiently long time (the "time constant") to warrant good statistics of the measurement. However, since nuclear reactions are random within certain probabilities, two consecutive runs with the same tool over the same interval will not provide identical answers. All oilfield gamma ray tools are referenced in the Houston test pit described earlier. Additionally, all service companies can provide on-site calibrations with a source of known strength, usually containing ^{226}Ra.

Natural gamma ray measurements are often described as having a spherical zone of investigation over which the measurement is averaged, but an ellipsoid might be a better approximation because of the long, narrow detector shape. The integrated geometrical factor reaches 95 % at 30 cm depth of penetration for a formation with density 2.0 g/cm^3, and practically 100 % for a density of 3.0 g/cm^3. The vertical resolution of gamma ray tools is generally about two to three times the detector length.

A hole correction factor has to be applied to gamma ray measurements to take into account the absence of rock material between the tool and the borehole wall. The corrections are different for centered and eccentered tools. There are two types of radioactive muds that have a detrimental effect on the gamma ray measurement. One is bentonite mud, which causes an overall shift in the measurement. The other is potassium chloride (KCl), which is a salt additive that has two effects on the gamma ray measurement. One is an overall increase caused by the potassium in the mud column. The other is a variable increase caused by the mud filtrate in the invaded zone, whereby the added radioactivity is generally higher in more permeable formations. If the gamma ray log is recorded behind casing or within tubing, there is generally a decrease in its readings depending on the material in the well. Correction curves are found in Hallenburg (1984) as well as in service company logging charts. However, under many logging conditions the corrections are often minor and not needed, unless the measurement is used for quantitative formation evaluation involving several wells.

Total gamma ray logging is not only done in wireline mode, but also during logging-while-drilling. It was one of the first measurements to be introduced and today is common in most LWD operations, used primarily as a lithology indicator in real-time and in subsequent formation evaluation. It is also one of the measurements in geosteering, where its response is continuously compared to a model of the geological layers, and the well trajectory is adjusted whenever significant differences occur (chapter 3.5).

Gamma ray logging is also done on cores and outcrops (e.g. Slatt et al., 1992), whereby a handheld instrument records the natural radioactivity at discrete intervals. Such studies help understand the relationship between natural gamma ray patterns and the various sedimentary units. They are also used to correlate cores with logs and put them on correct depth with respect to each other.

2.7.4 Natural Gamma Ray Spectroscopy Logging

Measurement Technique. The goal of natural gamma ray spectroscopy logging is to determine the concentrations of the three naturally radioactive elements thorium, uranium, and potassium. The tools record the energy spectrum of gamma rays with a large (typically 12 inch, or 30 cm long) NaI(Tl) scintillation detector. Due to the peak broadening in older scintillation detectors, and due to the small count rates at higher energies, only a small number of energy windows was used, typically three (Spectralog**) or five (Natural Gamma Ray Tool NGS*, and Spectral Gamma Ray***). The spectral windows are positioned such that they cover the characteristic energy peaks of the three reaction series (figure 2.7.2). The newer tools, however, have improved detectors and record the full gamma ray spectrum. The HNGS* (Hostile Environment Natural Gamma Ray Sonde), contains two cooled bismuth germanate (BGO) scintillation detectors, which are coupled with a stabilization source to prevent drifts. Since they are cooled they can be used in hot logging conditions, and their greater detector volume permits faster logging speeds. The Spectralog and the Compensated Spectral Gamma (CSNG***) tool use 256 and 768 channels respectively to record the full energy spectrum between 0 and 3 MeV. However, these spectra are still broad with few distinct peaks because of scattering interactions which degrade the higher energy spectrum (for an interesting account of the life of a gamma ray is consult Ellis, 1987, page 176–177). While logging speeds with the older tools were rather slow, the newer tools allow logging speeds of 1800 ft/hr because of their higher count rates.

In order to obtain the elemental concentrations from the count rates in three or five windows, a set of linear equations relating window count rates to elements was solved (Serra, 1984; Schlumberger, 1982). Additionally, to reduce spurious effects such as anticorrelations and negative readings, a Kalman filtering procedure was applied. In the newer tools, the spectrum is decomposed into a linear combination of the known spectral responses for each of the three elements. This is the same procedure as used in induced gamma ray spectroscopy discussed later in this chapter.

Figure 2.7.3 shows a natural gamma ray log with the three element concentrations (Th, U, K), the total gamma ray (GR or SGR), and the "compensated gamma ray" (CGR). The latter is obtained from the contributions of thorium and potassium only. It is thought to be a better representation of the radioactivity contributed from clays in argillaceous lithologies, where uranium is often asso-

---

* Mark of Schlumberger
** Mark of Baker-Hughes
*** Mark of Halliburton

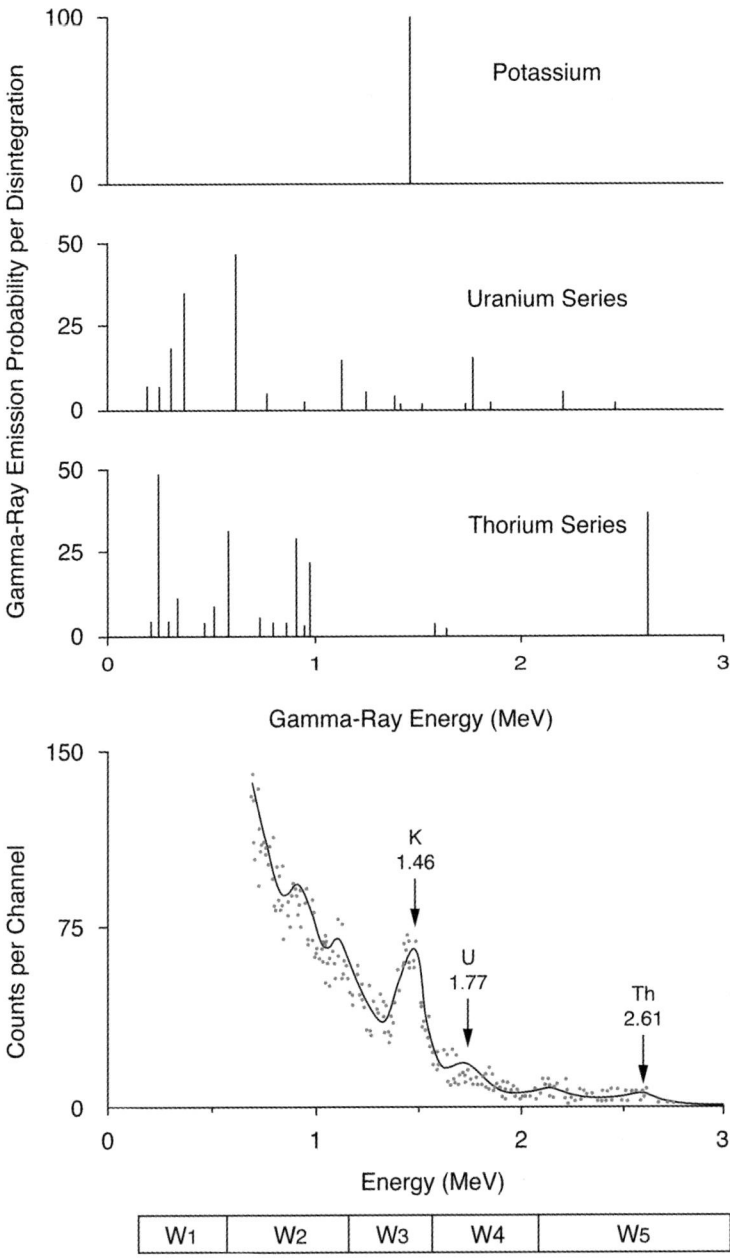

Figure 2.7.2 The gamma ray emission spectra of the three naturally radioactive elements. Potassium has a single peak from the ^{40}K to ^{40}Ar reaction. The thorium and uranium decay series are much more complicated, resulting in numerous peaks (top). The energy windows (W1–W5) of the NGS detector are positioned so that each of the three higher-energy windows is located on a characteristic peak of the three decay series (bottom). A typical spectrum is shown by way of illustration (after Tittman, 1986).

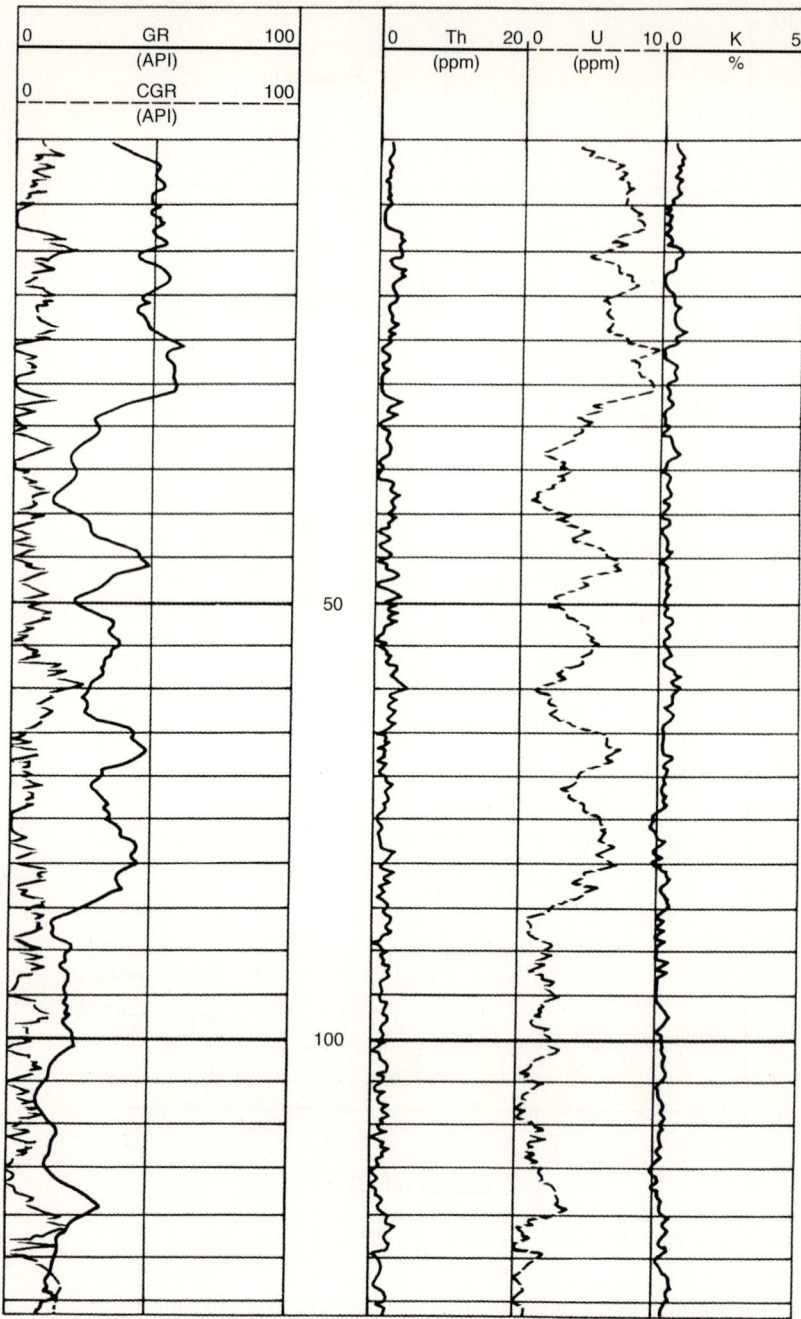

Figure 2.7.3 Results of a natural gamma ray spectroscopy log over a carbonate interval. While the total gamma ray is relatively high, the spectral analysis reveals that mostly uranium contributes to it. The compensated gamma ray (CGR) is accordingly very low and clay volumes are expected to be insignificant.

ciated with organic matter or minerals from outside the alumosilicate group. The total gamma ray count for the NGS can be approximated by the simple formula

$$\text{GR(API)} = 4 \cdot \text{Th(ppm)} + 8 \cdot \text{U(ppm)} + 16 \cdot \text{K(\%)} \qquad (2.7.5)$$

Therefore, the compensated gamma ray CGR is calculated as

$$\text{CGR(API)} = 4 \cdot \text{Th(ppm)} + 16 \cdot \text{K(\%)} \qquad (2.7.6)$$

2.7.5 Applications of Natural Gamma Ray Measurements

Volume of Clay

Gamma ray logs are mostly used as "shale indicators". Typically, a maximum and minimum value are determined on a gamma ray log, with the maximum (GR_{max}) corresponding to a shale and therefore the highest clay content, and the minimum (GR_{min}) to a clay-free ("clean") lithology such as a sandstone or a limestone. Intermediate values of the clay volume V_{cl} are calculated as

$$V_{cl} = c \cdot \frac{GR - GR_{min}}{GR_{max} - GR_{min}} \qquad (2.7.7)$$

where GR is the gamma ray reading at the depth of interest, and the coefficient c takes into account that a pure shale does not consist entirely of clay. Therefore, c is smaller than unity and for typical shales $c = 0.6$. Equation 2.7.7 is also sometimes written in terms of the volume of shale instead of the volume of clay, a somewhat unfortunate usage because shale is a lithological term, not a mineral, but in thin-bedded alternations of sands and shales this practice may have some justification. Equation 2.7.8 also implies that there is a linear relationship between radioactivity and volume of clay, which is not necessarily the case (Rider, 1996). Since uranium is often not associated with clay minerals, the use of CGR instead of the total gamma ray has to be preferred for the calculation of the clay volume:

$$V_{cl} = c \cdot \frac{CGR - CGR_{min}}{CGR_{max} - CGR_{min}} \qquad (2.7.8)$$

In the example shown in figure 2.7.3, which is from a carbonate sequence, the total gamma ray is seen to be quite high, while the CGR is low because most of the radioactivity is contributed from uranium. The clay volumes are accordingly expected to be very low.

Correlation and Pattern Analysis

Probably the most common use of gamma ray logs is for well-to-well correlation (Doveton, 1994) because of two main reasons: its strong lithological signal, and its wide availability. An overview of well correlation techniques is given in chapter 3.4, where also several examples are discussed. There are numerous well correlation examples in the literature, with particularly interesting examples to be found in Lake & Carroll (1986), Slatt et al. (1992), Doveton (1994) and Rider (1996). In many cases GR, SP and resistivity logs are used interchangeably and sometimes without being labeled. This illustrates that the most important feature is the *log shape*, not its readings.

Geologists have early recognized and classified certain typical log shapes: The bell, cylinder, and funnel shape, with either a smooth or a serrated expression (figure 2.7.4). These shapes are mostly used in clastic sequences and are thought to reflect characteristic grain size distributions, a supposition which Serra & Sulpice (1975) found true when comparing vertical grain size changes from cores with gamma ray responses. In fact, lithological columns computed from logs are now often displayed with a gamma ray log as a bounding curve, mimicking the weathering curve generally used for lithological columns obtained from outcrops. The gamma ray is thereby displayed such that low readings, corresponding to sand-rich intervals, show a thicker lithology column than the shales.

The basic log shapes are often combined into *sequences*, which can be related to standard depositional models. Figure 2.7.5 illustrates three sequences from three different depositional environments: a deltaic or fluvial channel bar, a prograding marine sand bar, and a prograding deep sea fan. Such type sequences have been described and put into a sequence stratigraphic concept by Van Wag-

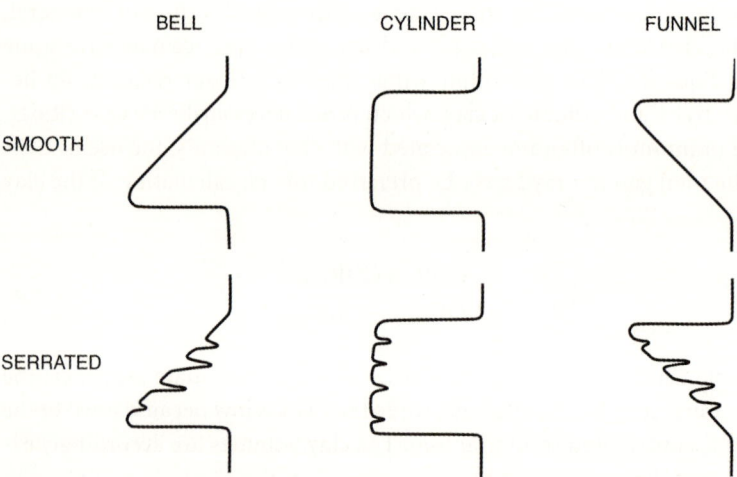

Figure 2.7.4 The three basic log shapes, bell, cylinder and funnel, in smooth and serrated expressions.

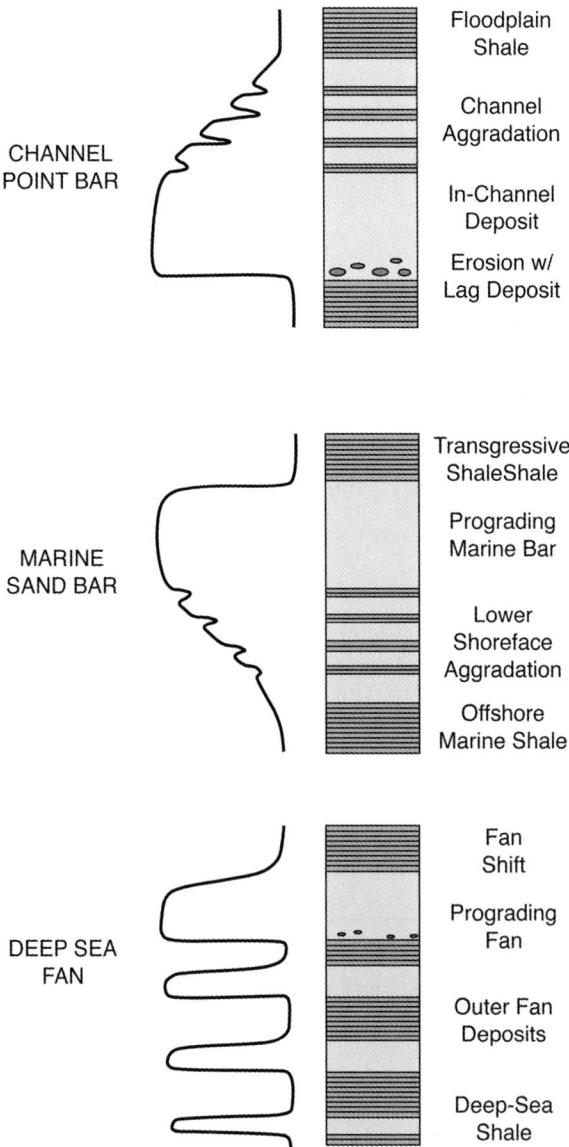

Figure 2.7.5 Example of three sequences, a channel point bar, a marine sand bar, and a deep sea fan, with their typical gamma ray expression on the left (after Rider, 1996).

oner et al. (1990) and Rider (1996). There is so far no strict formalism for either sequence stratigraphy or well correlation using well logs, but Rider (1996) outlines some useful rules. Generally, common geological sense has to prevail: Floodplain and transgressive shales (maximum flooding surfaces) are correlated first because they are likely to have large extents. The sands and other shales are then filled in whereby the depositional model is often continually adapted as the well correlation proceeds.

The parallelism between gamma ray value and grain size should be applied with great caution (Slatt et al., 1992). In fact, there is no inherent reason why there is a correlation between elemental composition and grain size. The two are completely independent from each other, and any correlation is entirely circumstantial: finer-grained sediments have generally more clay and heavy minerals which, in turn, have higher concentrations of the radioactive elements, principally thorium and potassium. There are too many exceptions to this rule to make it a working paradigm (the "hot sands", which are usually arkosic or micaceous sandstones, some fluvial shales with low radioactivities), and newer logging methods such a nuclear magnetic logging provide a more robust indication of grain size. In chapter 3.4 we illustrate how modern logs and new techniques can substantially facilitate sequence analysis and well correlation, thereby reducing some of the uncertainties inherent when using only gamma ray logs.

Mineral Identification

Natural gamma ray logging opens up the possibility of improving the lithological volume estimates from logs (Fertl, 1979; Hassan et al., 1976; Serra et al., 1980). The geochemistry of the radioactive elements was discussed above and summarized in table 2.7.3, but few of the minerals listed are major rock-forming minerals.

Table 2.7.4. Th, U, and K concentrations in common minerals, expressed as **average** and *range*. Absent values denote either low or unknown abundance (compiled from Schlumberger, 1982, Serra, 1984, and Edmundson & Raymer, 1979)

	Mineral	Th (ppm)	U (ppm)	K (%)
	Quartz	2	0.7	0.08
Feldspars	Orthoclase	5 *(3-7)*	*0.2 – 3*	14.0 *(10.9 – 16)*
	Albite	*0.5 – 3*	*0.2 – 5*	
	Anorthite	*0.5 – 3*	*0.2 – 5*	
Clays	Illite	*10 – 25*	*1.5 – 12.4*	6.7 *(3.5 – 8.3)*
	Kaolinite	*6 – 42*	*1.5 – 9.0*	0.35 *(0.0 – 0.5)*
	Smectite	*10 – 24*	*2.0 – 7.7*	1.6 *(0.0 – 4.9)*
Carbonates/Sulfates	Calcite			0.3 *(0.0 – 7.0)*
	Dolomite			0.07
	Anhydrite			
	Gypsum			
	Halite			

We therefore need to know how much of each radioactive element the common minerals contain. Table 2.7.4 is a compilation of published laboratory data. Not all minerals have entries, and the observed range is often very large, particularly evident in clays. Clays also are seen to have non-unique values when compared to other minerals, confirming that the correlation of natural radioactivity with grain size is circumstantial. However, there are some positive conclusions to be drawn from table 2.7.4. The first is the possibility of separating orthoclase from quartz, two minerals whose properties listed in previous tables of this chapter had been very close to each other. The other is the high level of potassium in illite, often a weathering product of K-feldspar, and its cousins in the mica group, biotite, muscovite, and glauconite, whose values are not shown in table 2.7.4 but which are similar to those of illite. In field studies in some North Sea reservoirs, Suau & Spurlin (1982) have found the NGS useful to distinguish kaolinite from illite. This distinction is important because illite reduces the permeability considerably more than kaolinite due to its fibrous texture.

Natural gamma ray logging is also useful in mineral exploration, for example for potassium salts or uranium ores, where often specially adapted tools are deployed (Hallenburg, 1992). Other fields where natural gamma ray logging is increasingly used are groundwater monitoring and some environmental applications.

In hydrocarbon reservoirs, however, natural gamma ray logging has been found to be of limited help in determining the abundance of commonly occurring minerals. Hurst (1990) and Humphreys & Lott (1990) conclude that other lithology-sensitive logs and a good *a priori* knowledge of the expected mineral assemblage are required for good mineral estimations. Schlumberger (1982) advocates the use of element ratios from natural gamma ray spectrometry, for example Th/K, in order to identify minerals. However, there is invariably one trace element involved in these ratios, and they are probably better avoided. On the other hand, there had been some success in identifying *unconformities* from these ratios.

2.7.6 Induced Gamma Ray Spectrometry

Measurement Principle. The purpose of induced gamma ray spectroscopy logging is to obtain elemental concentrations of the formation, a procedure sometimes referred to as "geochemical logging". This is achieved by emitting high-energy neutrons from a source on the tool, which then interact with the borehole mud and the formation. Only one type of interaction is of interest, the capture of thermal neutrons by formation nuclei during which characteristic gamma rays are emitted. The energy spectrum of these gamma rays is recorded, analyzed and concentrations of some elements in the formation are obtained. Other uses of gamma ray spectroscopy – mostly the determination of the C/O ratio from the inelastic spectra – were briefly discussed in the introductory section of

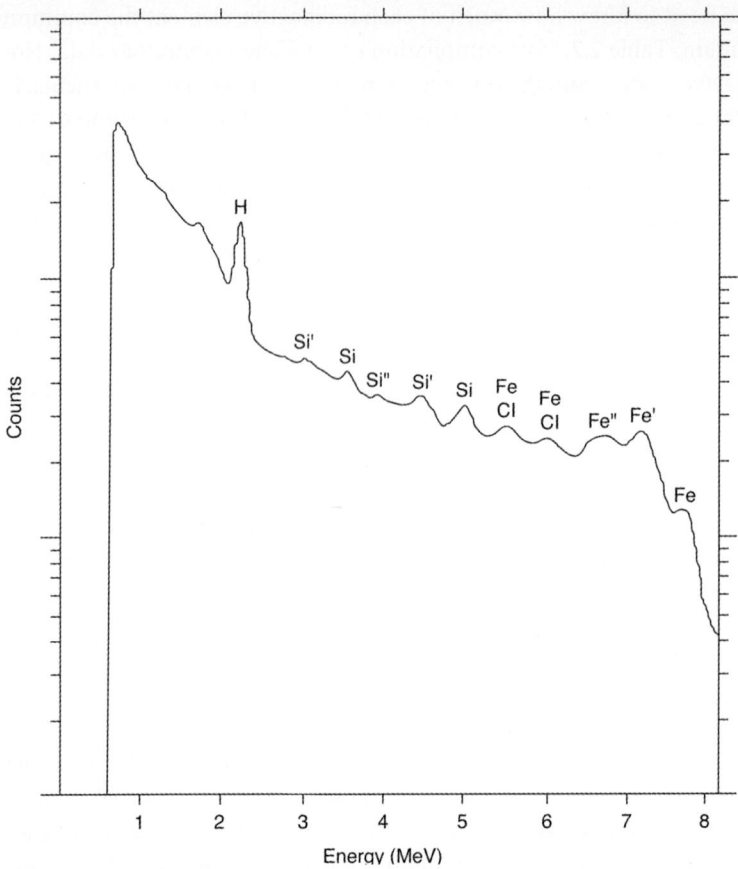

Figure 2.7.6. Example of a capture gamma ray spectrum. Some peaks are labelled with the elements which caused them (after Hertzog, 1978).

this chapter, and some reference papers can be found there. They are not discussed here as their major interest is for hydrocarbon saturation calculations.

A typical gamma ray capture spectrum is shown in figure 2.7.6. Notice how the counts decrease by more than an order of magnitude from lowest to highest energy, and how the peaks – whose elemental origins are indicated – are subtle. Both features are a result of the type of detector employed in the GST. This spectrum is decomposed into its components using standard spectra for each of the elements, some of which are shown in figure 2.7.7. At each depth, the linear combination of these standards is determined through a best-fit procedure, and *elemental yields* are obtained, the sum of which is 1 by definition. Elemental concentrations are obtained from the yields by rewriting each element as an oxide and putting their sums to unity in a so-called "closure model" (e.g. Schweitzer et al., 1988). Because of the complexities of nuclear interactions, and the imperfect resolution of commercially feasible detectors, there is only a limited number of

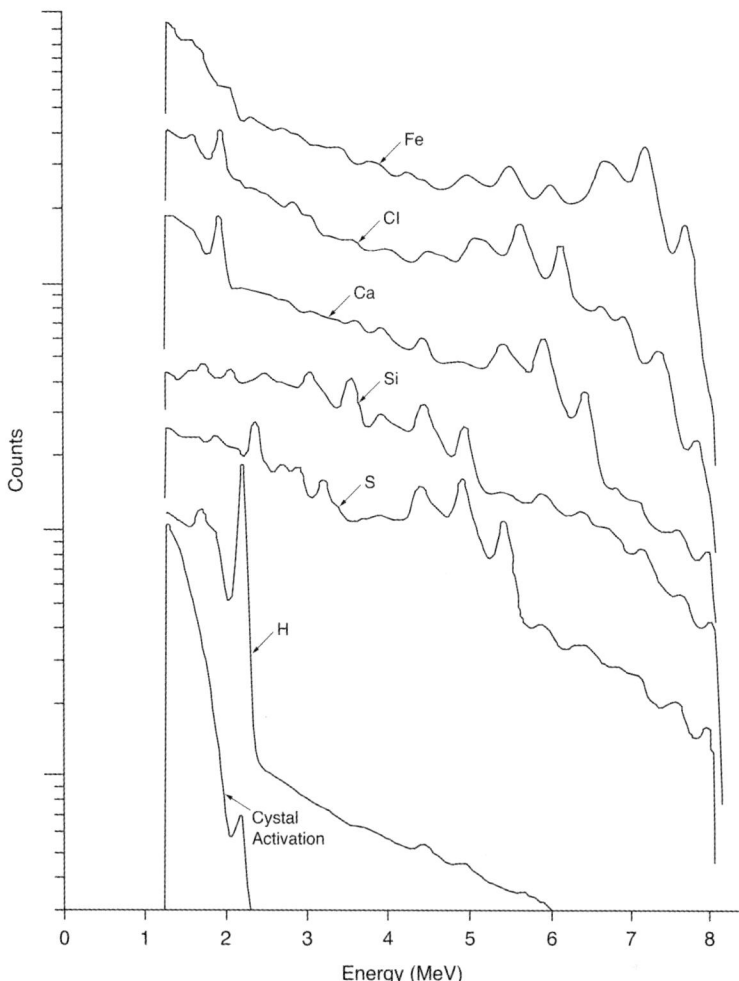

Figure 2.7.7. Capture spectra of some elements which are used in the decomposition of the total spectrum to obtain elemental yields (after Westaway et al., 1980).

elements which can be quantified in this manner. Furthermore, this list is different from the most common elements found in the earth's crust in general, and in sedimentary rocks in particular. The ease by which an element can be detected with neutron-induced gamma ray spectroscopy depends on three factors:

- the thermal absorption cross-section of the element, i.e. how large an obstacle the nucleus will present to an incoming neutron
- the gamma ray energy released during neutron capture, and
- the elemental abundance in the rock

Table 2.7.5 lists abundance and thermal cross-sections of all major and of some selected minor and trace elements in the earth's crust. It is seen that elements

Table 2.7.5. Crustal abundance and absorption cross-section of some major, minor, and trace elements (from Mason, 1966, and Ellis, 1987)

Element	Symbol	Atomic Number	Abundance in earth's crust[a]	Absorption cross-section[b]
Oxygen	O	8	46.6	0.000018
Silicon	Si	14	27.7	0.0061
Aluminum	Al	13	8.13	0.0091
Iron	Fe	26	5.00	0.0488
Calcium	Ca	20	3.63	0.0115
Sodium	Na	11	2.83	0.0246
Potassium	K	19	2.59	0.0573
Magnesium	Mg	12	2.09	0.00459
Titanium	Ti	22	0.44	0.136
Hydrogen	H	1	0.14	0.352
Phosphorous	P	15	0.11	0.0062
Manganese	Mn	25	0.095	0.259
Sulfur	S	16	0.026	0.0173
Carbon	C	6	0.020	0.00030
Chlorine	Cl	17	0.013	1.00
Nitrogen	N	7	0.002	0.141
Boron	B	5	0.001	75
Gadolinium	Gd	64	0.0005	333

[a] In weight percent
[b] Mass-normalized thermal absorption cross-sections (chlorine equivalents)

like gadolinium, chlorine, and boron are detectable because of their large cross-section, while silicon, iron, calcium, hydrogen, and titanium can be detected because of their abundance, although their cross-sections are smaller. Aluminum, sodium, and magnesium are difficult to identify because of their small cross-sections and their lack of a characteristic strong peak. Carbon and oxygen have very low thermal absorption cross-sections and can only be quantified through inelastic scattering.

The elemental capture spectroscopy sonde (ECS) has a standard americium-beryllium (AmBe) chemical source, which is also used in neutron porosity tools (figure 2.7.8). It emits neutrons at relatively low energies such that there are practically no inelastic interactions and thus the gamma ray spectrum is dominated by capture reactions. It is recorded by a cooled bismuth germanate (BGO) detector, which is the same as in the latest natural gamma ray spectroscopy tool (HNGS). The good efficiency of this detector allows a high logging speed of up to 1800 ft/hr, and the good spectral resolution provides the elements Si, Ca, S, Fe, Ti and Gd with good repeatability (Herron & Herron, 1996; Horkowitz & Cannon,

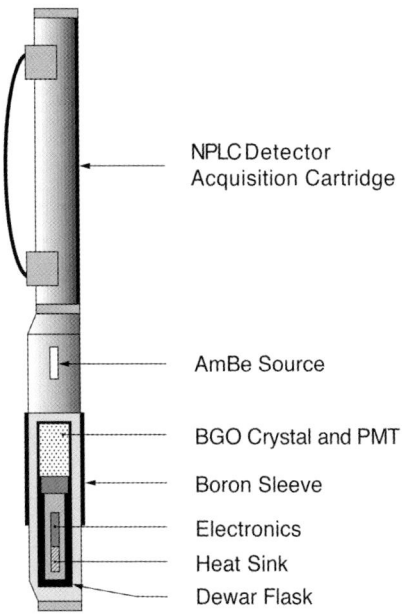

Figure 2.7.8 Sketch of Elemental Capture Spectroscopy sonde (ECS).

1997). A boron sleeve is mounted around the tool in order to reduce the thermal neutron flux impinging on the steel housing, which would otherwise contribute to an unacceptably high background or "tool signal". It also displaces borehole fluid and thus reduces the H and Cl signal from the mud. These two elements are not considered because they are primarily found in fluids, not solids.

Figure 2.7.9 compares elemental concentrations obtained from the ECS with chemical laboratory analysis on core samples. There is generally good agreement between log and core data. All scales are in weight percent. A pure quartz sandstone consists of 47 % Si, and a pure limestone of 40 % Ca; these two values define the scales chosen in the left two tracks. Several limestone beds can immediately be identified, with the most prominent one in the upper part and close to 20 feet thick. No sandstones seem to be present, at least no pure quartzitic ones. Since the iron concentrations obtained from the spectrum contain some contributions from aluminum, the corresponding core values for iron have been augmented by a factor of 0.14 of the core aluminum concentration, which gives the best match. The relatively high iron values suggest the remainder of the lithological column to consist mainly of shale.

This cursory discussion highlights how much lithological information can be directly drawn from such elemental concentration logs. The reservoir saturation tool, RST, can be used in a capture spectroscopy mode in cased holes to obtain very much the same information as the ECS in open holes (Horkowitz & Cannon, 1997). Because the tool is a pulsed neutron tool, and because of the difficult environment – tubing, fluid in annulus, casing, cement and formation all contri-

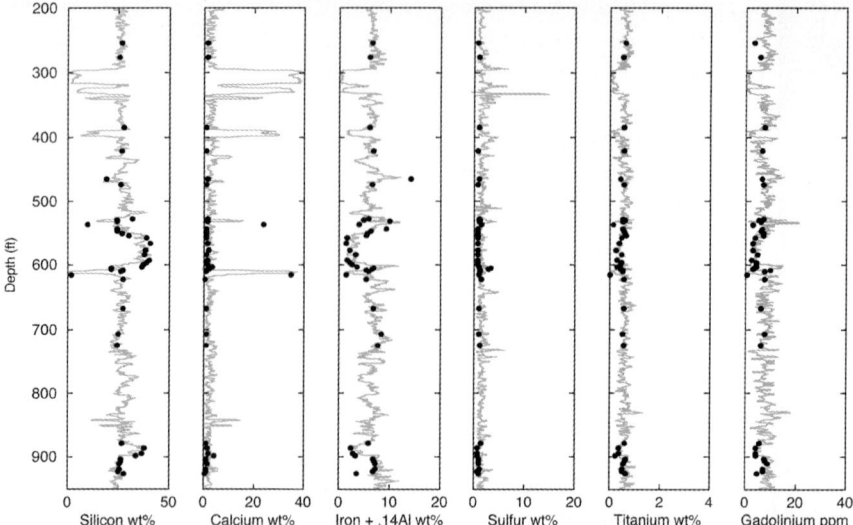

Figure 2.7.9 Si, Ca, Fe, S, Ti and Gd concentrations obtained from the ECS tool (curves) and from laboratory analysis on core plugs (dots) (from Herron & Herron, 1996).

bute to the signal – logging speeds are slow so that a sufficient formation signal can be obtained. At present, no logging-while-drilling spectroscopy tools are available.

Combining induced gamma ray spectroscopy data with other logs, for example natural gamma ray spectroscopy logs, or density and neutron porosity logs, can provide greatly improved estimation of lithological composition and other formation properties.

2.7.7 Applications of Spectroscopy Measurements

Mineral Concentrations

The principal use of elemental concentration logs is to transform them into quantitative logs of modal mineralogy. Numerous procedures have been proposed since the advent of nuclear spectroscopy, whereby initial enthusiasm slowly gave way to realism (Herron, 1986; Herron & Herron, 1990; van den Oord, 1991; Harvey & Lovell, 1992; Lofts et al., 1992, Herron & Herron, 1996; Horkowitz & Cannon, 1997). The reason for today's sober but realistic assessment lies in the fact that elemental compositions by themselves are difficult to interpret. To wit, imagine an feldspathic sandstone or an arkose undergoing various events of weathering and diagenetic alterations, after which it can be a fundamentally different rock: the feldspars might have turned into kaolinite, the micas into chlorite, calcium may have been released during some of these alterations and be precipitated as calcite, and quartz may may have been dissolved and been reprecipitated as overgrowth quartz in other places of the rock. And yet, the rock may have the same – or at least a very similar – elemental composition as it originally had.

The numerous alumosilicates such as feldspars, micas, and clays, are a major cause for this headache. Harvey & Lovell (1992) refer to it as the "problem of compositional colinearity": Given the concentrations of the major elements Si, Al, and K, it is impossible to determine the relative amounts of K-feldspar, illite, kaolinite, and muscovite, even if closure constraints are used (that the sum is 100 %). These minerals lie in the same compositional plane, or close to it, because their major elements are essentially the same (table 2.7.1). Similar situations can occur in the carbonate world, for example with calcite, high-Mg calcite, and dolomite, whose volumes cannot be determined from the concentrations of Mg and Ca.

In theory, however, it is straightforward to calculate minerals from elements: one takes the elemental concentrations at a particular measurement point and inverts it into a mineral assemblage, whereby the number of mineral cannot exceed the number of measurements by more than one (the closure condition being an additional equation). Herron (1986), for example, applied the approach of Pearson (1978) and extended it for non-clay phases. It relates, in matrix form, the mineral abundances M to the elemental abundances E by

$$E = CM \qquad (2.7.9)$$

where the coefficient matrix C contains matrix elements c_{ij} whose values are obtained from either the "pure" mineral compositions, or from test wells where geochemical analysis had been performed on rock samples. His element inputs include Al, Fe, and K, partly because they were measured by geochemical logs (at that time the GST), and partly because they are abundant and characteristic in the minerals present in his well data. In other words, there is no compositional colinearity. As minimum mineral suite he chose kaolinite, illite, and K-feldspar. Table 2.7.6 lists the c_{ij} coefficients for this particular data set. Subsequently, Herron & Herron (1990) extended this approach considerably by including more elements (Si, Ti, S, and Ca) as well as more minerals (quartz, calcite, smectite, pyrite, rutile and siderite). The corresponding coefficients of this extended

Table 2.7.6. Coefficients c_{ij} used to estimate weight percentages of kaolinite, illite and K-feldspar from concentrations of Al, Fe, and K (Herron, 1986)

Element	Kaolinite	Illite	K-feldspar
Al	19.0	9.2	10.0
Fe	0.14	10.6	0.05
K	0.35	4.0	12.0

matrix are given in their table 3. In a number of clastic field examples, Herron & Herron (1990) compared elemental concentrations (obtained with the GLT) to laboratory analysis on cores, normally using neutron activation analysis, and found good matches. They then compared their calculated mineralogy from the logs to XRD mineralogies obtained from core plugs in the laboratory and found equally good comparisons.

Van den Oord (1991) has a more reserved outlook on this. He finds elemental concentrations from these logs to be quite accurate, but the computed mineralogies to be strongly model-dependent and often at variance with core data. Lofts et al. (1992) tested a variety of mineral inversion models, including a set of simultaneous linear equations, a Euclidian distance model and a linear programming approach. They found that

"Perhaps the single largest problem is that of validation of the transform solution...This is especially so at low concentrations and remains a problem."

Their concern refers to the problem of compositional colinearity, which can only be avoided if the mineral suite is a priori known, if the minerals are not too similar in their geochemistry, and if sufficient characteristic elements are available from the logs. They also find that other lithology-sensitive logs help reducing errors, notably the lithodensity log, which provides the photoelectric factor measurement useful for estimating the amount of magnesium, and the natural gamma ray log, which provides potassium as a major rock-forming element.

Mineral Groups, not Minerals. With these concerns in mind, and with a robust tool (the ECS) at their disposal, Herron & Herron (1996) have proposed to calculate only four minerals, or rather mineral groups, namely

1. clays
2. quartz-feldspars-mica
3. carbonates (calcite and dolomite)
4. evaporites (anhydrite and gypsum)

As inputs they use the major rock-forming elements measured with the ECS, Si, Ca, Fe and S. Magnesium can be calculated from an estimation of dolomite using other open-hole logs, notably the photoelectric factor. Since no aluminum measurement is available, they propose to compute the Al concentration from a combination of measurable elements which correlate or anticorrelate with Al. For this, Herron & Herron (1996) propose the empirical relationship

$$Al = 0.34(100 - SiO_2 - CaCO_3 - MgCO_3 - 1.99Fe) \qquad (2.7.10)$$

which was obtained from a data bank of 12 wells. The correlation coefficient with core-derived aluminum obtained with equation 2.7.10 is reported to be 0.99. This relationship in essence states that whatever is not tied to silicon oxide, carbo-

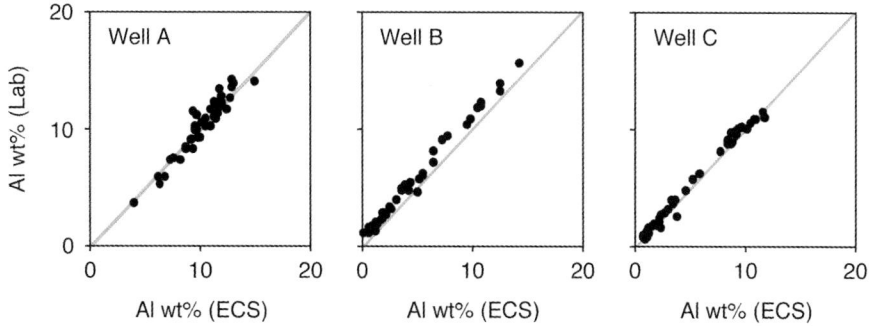

Figure 2.7.10 Aluminum concentrations estimated from Si, Ca, Fe and Mg concentrations obtained with the ECS compared to laboratory measurements on core plugs (from Herron & Herron, 1990).

nates, or iron, must correlate with clays and hence with aluminum. Figure 2.7.10 shows comparisons for three wells of the Al concentration measured on cores using neutron activation analysis with the concentrations obtained using this equation. Visibly, the correlations are very good and seem to suggest that calculating Al rather than measuring it in an expensive and complicated procedure may be justified. Again using their data bank, Herron & Herron (1996) then compute the amount of clay by multiplying Al with a factor of about 5, since they found that average clays contain 20 % aluminum. They subsequently calculate carbonate using Ca and Mg, whereby the possible contributions by plagioclase feldspars are accounted for. Finally, they attribute the remainder to the quartz-feldspar-mica group of minerals. If anhydrite or gypsum is present, it is calculated from the amount of sulfur and calcium prior to any other mineral.

Unlike the previously discussed approaches, this one is sequential and relies on the relationship of the major measurable elements with the major rock-forming mineral groups found in typical reservoirs. Figure 2.7.11 shows the results of such a computation in a lithological sequence that involves all four mineral groups. It is particularly interesting to compare the mineral concentration column to the other logs, which even to an experienced interpreter are difficult to analyze. The information contained in the mineralogical column, on the other hand, is direct and can be readily interpreted in terms of depositional or diagenetic history, reservoir potential, or sequence stratigraphy.

In summary, mineral concentrations are clearly the major objective of induced (and natural) gamma ray spectroscopy logging. In certain cases, where the mineral suite in the rocks of interest is known, sufficient constraints can be applied to the element/mineral inversion that accurate mineral concentrations are obtained. In the majority of cases, however, it might be wiser to limit the inversion to a select group of rock-forming minerals. In either case, the resulting mineral volumes are likely to be better estimates than those obtained from standard open hole logs alone.

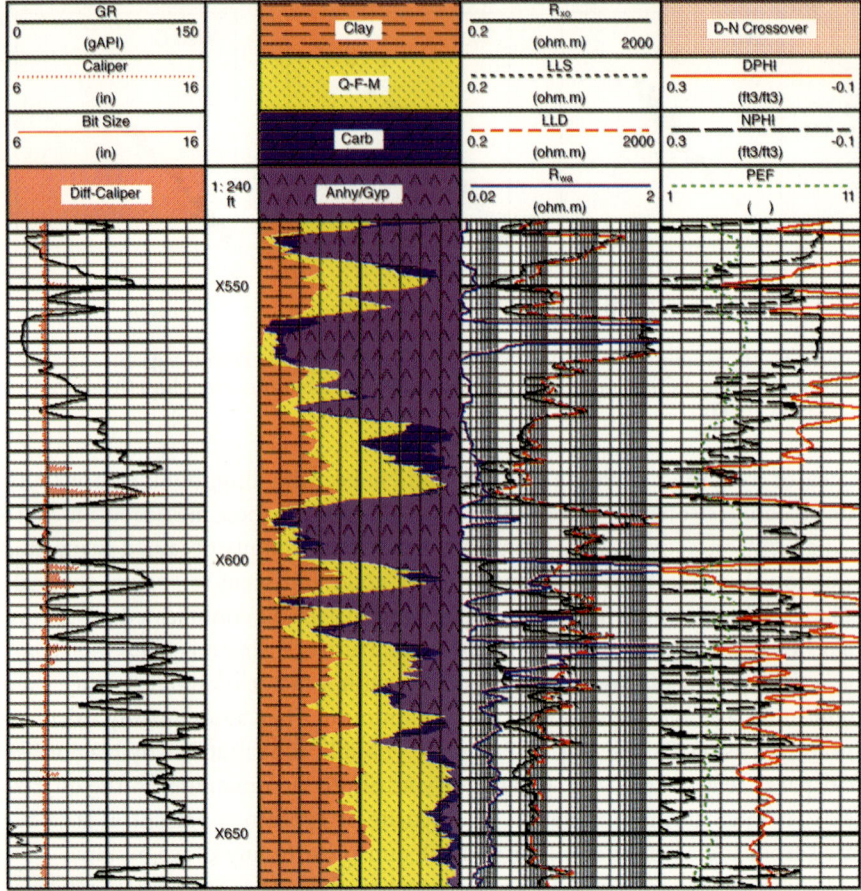

Figure 2.7.11 Example of mineral volumes calculated from elemental concentrations from the ECS tool. All four mineral groups are present: Clay, quartz/feldspars/mica, carbonates, anhydrite/gypsum. Volumes shown are in solid weight percentages.

Permeability

Herron (1987) has proposed to estimate the intrinsic permeability of clastic sediments from downhole geochemical data using multivariate statistics. This approach was recently extended and put on a more physical basis (Herron et al., 1998).

Estimating permeability from porosity is known to be inadequate unless some information on the geometry of the pore space is included. Johnson et al, (1986) have proposed using a parameter Λ (or just simply called "lambda"), which is inversely proportional to the ratio of surface area over pore volume. For simple pore geometries, such as regular packings of spheres, they found $k \propto \Lambda^2 \phi^m$

where m is the same exponent as used in Archie's Law (equation 1.3.3). From this, a permeability estimate k_Λ can be made of the form

$$k_\Lambda = \frac{Z \cdot \phi^m}{\left(\dfrac{S}{V_P}\right)^2} \qquad (2.7.11)$$

where Z is a proportionality constant. One notices that the same players as in the NMR-derived permeability estimates are involved (equations 2.6.6 and 2.6.7), since the surface relaxation rate T_2 is inversely proportional to S/V. Herron et al. (1998) realized from fits with core data that the exponents in equation 2.7.11 have to be increased significantly for permeabilities below 100 mD, typically by factor of 1.7, and that Z is also a function of the permeability. They propose to adapt equation 2.7.11 such that the surface area/pore volume ratio is estimated from the mineral composition. This can be justified by arguing that every mineral has a typical shape, and since the pore space is the mold of the grain space, every mineral has a typical pore shape. Although a disputable assumption, it essentially states that clays have a larger specific surface area than framework minerals such as quartz or feldspars[3]. The first step is to estimate the surface area/pore volume ratio S/V_p from porosity-independent variables, the specific surface area S_o, and the matrix density ρ_m as

$$\frac{S}{V_p} = S_o \rho_m \frac{1-\phi}{\phi} \qquad (2.7.12)$$

The total specific surface area of a sedimentary rock is then computed as the sum of its components, or

$$S_o = \sum M_i S_{oi} \qquad (2.7.13)$$

where M_i is the mass fraction of component i. These values are approximately known from core data and can be taken as 60, 2 and 0.22 for clay, carbonate and quartz respectively (Herron et al., 1998). The permeability is then estimated as

$$k_\Lambda = \frac{Z \cdot \phi^{m+2}}{\left(\sum M_i S_{oi}\right)^2 \rho_m^2 (1-\phi)^2} \qquad (2.7.14)$$

Figure 2.7.12 shows an example of such a k-lambda permeability estimation, together with permeabilities measured on core plugs. According to Herron et al. (1998), the correlation coefficient in this case is 0.84. Although promising, this

---

[3] Permeability estimates are often considered acceptable if they lie within an order of magnitude of the actual value

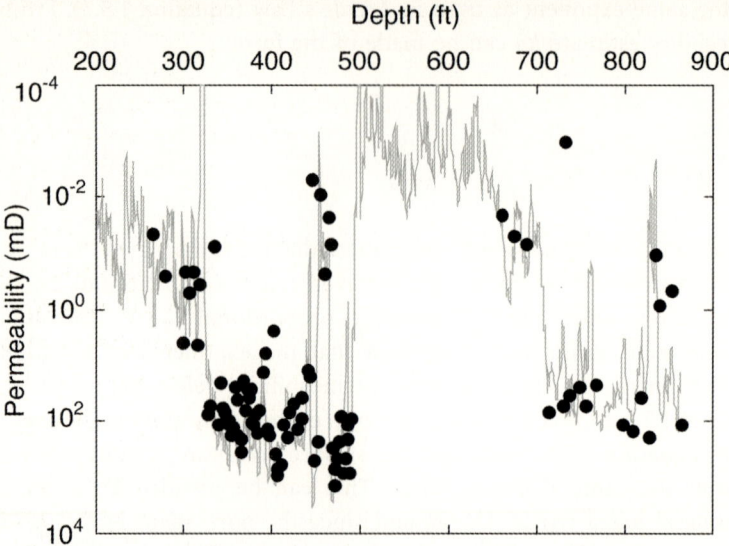

Figure 2.7.12. Permeability estimated from mineral volumes using the k-lambda concept, compared to core permeabilities (from Herron et al., 1998).

approach needs better validation and perhaps integration with nuclear magnetic resonance results, which, after all, provide a good measurement of S/V_p.

Other Applications

Organic Carbon Estimation. S. Herron (1987) has proposed a method to estimate total organic carbon(TOC) from geochemical logs, with the objective of spotting potential source rocks. Her approach is based on a measurement of the carbon/oxygen ratio using inelastic scattering spectra such as obtained from the GST or the RST. The formation oxygen is then estimated by modelling the formation as a two-component system, a solid matrix filled with water. For each of the components, the oxygen is estimated[4] and from the sum of the two and the C/O ratio the total carbon is calculated. A more elaborate model considers also the organic carbon with its own oxygen weight percentage. The inorganic carbon is obtained from a log interpretation that preferably includes spectroscopy data as in the example shown in figure 2.7.11. Total organic carbon is then computed by subtracting the inorganic carbon from the total carbon. S. Herron (1987) reports some success of this method in a shaly sequence, but her approach has the disadvantage of featuring the subtraction of two similar numbers (the total carbon, and the inorganic carbon) and the error, therefore, is expected to be large – unless very organic-rich sediments are present. Additionally, *a priori*

---

[4] Most common minerals have an oxygen weight percentage of close to 50 %.

estimates have to be made to account for the composition of the organic carbon in the rock, and the presence of residual oil might interfere. Myers & Jenkyns (1992) found the method less precise than using conventional density logs, but they point out that under some conditions such as overpressuring, or the presence of certain kerogens or minerals, their method may be at a disadvantage. In none of the two cases can an assessment of source rock maturity be made.

Well Correlation. The lithologies obtained from inverting elemental logs provide excellent data for lithostratigraphic correlations. An example is discussed in chapter 3.4.3, but other examples can be found in Herron & Herron (1990). Well correlation can of course also be done with the elemental concentration logs alone.

2.7.8 Summary

The biggest contribution of natural and induced gamma ray spectroscopy is that they are mostly sensitive to the rock composition. Therefore, they offer numerous possibilities for a better geological characterization of rocks in wellbores. These include:

- Mineral volume calculations from elemental concentration logs
- Well correlation and sequence analysis with total gamma ray, elemental logs, or mineral volume logs
- Permeability estimations from mineral volumes
- Organic carbon estimates

These logs are also very useful for formation evaluation. The volumes of clays, for example, can be significantly more accurate when nuclear spectroscopy information is included. Consequently the estimation of fluid saturation is also improved.

References

Adams JA, Weaver CE (1958) Thorium to uranium ratios as indicators of sedimentary processes: example of concept of geochemical facies. Am Assoc Petrol Geol Bull 42:

Belknap WB, Dewan JT, Kirkpatrick CV, Mott WE, Pearson AJ, Rabson WR (1959) Calibration facility for nuclear logs. Drill and Prod Prac API: 289–317.

Burke JA, Campbell RL, Schmidt AW (1969) The litho-porosity crossplot. The Log Analyst (Nov–Dec).

Clavier C, Rust DH (1976) MID-plot: a new lithology technique. The Log Analyst, 17: 6.

Dewan JT (1983) Essentials of modern open hole log interpretation. PennWell, Tulsa.

Doveton JD (1994) Geologic log analysis using computer methods. Am Assoc Petrol Geol Computer Applications in Geology 2.

Edmundson HN, Raymer LL (1979) Table of radioactive logging parameters of common minerals. Trans 6th Europ Symp Soc Prof Well Log Analysts: Paper BB.

Ellis DV, Flaum C, Roulet C, Marienbach E, Seeman B (1983) The litho-density tool calibration. 58th Ann Conf Soc Petr Eng: Paper 12048.

Ellis DV (1987) Well-logging for earth scientists. Elsevier, Amsterdam.
Ellis DV (1990) Neutron and gamma ray scattering measurements for subsurface geochemistry. Science 250, 82–87.
Fertl HW (1979) Gamma ray spectral data assists in complex formation evaluation. Trans 6th Europ Symp Soc Prof Well Log Analysts: Paper Q.
Flaum C, Pirie G (1981) Determination of lithology from induced gamma ray spectroscopy. Trans 22nd Symp Soc Prof Well Log Analysts: Paper H.
Gardner JS, Dumanoir JL (1980) Litho-density log interpretation. Trans 21^{rst} Symp Soc Prof Well Log Analysts: Paper
Gilchrist WA Jr, Quirein JA, Boutemy YL, Tabanou JR (1982) Application of gamma ray spectroscopy to formation evaluation. Trans 23rd Symp Soc Prof Well Log Analysts: Paper
Hallenburg JK (1984) Geophysical logging for mineral and engineering applications. PennWell Books, Tulsa, Oklahoma.
Hallenburg JK (1992) Nonhydrocarbon logging. The Log Analyst 33: 259–269.
Harvey PK, Lovell MA (1992) Downhole mineralogy logs: mineral inversion methods and the problem of compositional colinearity. In: Hurst A, Griffiths, CM, Worthington PF (eds): Geological Applications of Wireline Logs. Geol Soc Spec Publ 65: 361–368.
Hassan M, Hossin A, Combaz A (1976) Fundamentals of the differential gamma ray log interpretation technique. Trans 17th Symp Soc Prof Well Log Analysts: Paper H.
Herron MM (1986) Mineralogy from geochemical logging. Clays and Clay Minerals 34, 204–213.
Herron MM (1987) Estimating the intrinsic permeability of clastic sediments from geochemical data. Trans 28th Symp Soc Prof Well Log Analysts: Paper HH.
Herron SL (1987) A total organic carbon log for source rock evaluation. The Log Analyst 28: 520–527.
Herron MM, Herron SL (1990) Geochemical logging applications. In: Hurst A, Lovell MA, Morton A (eds) Geological Applications of Wireline Logs. Geol Soc Spec Publ 48: 165–175.
Herron MM, Herron SL (1996) Quantitative lithology: An application for open and cased hole spectroscopy. Trans 37th Symp Soc Prof Well Log Analysts: Paper E.
Herron MM, Johnson DL, Schwartz LM (1998) A robust permeability estimator for siliciclastics. 73rd Ann Conf Soc Petr Eng: Paper 49301.
Hertzog RC (1978) Laboratory and field evaluation of an inelastic neutron-scattering and capture gamma ray spectroscopy tool. 53rd Ann Conf Soc Petr Eng: Paper 7430.
Hertzog RC, Colson L, Seeman B, O'Brien M, Scott H, McKeon D, Wraight P, Grau JA, Ellis DV, Schweitzer JS, Herron MM (1987) Geochemical logging with spectrometry tools. 62nd Ann Conf Soc Petr Eng: Paper 16792.
Horkowitz JP, Cannon DE (1997) Complex reservoir evaluation in open and cased wells. Trans 38th Symp Soc Prof Well Log Analysts: Paper W.
Humphreys B, Lott GK (1990) An investigation into nuclear log responses of North Sea Jurassic sandstones using mineralogical analysis. In: Hurst A, Lovell MA, Morton A (eds) Geological Applications of Wireline Logs. Geol Soc Spec Publ 48: 223–240.
Hurst A (1990) Natural gamma-ray spectrometry in hydrocarbon-bearing sandstones from the Norwegian continental shelf. In: Hurst A, Lovell MA, Morton A (eds) Geological Applications of Wireline Logs. Geol Soc Spec Publ 48: 211–222.
Johnson DL, Koplik J, Schwartz LM (1986) New pore-size parameter characterizing transport in porous media. Phys Rev Let 57: 2564–2567.
Lake LW, Carroll HB (1986) Reservoir Characterization. Academic Press.
Lofts JC, Harvey PK, Lovell MA, Locke J (1992) The application of induced gamma-ray spectroscopy measurements: characterisation of a North Sea reservoir. Conference Record, 1992 IEEE Nuclear Science Symposium and Medical Imaging Conference: volume 1.
Mason B (1966) Principles of geochemistry. J. Wiley Sons, New York.
Mayer C Sibbit AM (1980) GLOBAL, A new approach to computer processed log interpretation. 55th Ann Conf Soc Petr Eng: Paper 9341.

Myers KJ, Jenkyns KF (1992) Determining total organic carbon contents from well logs: an intercomparison of GST data and a new density log method. In: Hurst A, Griffiths, CM, Worthington PF (eds): Geological Applications of Wireline Logs. Geol Soc Spec Publ 65: 369–376.

Pearson MJ (1978) Quantitative clay mineralogical analysis from the bulk chemistry of sedimentary rocks. Clays and Clay Minerals 26: 423–433.

Poupon A, Hoyle WR, Schmidt AW (1971) Log analysis in formations with complex lithologies. 45th Ann Conf Soc Petr Eng: Paper 2925.

Rider M (1996) The geological interpretation of well logs. Gulf Publishing Co.

Schlumberger (1982) Essentials of NGS Interpretation.

Schlumberger (1989) Log interpretation charts. Schlumberger Educational Services, Houston.

Schmidt AW, Land AG, Yunker JD, Kilgore EC (1971) Applications of the CORIBAND technique to complex lithologies. Trans 12th Symp Soc Prof Well Log Analysts: Paper AA.

Schweitzer JS, Ellis DV, Grau JA, Hertzog RC (1988) Elemental concentrations from gamm-ray spectroscopy logs. Nuclear Geophysics 2: 175–181.

Serra O (1984) Fundamentals of well log interpretation. 1. The acquistion of logging data. Developments in Petroleum Science 15A, Elsevier, Amsterdam.

Serra O, Sulpice L (1975) Sedimentological analysis from sand shale series from well logs. Trans 16th Symp Soc Prof Well Log Analysts: Paper W.

Serra O, Baldwin J, Quirein J (1980) Theory, interpretation and practical applications of natural gamm-ray logging. Trans 21st Symp Soc Prof Well Log Analysts: Paper Q.

Slatt RM, Jordan DW, D'Agostino AE, Gillespie RH (1992) Outcrop gamma-ray logging to improve understanding of subsurface well log correlation. In: Hurst A, Griffiths, CM, Worthington PF (eds): Geological Applications of Wireline Logs. Geol Soc Spec Publ 65: 3–19.

Suau J, Spurlin J (1982) Interpretation of micaceous sandstones in the North Sea. Trans 23rd Symp Soc Prof Well Log Analysts: Paper G.

Tittman J (1986) Geophysical well logging. Academic Press, Orlando.

Tittman J, Wahl JS (1965) The physical foundations of formation density logging (gamma-gamma). Geophysics 30: 284–294.

Van den Oord RJ (1991) Evaluation of geochemical logging. The Log Analyst 32: 1–12.

Van Wagoner JC, Mitchum RM, Campion KM, Rahmanian VD (1990) Silicicalstic sequence stratigraphy in well logs, outcrops and cores. Am Assoc Petrol Geol Methods in Exploration Series 7.

Westaway P, Hertzog RC, Plasek RE (1980) The gamma ray spectroscopy tool, inelastic and capture gamma ray spectroscopy for reservoir analysis. 55th Ann Conf Soc Petr Eng: Paper 9461.

2.8 Paleomagnetic Logging

2.8.1 Basics of Paleomagnetism

Paleomagnetism is the science of the earth's magnetic record of the past. One of its major interest lies in the fact that the polarity of the earth's magnetic field has changed at seemingly random intervals over geologic time. Thus, periods of normal polarity alternate with periods of reverse polarity, with the intervals lasting between 10^4 to 10^8 years. This sequence of polarity changes has been the subject of much research over the past few decades in order to establish a reference scale that can be used for geochronologic applications. A particularly good source has been the polarity sequences found on spreading seafloors (Vine & Matthews, 1963) which were complemented by studies on outcrops. The resulting *Geomagnetic Polarity Time Scale* (GPTS) had a range initially confined to the Pliocence-Pleistocene (Cox at al, 1963) but was subsequently extended to all of the Tertiary and the Late Cretaceous (Heirtzler et al., 1968; Cox, 1982; figure 2.8.1) as well as over the Mesozoic (Haq et al., 1987). An interesting observation was the geomagnetic normal "superchron" from the Aptian to the Santonian, lasting from 118 Ma to 83 Ma, or a total of 35 million years without any geomagnetic reversal. Knowledge of the paleomagnetic record for the Late Cretaceous and Tertiary is significantly better than for older periods, notably because of the absence of any oceanic crust older than Late Jurassic in the present ocean basins. Therefore, the GPTS for these ages must be determined from exposed stratigraphic sections on land, which are generally less magnetized, often diagenetically altered, and whose ages are not as accurately defined as for oceanic crust. Relatively vague paleomagnetic stratigraphic sequences have been proposed for the Paleozoic (Cox, 1982), but at present they are not accurate enough to be used for geochronology. However, significant improvements have been made for the younger periods (Cande & Kent, 1992; 1995) and today the accuracy of paleomagnetic age determination is often comparable to biostratigraphic methods.

It is not clear why the earth's magnetic field changes its polarity. The field is described as a geocentric axial dipole (Butler, 1992), i.e. the magnetic field is thought to be produced by a single magnetic dipole at the center of the earth and aligned with its axis of rotation. The geomagnetic field *H* at the earth's surfaces is generally illustrated by isomagnetic charts, which show the magnetic field properties on a world map. Among the field properties are the magnitude of the field *H*, its inclination *I* and its declination *D*. The inclination *I* is defined as the vertical angle between the horizontal and *H*, while the declination *D* is the azimuthal angle between the horizontal component of *H* and geographic north. *I* and *D* are exactly identical to the definitions of dip and azimuth of a plane shown in figure 2.1.5 if the line of steepest descent on the plane is compared to the direction of *H*.

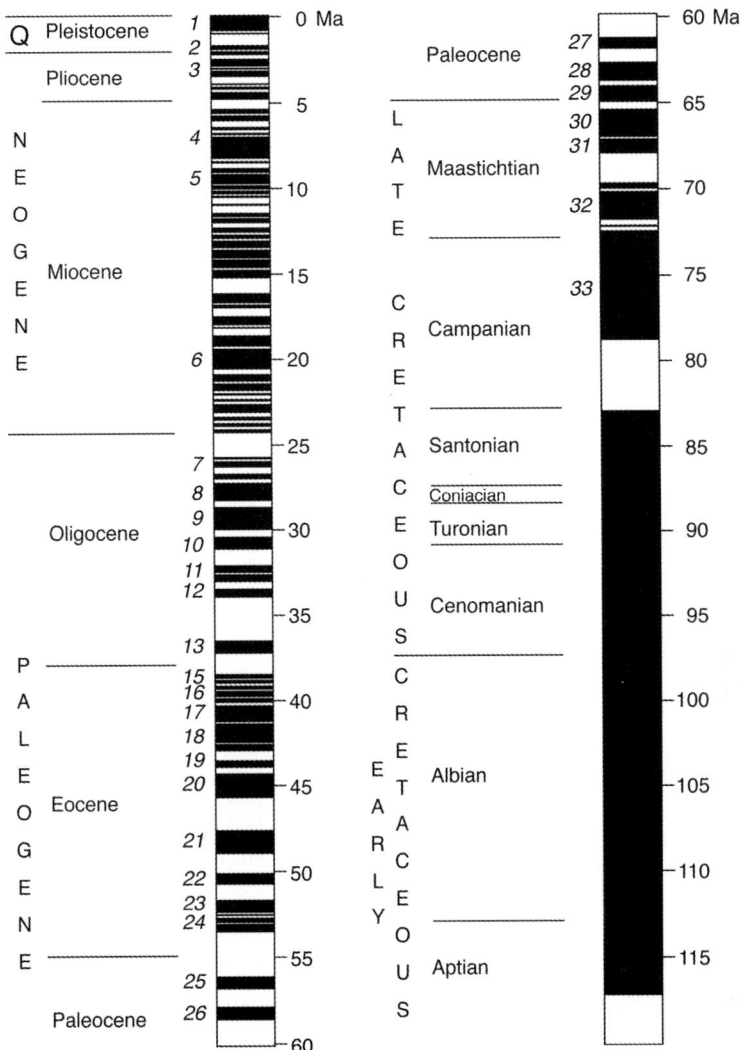

Figure 2.8.1. The geomagnetic polarity time scale (GPTS) from the recent to the upper Aptian (from Butler, 1992, after Cox, 1982). Black are normal polarities, white reversals. Notice the long normal magnetic period in the middle to late Cretaceous, during which magnetostratigraphy is not possible.

In the following we define the basic relationships used in paleomagnetism, with their measurement units defined in table 2.8.1.

A magnetic dipole moment M is defined as a pair of magnetic charges some distance separated from each other. It can alternatively be viewed as a loop of electrical current. The magnetic field H is defined as the force experienced by a unit of positive magnetic charge placed in the region influenced by the magnetic

217

Table 2.8.1. Quantities and SI units used in magnetism

	Symbol	Fundamental Units	Unit
Magnetic Induction	B	kg s^{-1} C^{-1}	tesla (T)
Magnetic Field	H	C s^{-1} m^{-1}	ampere m^{-1} (A/m)
Magnetization	J	C s^{-1} m^{-1}	ampere m^{-1} (A/m)
Magnetic Moment	M	C s^{-1} m^{2}	ampere m^{2} (Am^{2})
Magnetic Susceptibility	χ	dimensionless	–
Magnetic Permeability	μ	m kg C^{-2}	henry m^{-1} (H/m)

Note: In the cgs system, B, H, and J all have the same fundamental units, but are expressed in gauss *(B, J)* and oersted *(H)*.

dipole. A magnetic dipole moment free to rotate will align itself with the magnetic field, the torque of alignment being proportional to the magnetic dipole moment and the magnetic field. A material within a magnetic field acquires an *induced magnetization J* related to the external applied field as

$$J = \chi H \qquad (2.8.1)$$

where χ is the magnetic susceptibility of the material. Although many materials show isotropic magnetic susceptibility, some do not (requiring a vector notation of χ), which may result in the induced magnetization not being aligned with the magnetic field.

The magnetic induction *B* is defined as

$$B = \mu_o(H + J) \qquad (2.8.2)$$

which is the vector sum of the magnetic field and the induced magnetization (times the permeability of free space μ_o). Combining equations 2.8.1 and 2.8.2 yields

$$B = \mu_o(H + \chi H) = \mu_o(1 + \chi)H = \mu H \qquad (2.8.3)$$

where μ is the magnetic permeability of the material.

Materials may also possess a *remanent magnetization*. This is a property the material acquired in the past and which it will have even if no external magnetic field is present. The remanent magnetism in rocks is generally caused by finely distributed ferromagnetic minerals, of which the iron-titanium oxides are by far the most important ones although usually representing only a minor volume of the total rock. The most common ferromagnetic minerals are titanomagnetites, which have compositions between magnetite (Fe_3O_4) and ulvospinel (Fe_2TiO_4). Much less common although locally abundant are titanohematites, goethite, and the sulfide pyrrhotite. The latter is ferrimagnetic, i.e. the magnetic moments in and between the mineral layers are partly antiparallel, as opposed to ferromagnetic minerals, where they are parallel throughout a domain of the mineral.

There are three major types of remanent magnetism:

- *Thermoremanent magnetism*, acquired during cooling in a magnetic field from a high temperature to below the blocking temperature (which is somewhere near but below the Curie temperature). In this process, a bias in the distribution of magnetic moments at higher temperature freezes when the material drops below the critical temperature, producing remanent magnetization.
- *Chemical remanent magnetism*, in which a material, at a temperature below the blocking temperature of thermoremanent magnetism, is chemically altered such that a ferromagnetic mineral is produced. Alternatively, a ferromagnetic mineral might be precipitated from the pore water.
- *Detrital remanent magnetism*, which is acquired during deposition, generally through alignment of ferromagnetic grains with the earth's magnetic field. Sedimentary particles, for example, can be aligned this way while they settle in water. This type of remanent magnetism may be affected by postdepositional reorientation, for example through bioturbation or compaction.

Other forms of natural remanent magnetism are unsuitable for geochronological applications because their time of acquisition is often completely unrelated to the time of deposition of the sediment. These include viscous remanent magnetism, which is slowly acquired during exposure to weak magnetic fields such as the earth's field, and isothermal remanent magnetism, which can be caused by exposure to very strong magnetic fields such the ones occurring during lightning. While the latter seems to affect a significant number of the rocks at the earth's surface, it seems to be relatively unimportant in the subsurface.

In igneous rocks, thermoremanent magnetism is most important, while in sedimentary rocks detrital and chemical remanent magnetism are most common. Biochemical processes have also been cited as possible sources of remanent magnetism in sediments. Particularly interesting are magnetotactic bacteria which contain magnetite crystals arranged in chains, and which have been found as fossils in rocks up to 700 million years old. They seem to be more abundant in fine-grained carbonates and may account for a significant part of stable detrital remanent magnetism in these rocks. Apparently, most sedimentary rocks with biogenic magnetite also contain significant amounts of primary detrital remanent magnetism (Butler, 1992). It is not clear, however, how much magnetotactic bacteria found in oil zones perturb the detrital paleomagnetic signal, but according to Machel (1995) the effect might be significant.

In the presence of rocks with natural remanent magnetism, the total magnetic induction can be written as

$$B = \mu_o(H + J) + R \qquad (2.8.4)$$

where R represents the contribution due to the remanent magnetization of the rock. Table 2.8.2 contains some typical values for the quantities contained in equation 2.8.4.

By measuring the direction of remanent magnetism in sedimentary rocks, one hopes to obtain the direction of the earth's magnetic field at the time of deposition, or shortly thereafter. Comparing a sequence of magnetic directions – or rather, polarities – with the GPTS standard can then provide a sequence of absolute, geochronological ages. The principles of magnetic stratigraphy can be found in Hailwood (1989) and Opdyke & Channell (1996), while field studies are for example discussed in Lowrie & Alvarez (1977), Tauxe & Opdyke (1982) and in several articles in the volume edited by Turner & Turner (1995). A paleomagnetic borehole measurement has often been suggested, but the challenge of measuring a signal roughly one to ten million times smaller than the earth's magnetic field (table 2.8.2) has been formidable, particularly when considering that conditions in the borehole are substantially more difficult than in the laboratory. The actions of the ferromagnetic drill bit and the entire drill string as well as the composition of the mud, which can contain ferro- and ferrimagnetic particles, may strongly influence the measurement. Furthermore, the layers may be structurally distorted, the wells may be deviated, and significant plate tectonic drifts may have occurred since deposition of the rocks of interest. And, as if all of this was not challenging enough, the relative directions of the earth's magnetic field, the susceptibility and the well trajectory may combine to produce a "blind spot", i.e. a configuration where reversals are not detectable by a downhole measurement because the inductions J and R project orthogonally onto the earth's magnetic field induction B. For a vertical well with horizontal layers, this condition occurs at an inclination of the earth's magnetic field of

Table 2.8.2. Some values of relevance in paleomagnetism. Figures for natural remanent magnetism in rocks are approximate only.

Magnetic Permeability of Vacuum μ_o	$1.26 \cdot 10^{-6}$ H/m
Magnetic Moment of Earth	$8 \cdot 10^{22}$ Am^2
Earth's Magnetic Field at Surface	24,000 – 66,000 nT
Natural Remanent Magnetism:	
Basalt	1,000 nT
Granite	100 nT
Siltstone	10 nT
Limestone	0.1 nT

35.3°, which is typically found at latitudes around 10° to 30° (north and south). In the equatorial band between these two critical zones the magnetic induction caused by the susceptibility of the rock increases the total induction, while at higher latitudes it reduces it. Specifically, the magnitude of the induced magnetization projected onto the earth's magnetic field (J') varies with the inclination of the earth's magnetic (I) field as

$$J' \propto [1 - 3 \sin^2 I] \qquad (2.8.5)$$

This relationship is depicted in figure 2.8.2.

Similarly, the magnitude of the remanent magnetization projected onto the earth's magnetic field (R') is a function of the present-day inclination (I) and declination (D) of the earth's magnetic field, and additionally of the paleoinclination (i') and paleodeclination (d') of the earth's magnetic field at the time when the magnetization was acquired. It follows

$$R' \propto \left[\cos I \, \cos i \, \cos(D - d) - 2 \sin I \sin i \right] \qquad (2.8.6)$$

Here also, there is a blind spot but it is more difficult to determine as it involves more variables. Prior to considering paleomagnetic logging, therefore, it is necessary to perform an assessment of whether the blind spots are avoided. For non-vertical wells and dipping layers, additional complications occur within equations 2.8.5 and 2.8.6.

In the preceding discussion we omitted some aspects of paleomagnetism considered less relevant in the context of this book: The magnetic field intensities, the detailed nature of ferromagnetism, or the laboratory measurement of natural remanent magnetism and its stability. Instead, focus was laid on

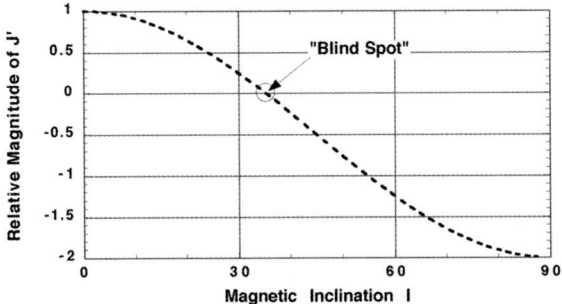

Figure 2.8.2. The projection of the induced magnetization onto the earth's magnetic field varies strongly with field inclination and thus with latitude. It goes through a blind spot at a field inclination of 35.3°, which typically occurs at latitudes between 10° and 30° north and south. Therefore, before planning paleomagnetic logging, the local field inclination needs to be checked. A few degrees north and south of the blind spot the magnitudes are already large enough again. Curve shown is only valid for a vertical well with horizontal layers.

the basic notions needed to understand the paleomagnetic measurement in boreholes. A fuller picture of paleomagnetism can be obtained from textbooks such as Butler (1992).

2.8.2 Measurement Principle

The idea of a magnetic downhole measurement dates back to the early 1950s, when the Magnolia Petroleum Company (now Mobil) developed a logging tool which featured a conventional airborne magnetometer modified for downhole use, and a susceptometer which measured the self-inductance of a solenoid caused by varying susceptibilities of the rocks (Broding et al, 1952). The main purpose of this tool was to measure the downhole magnetization in order to improve interpretations of surface magnetic surveys[1]. In the late 1980s, the French oil company TOTAL combined forces with the Centre National de la Recherche Scientifique (CNRS)[2] and later with the Laboratoire d'Electronique de Technologie et d'Instrumentation (LETI)[3] to develop a high-precision total magnetic induction and susceptibility measurement for borehole applications (Pozzi et al., 1988; Pozzi et al., 1993; Thibal, 1995; Vibert-Charbonnel, 1996). These two measurements were packaged by Schlumberger into one tool (figure 2.8.3; Schlumberger, 1993) and equipped with the necessary electronics to be used as a research and commercial tool under the name Geological High-Resolution Magnetic Tool (GHMT*).

Measurement of the magnitude and direction of R is the most direct approach to determine paleomagnetic reversals, but this cannot be done downhole. Instead, an indirect approach based largely on equation 2.8.4 has to be taken. It involves measuring the total induction B as well as the rock susceptibility at a given depth downhole. This allows determining the induced magnetization in the rock following equation 2.8.1. Additionally, the earth's magnetic field H is determined, either from a surface measurement while the logging is performed, or from nearby geophysical observatories. The decrease of the earth's magnetic field with depth is quite well known and has to be taken into account. From these three quantities, the projection of R onto the direction of B can be estimated, and magnetic reversals in the rock can thus be identified.

The required sensor accuracies are very high. Rock susceptibilities vary from 10^{-2} to 10^{-6}, implying that a sensor should have a range equivalent to the first figure, and an accuracy equivalent to the second. From the relative values of the earth's magnetic field and the typical values of natural remanent magnetism

---

[1] Just like downhole resistivity measurements had initially been developed as an aid for interpreting surface resistivity surveys (chapter 1.2)!

[2] The French National Center for Scientific Research

[3] A research laboratory of the French Atomic Energy Commission (CEA).

* Mark of Schlumberger

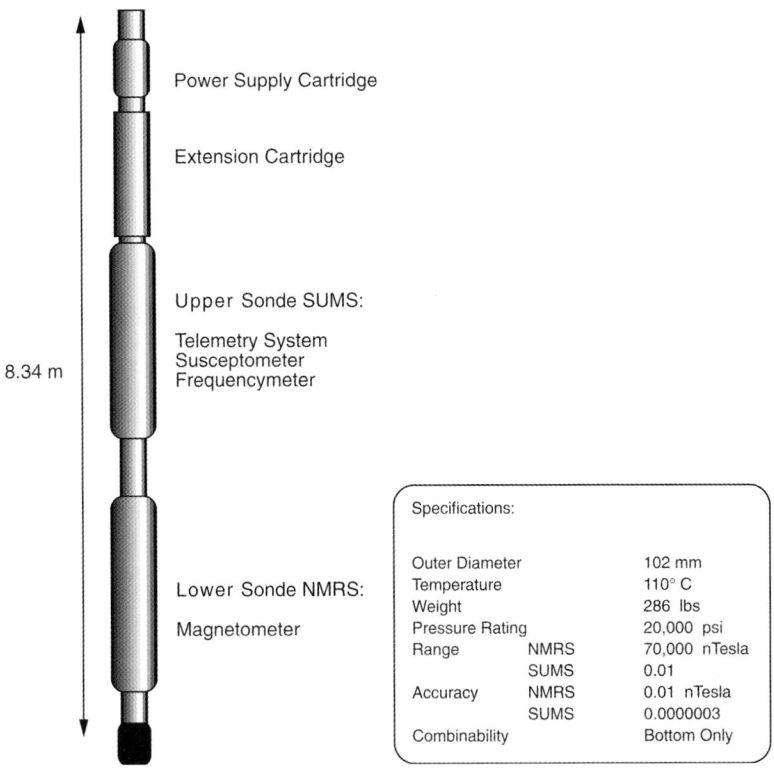

Figure 2.8.3. Tool sketch of the geologic high-resolution magnetic tool (GHMT), with the two main components, the susceptometer and the magnetometer.

found in sedimentary rocks (table 2.8.2), the total magnetic field measurement needs to have a range of about 70,000 nT and accuracy better than 0.1 nT.

The susceptibility sensor of the GHMT consists of a pair of induction coils with which the mutual inductance is measured similar to an induction logging tool. After elimination of direct mutual coupling, the in-phase receiver voltage is approximately linear with susceptibility, while the out-of-phase voltage gives an induction reading related to the formation conductivity. The two coils are separated by 80 cm and operate at 220 Hz, which is a frequency about 100 times lower than induction tools. This was shown to be an optimal frequency to boost the in-phase signal and reduce the out-of-phase signal. Comparison of this measurement with susceptibilities measured on cores at the surface showed that the sensor meets the required specifications mentioned above (figure 2.8.4).

The total magnetic induction B is measured by a nuclear magnetic resonance (NMR) device (see chapter 2.6). It takes advantage of the fact that protons in a magnetic field B precess at a frequency

$$f_L = \gamma \cdot B \tag{2.8.7}$$

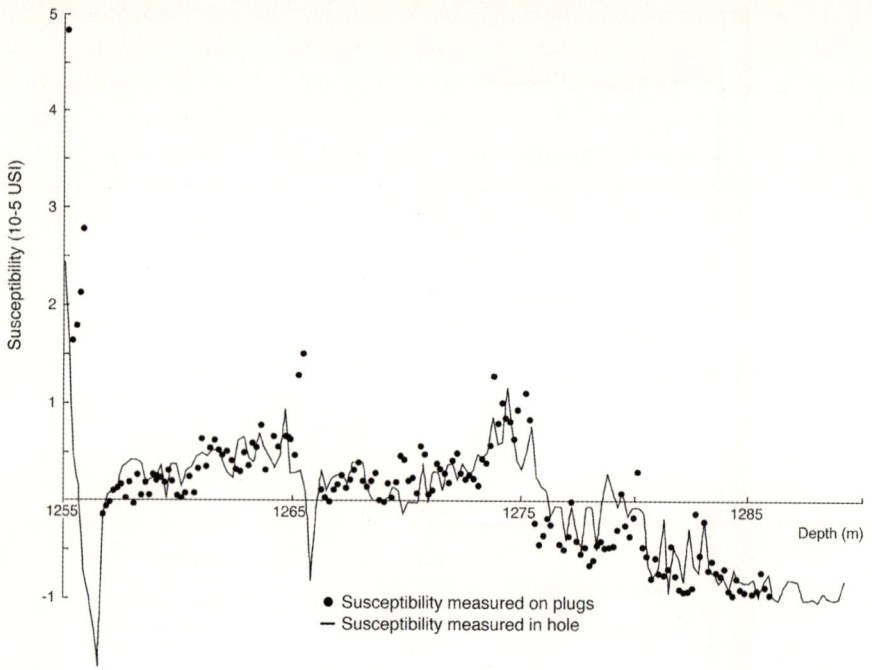

Figure 2.8.4. A comparison of susceptibilities measured in the borehole and on core. Jurassic limestones, Les Bagneaux well 4 (c. 100 km SE of Paris, France). After Pozzi et al. (1993).

where $\gamma = 4.2576076 \cdot 10^{-7}$ Hz/T is the temperature-independent Larmor precession constant. The measurement is therefore a high-precision frequency determination of the proton spin, which is achieved by a so-called electronic pumping NMR device, which has a sensitivity of 10^{-2} nT. Its range goes up to 70,000 nT in 16 intervals, each covering from 2,000 to 10,000 nT of the total range. The appropriate interval has to be selected such that the expected total magnetic field at the measurement location is centrally covered by the interval. Only with this offset-control of the magnetic measurement can the exceptionally high accuracy be achieved. Ideally, with this measurement a field of 50,000.01 nT can be distinguished from one of 50,000.00 nT.

2.8.3 Processing

Despite the high sensor accuracies, the measurement with this tool is still looking for a very small signal in a difficult environment. The initial processing of the data proposed by Pozzi et al. (1993) has been somewhat modified and the presently used approach contains a strong statistical component. If the earth's magnetic field (or actually, its induction) is subtracted from the total field measurement, the remainder contains the induction caused by the magnetic susceptibility of the rock, plus the remanent magnetic component. The induction, however, can be computed from the susceptibility measurement and the earth's magnetic field using equation 2.8.1. With this value of J (which is a scalar at this point), an estimate of R is possible from equation 2.8.4 by taking into account the values of B, H and J. In a simple geometric scenario[4], the value of R should increase the total

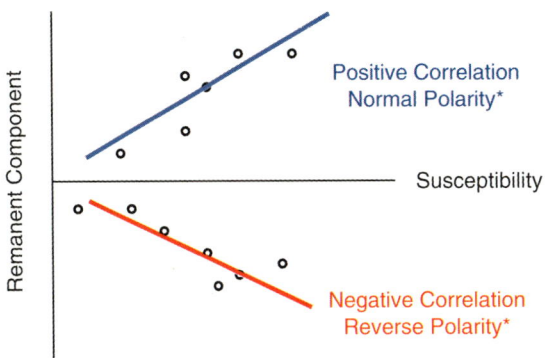

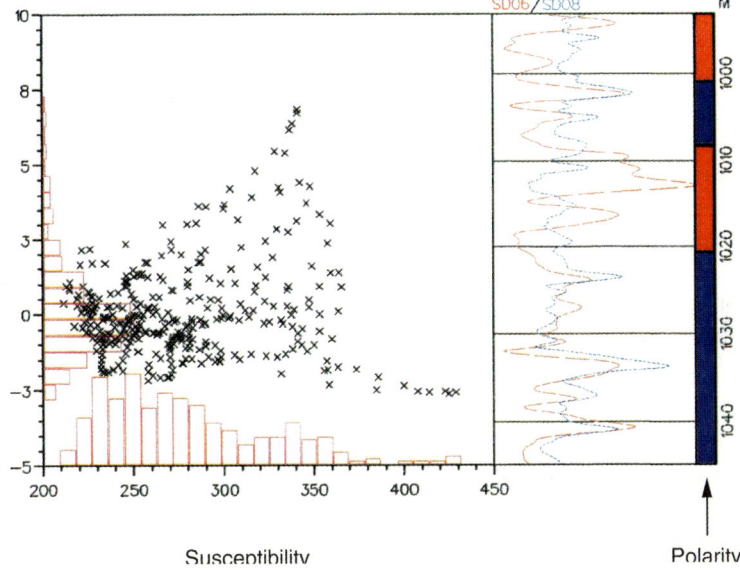

Figure 2.8.5. Correlation of the remanent component R with the susceptibility χ (top). If successive points show a positive correlation, a normal polarity is deduced assuming the simple scenario discussed in the text. Conversely, a negative correlation indicates reverse polarity. These correlations can also be observed when the logs are plotted versus depth (bottom): The synchronized behaviour of the two curves in the lower half indicates normal polarity, while for much of the upper half the polarity is reverse. Lower Jurassic, Paris Basin.

225

magnetic field B if its polarity is normal, and reduce it if it is reversed. Since the induction J caused by the magnetic susceptibility of the rock points in the direction of the earth's magnetic field – which is normal at present – the remanent magnetization R is parallel with J if the remanent signal is normal, and opposite if the remanent signal is reversed. The procedure then consists of cross-plotting several successive values of R and the susceptibility χ, and of calculating the linear slope using a least-mean squares fit (figure 2.8.5). By doing this repeatedly over successive windows, a sequence of slopes is obtained whose signs (negative or positive) should reflect the polarity of the remanent magnetization of the rocks.

This cross-plotting technique has the advantage that it reduces the noise by averaging several measurement levels. On the other hand, it diminished the vertical resolution and therefore the ability to detect small reversals. The slope of correlation between R and χ is known as Koenigsberger's ratio and is a function of the type and amount of magnetic substance in the rock, the grain size, the magnetic domains, the stress and the thermal history. If a lithological boundary is crossed within one correlation interval, this ratio may therefore change and so will the slope of the linear correlation. However, the sign of the correlation (positive or negative) should not change if everything else remains equal.

Figure 2.8.6 shows an example of such a correlation, using several windows of different correlation lengths. The polarity at a given depth is only indicated if the sense of correlation remains the same throughout the entire range of window sizes. If it is not, no polarity is assigned and the polarity column is left blank.

In practically all wells, the processing procedure described here has to be significantly expanded in order to take into account the non-ideal conditions under which the measurements were recorded. An important correction to be made is the change (decrease) of the earth's magnetic field with depth. The amount of this change – a relatively constant gradient – is generally known and can be applied to the measurement of the earth's field at the surface in order to provide the in-situ value. If this correction is not properly done, the value for the remanent rock magnetization obtained through equation 2.8.4 may be erroneous. A similar correction has to be made for regional magnetic anomalies. The magnitude and direction of these in the wellbore have to be obtained through modeling calculations.

Layer dip and well deviation constitute other factors that have to be taken into account during processing. Their effects are known and can be compensated for. While well deviation is generally known, layer dips and azimuths are best obtained either from a dipmeter survey (chapter 2.1), from borehole imagery (chapters 2.2 – 2.5) or, if nothing else is available, from seismic surveys. It is recommended to perform a modeling calculation prior to logging the tool to

---

[4] No borehole deviation, no layer dip, no plate movements since deposition of the layer and a near-equatorial latitude.

ascertain that well deviation, layer dip and the earth's magnetic field do not combine to form a blind spot, whereby the induced component is at a right angle to the earth's magnetic field and thus cannot be detected (similar to the blind spot for a vertical well with horizontal layers discussed above and illustrated in

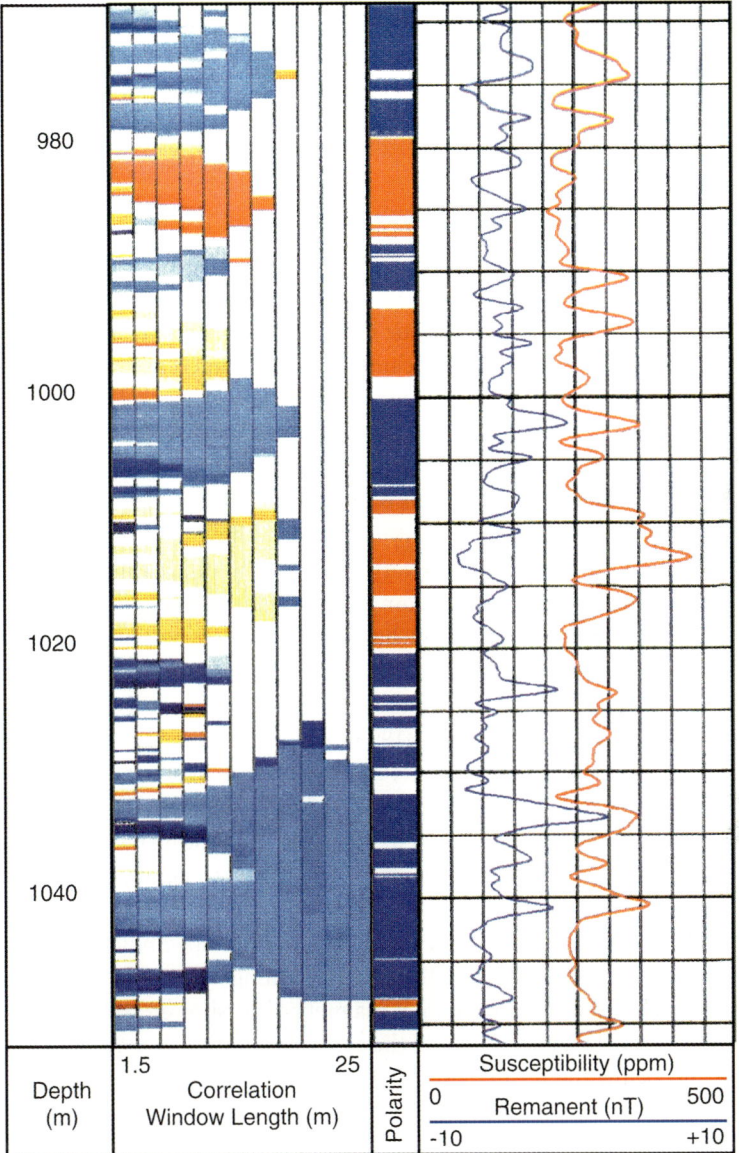

Figure 2.8.6. Paleomagnetic results obtained in a well in the Paris Basin. Polarities (middle track) are determined from the correlations using variable window lengths (left track). The susceptibility and remanent component are shown on the right.

227

figure 2.8.2). Modeling can also be done for the remanent magnetization, whereby the paleolatitude and paleolongitude of the location at the time of acquisition of the remanent magnetization have to be known. For a vertical well with horizontal layers, equation 2.8.6 applies.

Finally, there are many other factors, which can perturb borehole magnetic measurements. Among these are the surface equipment (the rig, the generators etc.), the casing which is generally made of steel, pieces of the drill string etc. which have fallen into the well, or magnetic particles in the mud. Corrections for many of these are possible and often involve empirical signal-processing methods, which may be easier to implement than a correction based on a physical understanding of the disturbance.

2.8.4 Interpretation

There are at present only few paleomagnetic borehole data sets available, mostly because it is a new technique, but also because age dating in oil and gas wells is classically and still preferably done with biostratigraphy.

Pozzi et al. (1993) have compared downhole paleomagnetic measurements in weakly magnetized Jurassic rocks from the Paris basin in France to laboratory measurements of rock magnetism on cores of the same well. They found that the downhole susceptibility measurements very accurately follow the core measurements although the values were generally very small (below 10^{-5}, see figure 2.8.4). They also found a secondary remanent component in the cores, which they attribute to the coring process. This effect has been modeled by Shi & Tarling (1995), who found that the magnetic field associated with a core barrel is strongest at the very tip of the device, while inside the string and on the outside the field is relatively weak. Pozzi et al. (1993) argue that this is a significant advantage of paleomagnetic measurements in the borehole over measurements on cores. Additional arguments in favor of this are the larger volume investigated by the downhole measurement compared to cores, and the generally lower diagenetic overprint of rocks at depth when compared to outcropping formations.

Dubuisson et al. (1995) and Thibal et al. (1995) report downhole magnetic measurements in young sediments of the Ocean Drilling Program Leg 145 in the North Pacific. They found that the total magnetic field showed strong variations around the average magnetic field of the earth, amounting to between 20 to 100 nT above and below the mean value. These variations can readily be correlated with the GPTS of the Pliocene and the Quaternary (figure 2.8.7). Furthermore, comparison of the polarity sequence thus obtained compares very well with those obtained from magnetic measurements on cores. This well represents, therefore, a simple but important step in the validation process of this technique. It is unique in that the remanent signal is very strong, dominating the induced signal because of abundant fine-grained and well-oriented magnetite in these pelagic sediments (Thibal et al., 1995). However, the well lacks many of the complicating factors one might typically encounter in wells of the oil and gas industry (see chapter 2.8.1). Furthermore, since the sedimentation rate appears to have been fairly constant, and an unequivocal marker – the ocean surface representing the present – is available, correlation with the GPTS is greatly facilitated and therefore age dating is easy and accurate.

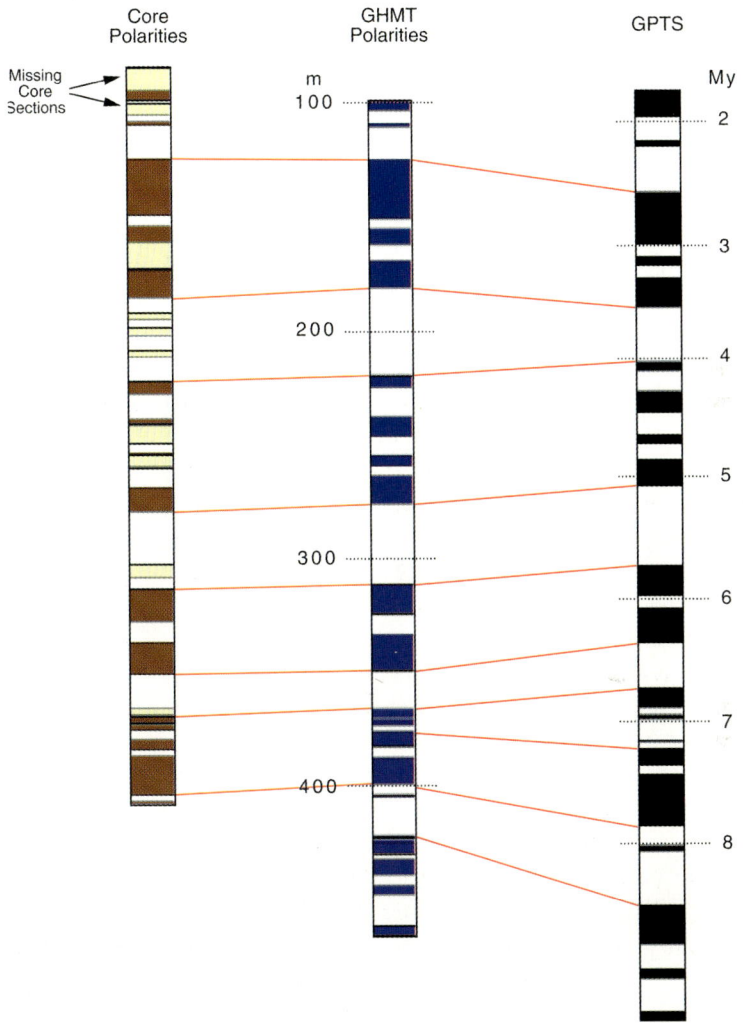

Figure 2.8.7. Polarity sequence in ODP Leg 145, hole 884E from downhole paleomagnetic measurements (left) and from cores (right). Correlation with the geomagnetic polarity time scale of Cande & Kent (1992) for the Plio/Pleistocene is straightforward (after Thibal et al., 1995).

Thibal (1995) discusses a relatively complex field example, which might be more representative of wells in the oil and gas industry. Two nearby wells in the Aquitaine basin in France had been logged with the commercial version of the paleomagnetic tool (GHMT). The processing steps outlined in chapter 2.8.3 resulted in two polarity sequences, of which the intervals around the Eocene

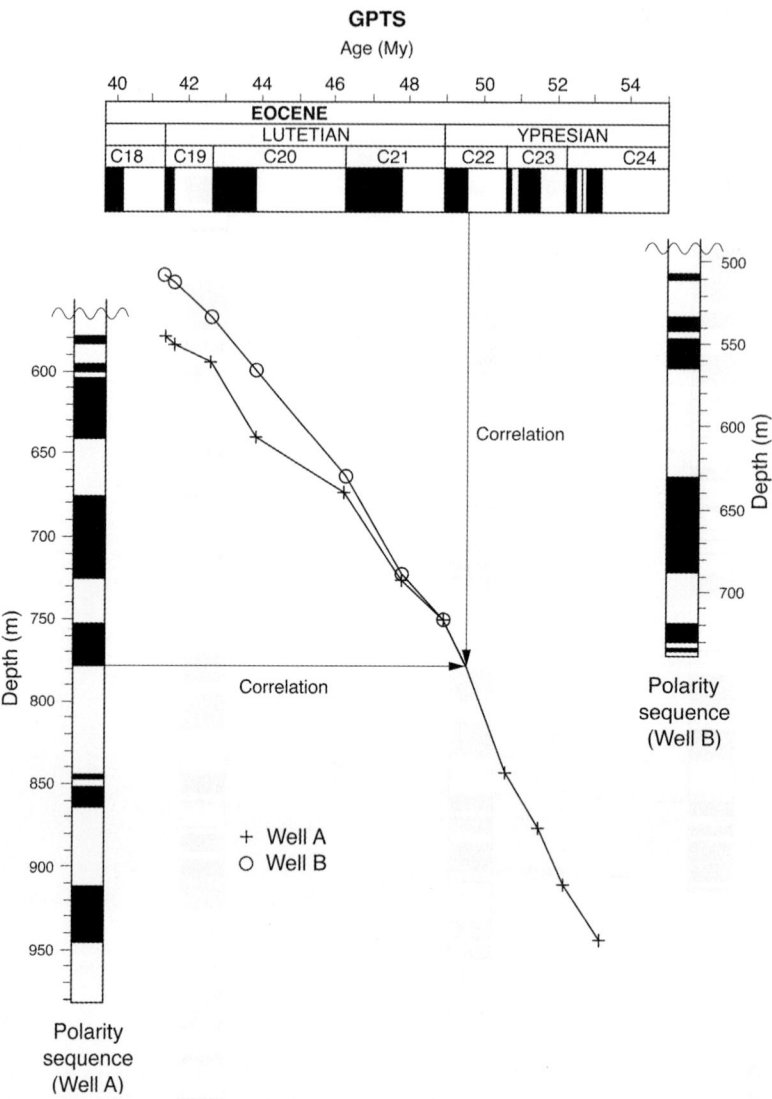

Figure 2.8.8. Example of polarity sequences obtained from the GHMT over Eocene carbonate sections of two nearby wells in the Aquitaine basin in southern France (Thibal, 1995). Correlation with the GPTS (top) is made for both wells separately and results in the two curves shown in the middle. See more detailed discussion in the text (from Thibal et al. 1999).

are shown in figure 2.8.7. The correlation with the GPTS is made difficult by the presence of an unconformity spanning several million years as well as different sedimentation rates in the two wells. However, after careful analysis, Thibal

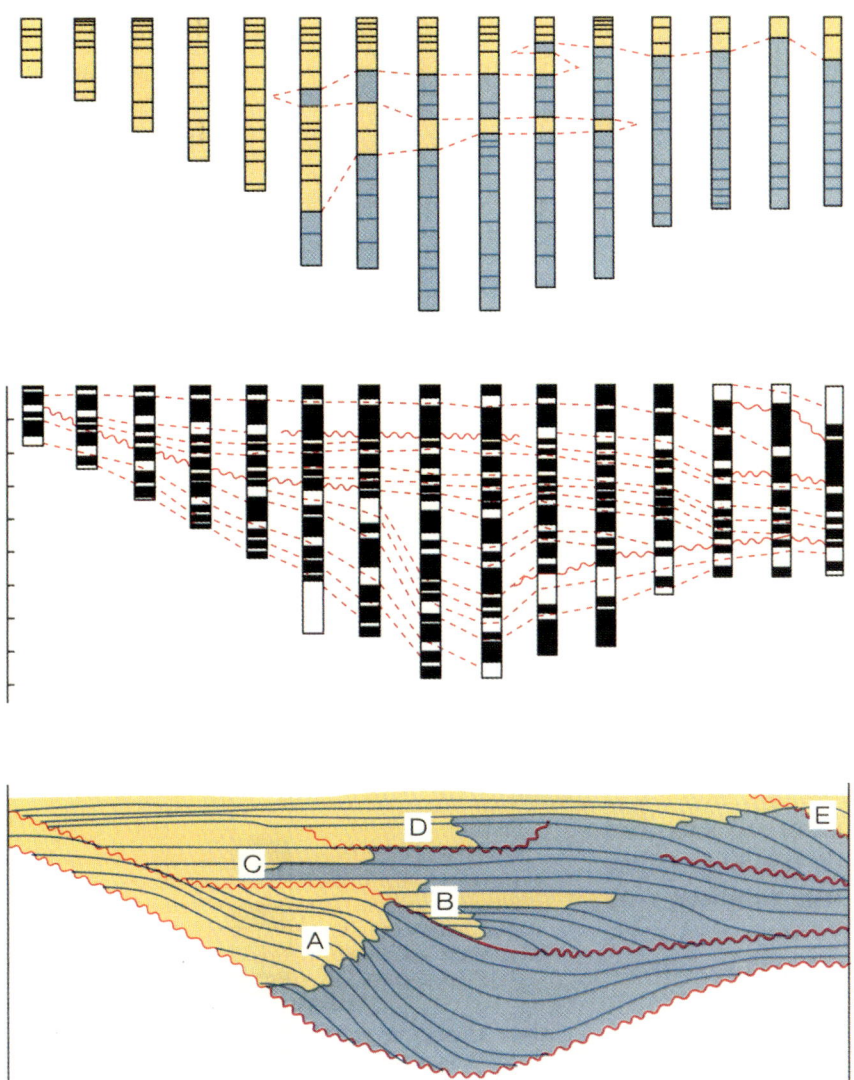

Figure 2.8.9. A synthetic correlation example of several wells in a clastic passive margin setting. The lithological correlation (top) distinguishes continental/transitional lithofacies in yellow, and fully marine facies in blue. The chronostratigraphic correlation (middle) gives a different picture and allows identification of unconformities. Combining the two correlations, however, tells a more complete story. It reveals five depositional tracts in a generally prograding setting interrupted by rapid sea level changes (from Schlumberger, 1993).

(1995) proposes the correlation shown in figure 2.8.8. The polarity sequences in this figure have been plotted vertically and correlated to the GPTS for the relevant time interval, which is displayed horizontally. The resulting correlation, therefore, has a slope of depth/time, which represents the sedimentation rate after compaction. It can be seen that well A, on the left, has a rather constant sedimentation rate in the lower Eocene before decreasing steadily towards the unconformity. Well B is not as deep but shows a similar development although later

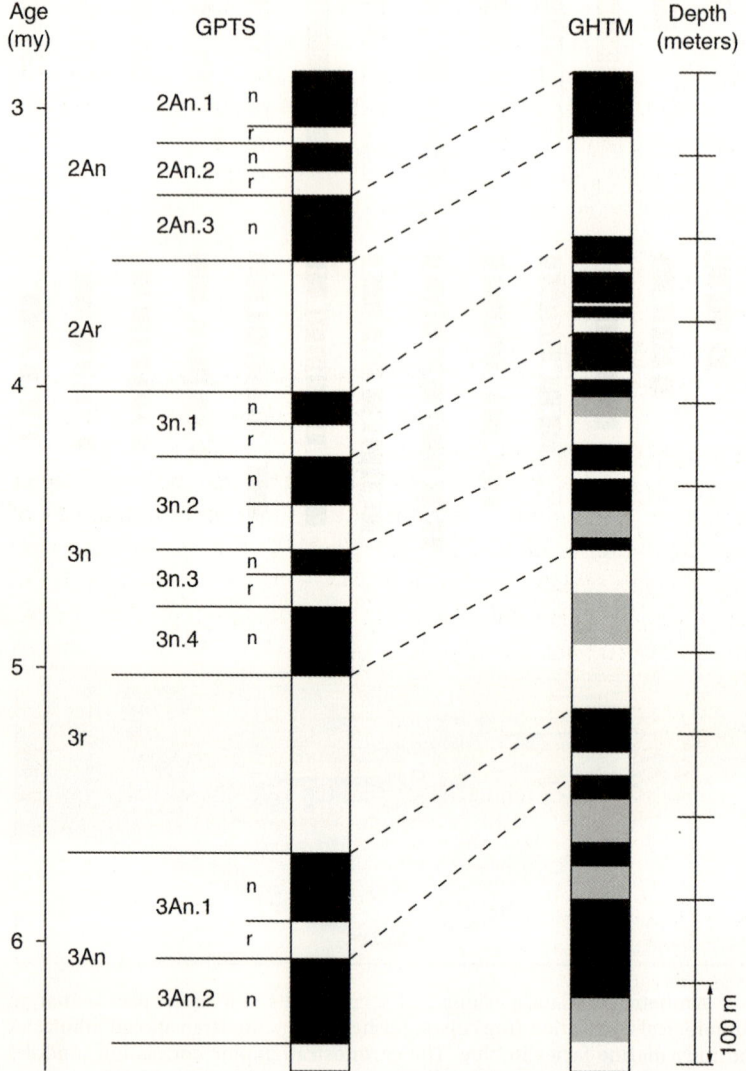

Figure 2.8.10. An example of age dating by matching the paleomagnetic logging results from the GHMT (right) with the global polarity time scale GPTS for the Pliocene (Cande & Kent, 1995; left). The results suggest a very high and slightly variable sedimentation rate.

in time. A possible explanation for this decreasing sedimentation rate is an increasing starvation in carbonate supply of this area. This example illustrates that both lithostratigraphic and chronostratigraphic correlation appear now to be possible with well logs: lithostratigraphy with the standard, rock-sensitive logs such as the gamma ray or more advance nuclear spectroscopy logs (chapter 2.7), and chronostratigraphy with the paleomagnetic measurement discussed here. See also chapter 3.4.3 for a further discussion of this topic. The rock magnetic susceptibility by itself, which is part of the downhole paleomagnetic measurement, has been proposed by various authors as a mineralogical indicator and a lithological correlation aid (Robinson, 1992; Lovlie & Van Veen, 1995) that could help identifying stratigraphic sequence boundaries. This possible application remains so far largely untested in hydrocarbon exploration (Schumacher, 1990), but there is well-founded hope that sequence stratigraphy will soon be possible on a field-wide and regional basis primarily using borehole measurements. The conceptual sketch in figure 2.8.9 shows how a combination of lithological and chronological correlation can result in the correct identification of unconformities and the proper distinction of system tracts. This leads to a genetically correct subdivision of the geological units in a field, which in turn will also lead to a better reservoir model.

An example of age dating in a field under development is shown in Figure 2.8.10. The well penetrates young Tertiary clastics in West Africa. Comparison of the polarities obtained from the GHMT survey with the GPTS from Cande & Kent (1995) suggests most of the sediments to be of Pliocene age with a rather high and slightly variable sedimentation rate. Grey zones indicate intervals of uncertain polarities. Alternative matches with the GPTS are possible, but they all would suggest highly variable sedimentation rates. The interpretation shown here is therefore the most plausible one from a sedimentological point of view, but alternative interpretations are possible.

2.8.5. Summary

In summary, paleomagnetic borehole logging has the following possible applications:

- Age dating
- Identification of unconformities
- Chronostratigraphic well-to-well correlation
- Sequence stratigraphy

References

Broding RA, Zimmermann CW, Somers EV, Wilhelm ES, Stripling AA (1952) Magnetic well logging. Geophysics 17: 1–26.

Butler R (1992) Paleomagnetism. Blackwell, Boston.

Cande SC, Kent DV (1992) A new geomagnetic polarity time scale for the late Cretaceous and Cenozoic. J. Geophys Res 97, B10: 13917–13951.

Cande SC, Kent DV (1995) Revised calibration of the geomagnetic polarity time scale for the late Cretaceous and Cenozoic. J. Geophys Res 100, B4: 6093–6095.

Cox A, Doell RR, Dalrymple GB (1963) Geomagnetic polarity epochs and Pleistocene geochronometry. Nature 198: 1049–1051.

Cox A (1982) Magnetostratigraphic time scale. In: Harland WB et al (ed) A geologic time scale, Cambridge University Press, pp 63–84.

Dubuisson G, Thibal V, Barthes J, Pocachard J, Pozzi JP (1995) Downhole magnetic logging in sediments during leg 145 – usefulness and magnetostratigraphic interpretation of the logs at site 884. In: Rea DK, Basov IA, Scholl DW, Allen JF (eds) Proceedings of the Ocean Drilling Program, Scientific Results, 145, Texas A&M University, pp 677–688.

Hailwood EA (1989) Magnetostratigraphy. Geol. Soc. London, Spec. Report, 19,

Haq BU, Hardenbol J, Vail PR (1987) Chronology of the fluctuating sea levels since the Triassic. Science 235: 1156–1167.

Heirtzler JR, Dickson GO, Herron EM, Pitman WC III, Le Pichon X (1968) Marine magnetic anomalies, geomagnetic field reversals, and motions of the ocean floor and continents. J Geophys Res 73: 2119–2136.

Lovlie R, Van Veen P (1995) Magnetic susceptibility of a 180 m core: reliability of incremental sampling and evidence for a relationship between susceptiblity and gamma ray activity. In Turner P, Turner A (eds) Paleomagnetic applications in hydrocarbon exploration and production. Geol Soc Spec Publ 98, pp 259–266.

Lowrie W, Alvarez W (1977) Upper Cretaceous-Paleocene magnetic stratigraphy at Gubbio, Italy, III. Upper Cretaceous magnetic stratigraphy. Geol Soc Am Bull 88: 374–377.

Machel HG (1995) Magnetic mineral assemblages and magnetic contrasts in diagenetic environments – with implications for studies of paleomagnetism, hydrocarbon migration and exploration In Turner P, Turner A (eds) Paleomagnetic applications in hydrocarbon exploration and production. Geol Soc Spec Publ 98, pp 9–29.

Opdyke ND, Channell JET (1996) Magnetic Stratigraphy. Academic Press.

Pozzi JP, Martin JP, Pocachard J, Feinberg H, Galdeano A (1988) In–situ magnetostratigraphy: Interpretation of magnetic logging in sediments. Earth Planet Sci Lett 88: 357–373.

Pozzi JP, Barthes V, Thibal J, Pocachard J, Lim M, Thomas T, Pages P (1993) Downhole magnetostratigraphy in sediments: Comparison with the paleomagnetism of a core. J Geophys Res 98: 1939–1957.

Robinson SG (1990) Applications for whole-core magnetic susceptibility measurements of deep-sea sediments: Leg 115 results. Proceedings of the Ocean Drilling Program, Scientific Results, 115, Texas A&M University: 737–770.

Schlumberger (1993) Harnessing paleomagnetics for logging. Oilfield Review, v 5, n 4: 4–13.

Schumacher D (1990) Proposal for magnetostratigraphic dating of Plio-Pleistocene seismic sequence boundaries, Gulf of Mexico. GCSSEPM Foundation 11th Ann Res Conf: 329–333.

Shi H, Tarling DH (1995) Magnetic field of a core barrel. In: Turner P, Turner A (eds): Paleomagnetic applications in hydrocarbon exploration and production. Geol Soc Spec Publ 98, pp 267–272.

Tauxe L, Opdyke ND (1982) A time framework based on magnetostratigraphy for the Siwalik sediments of the Khaur area, northern Pakistan. Paleogeogr Paleoclimat Paleoecol 37: 13–61.

Thibal J (1995) Analyse de l'aimantation des sédiments par diagraphies magnétiques; magnetostratigraphie, cyclicités climatiques et intensité du champ géomagnétique. Ph. D. thesis, Univ. Paris XI Orsay, France.

Thibal J, Pozzi JP, Barthes V, Dubuisson G (1995) Continuous record of geomagnetic field intensity between 4.7 and 2.7 Ma from downhole measurements. Earth Planet Sci Lett 136: 541–550.

Thibal J, Etchecopar A, Pozzi Jp, Barthès V, Pocachard J (1999) Comparison of magnetic and ray logging for correlations in chronology and lithology: example from the Aquitaine Basin (France). Geophys J Int 137: 839–847.

Turner P, Turner A (1995) Paleomagnetic applications in hydrocarbon exploration and production. Geol Soc Spec Publ 98.

Vibert-Charbonnel P (1996) Méthodes de traitement des mesures magnétiques en forage pour la datation haute-résolution des séries sédimentaires. Ph.D. thesis, Inst. National Polytechnique de Grenoble, France.

Vine FJ, Matthews DH (1963) Magnetic anomalies over ocean ridges. Nature, 199: 947–949.

2.9 Core Sampling

2.9.1 Introduction

Core sampling is to geologists what well testing is to reservoir engineers: it provides them with a physical specimen, rather than a downhole measurement. Fluid and rock samples brought to surface have the advantage that they can be subjected to tests and measurements which are not possible in the wellbore.

The are two types of coring: drill-string, or conventional coring, and wireline coring (figure 2.9.1). *Conventional coring* is quite commonly done in exploration wells and in selected development wells. It retrieves a core from the bottom of the borehole, where a special coring bit cuts out a cylindrical portion of the rock in front of it. This is then pushed into the inner core barrel, which remains stationary while the drill bit and its connection to the drill string (the outer core barrel) rotate as the drilling proceeds. At convenient intervals, the core is retrieved to the surface, whereby a core catcher prevents it from slipping out of the barrel. Typically, core barrels are 9 meters (30 feet) long, and core diameters can vary but are often around 10 centimeters (4 inches). Special techniques have been developed to minimize the influence of the drilling process on the core, for example to reduce the mechanical action so that an optimum core recovery is achieved, or to reduce the invasion of drill mud so that the core still contains the original fluids (figure 2.9.2)

Wireline coring is also referred to a sidewall coring and is done with a device mounted on a wireline cable after the well has been drilled. A correlation measurement such as a gamma ray tool is used to put the coring tool at the desired depth. Three types of sidewall coring exist, but only two are commonly used:

- percussion coring
- core slicing, and
- sidewall rotary coring.

The *core slicer*, developed, among others, by Schlumberger in the 1960s, consisted of two motorized saw blades oriented at 60° to each other which were able to cut up to four triangular pieces, each one meter long, out of the borehole wall. This mechanically very demanding design was abandoned in the 1990s in favor of sidewall rotary coring.

Wireline coring is often used to sample zones that appear of interest after drilling and logging has been done. In many places it is also used to take shale samples for biostratigraphic analysis. In this chapter we only discuss the two common types of wireline core sampling.

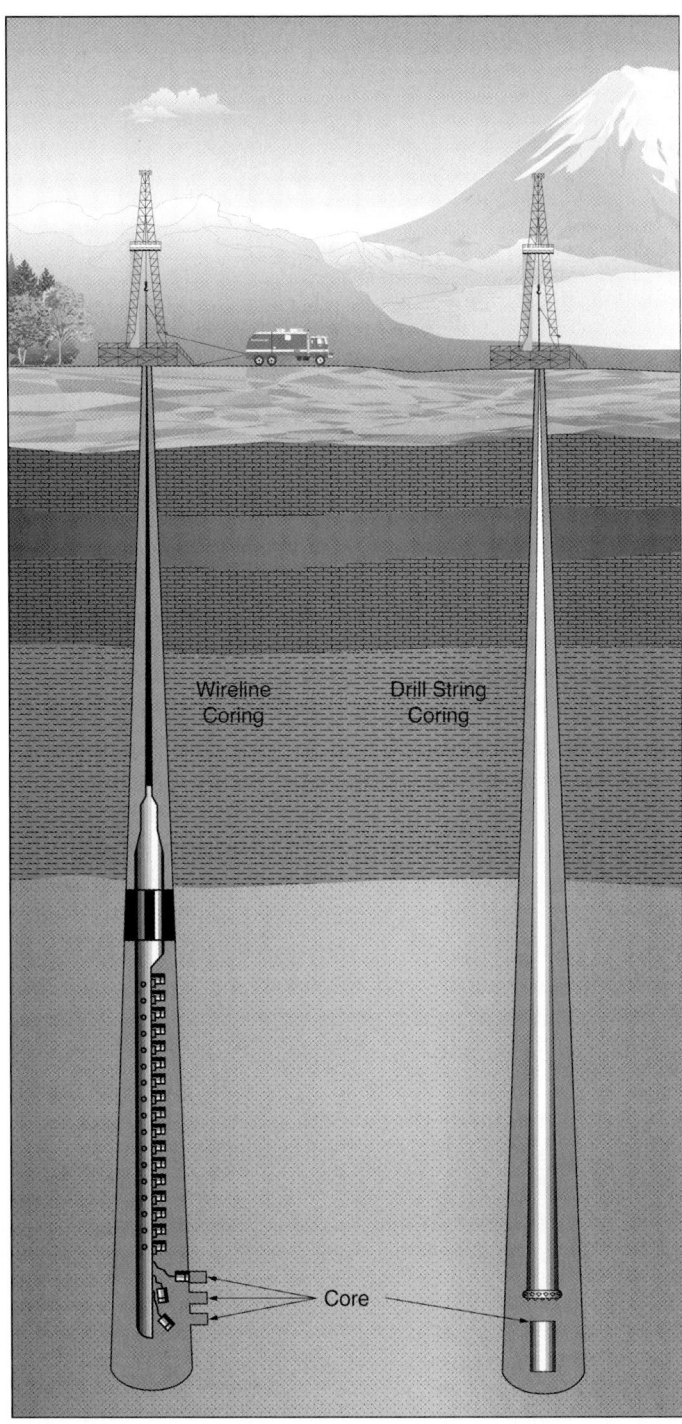

Figure 2.9.1. Conventional or drill-string coring (right), and wireline coring (left).

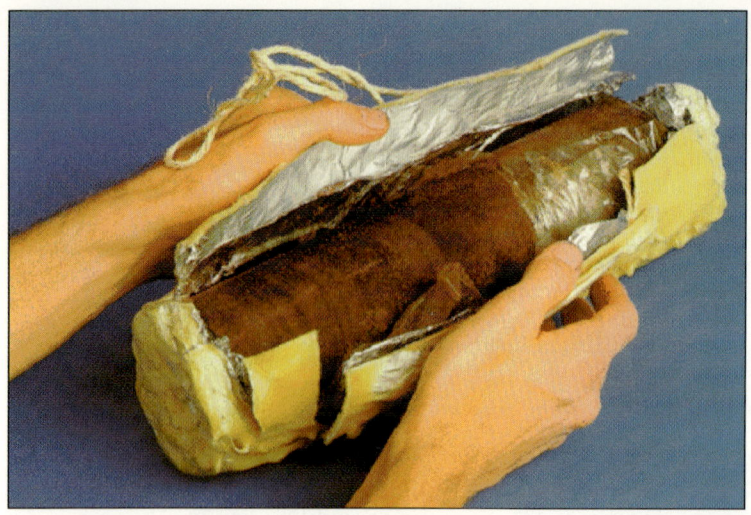

Figure 2.9.2 Photograph of a preserved core obtained from conventional drill-string coring. The brown color indicates that the original oil saturation is little affected by the drilling process.

2.9.2 Percussion Coring

Developed in the 1930s and little changed since, percussion coring is still the wireline coring service of choice, particularly in softer lithologies where the recovery rates are higher than in harder rocks[1]. Sometime only referred to *as sidewall coring,* it has a relatively simple operating principle. A hollow cylinder such as the one in figure 2.9.3 is shot into the borehole wall by a powder charge ignited with an electric current. The bullet is then retrieved by means of a steel cable which is attached to the main tool. This is simply achieved by pulling the tool upwards. The coring guns contain up to 50 bullets (figure 2.9.4), but only one is fired at a time. The cores remain in the bullets until the tool reaches the surface, at which point they are retrieved. Sample cores are shown in figure 2.9.5.

The powder charges and particularly the bullet shapes are critical factors influencing recovery rates. For softer lithologies such as shales, a relatively long and wide bullet is used, while for harder rocks such as limestones a short and narrow shape like the one shown in figure 2.9.3 is preferred (Desbrandes, 1985, p. 338). Typically, the core barrel length ranges from 4 to 6.5 cm, and the diameter from 1.75 to 2.5 cm. Holes on the sides of the bullets allow the mud to escape while the bullet penetrates the rock. What bullet gives the best result is

---

[1] Current tools on the market include the Chronological Sample Taker CST (Schlumberger), and the Sidewall Corgun SWC (Baker-Hughes).

Figure 2.9.3 The core barrel of a percussion or sidewall coring tool. The shape of these hollow bullets is adapted to the different rock types. The one in this photograph is designed for hard rocks such as limestones (Baker-Hughes document).

Figure 2.9.4 Two gun barrels for sidewall coring which can be mounted in tandem, each containing 25 bullets (Baker-Hughes document).

Figure 2.9.5 Two sidewall cores retrieved with a percussion coring device.

often only known after trial and error. Additionally, the engineer's skill in accurate positioning and careful tool movement is equally important for guaranteeing sucess..

Applications. Sidewall cores obtained from percussion coring have generally undergone a serious shock from the bullet impact, causing mechanical damage in the form of fracturing and deformation. They can rarely be used for accurate petrophysical analysis, although in favorable cases good porosity measurements can be performed on them. Permeabilities often cannot be measured at all because not enough material may be available, or the core may be fractured. In reservoir rocks, fluid saturations are usually altered because the cores come from the invaded zones, but oil shows can still be seen.

On the other hand, percussion sidewall cores can provide excellent information on lithology and biostratigraphy, and they have been successfully used for source rock analysis.

2.9.3 Rotary Coring

In hard rocks, the alternative to percussion coring is rotary coring. The technique was developed by Gearhart Industries in the 1980s (Hashmy et al., 1985) and today two tools are on the market[2]. Their operating principles are very similar: A tiltable unit contains a small drill bit which is driven by a downhole motor (figure 2.9.6). When the tool is at the desired depth, it is locked in place with an anchor pressed against the borehole wall. The drill unit is then tilted from its resting position, which is parallel to the tool axis, to a position perpendicular to the borehole wall (figure 2.9.7). The coring process is then started and con-

---

[2] The Mechanical Sidewall Coring Tool MSCT (Schlumberger), the Rotary Sidewall Coring Tool, named RCOR (Baker-Hughes) and RSCT (Halliburton) respectively.

Figure 2.9.6 The drill bit of the Rotary Sidewall Coring Tool (RCOR) in drilling position (Baker-Hughes document).

tinuously monitored from a surface control system, and some parameters such as the rotating plate pressure can be adjusted. Once the drill bit has penetrated deep enough, it is retracted, breaking the core from the wall, and repositioned into its original orientation. A rod then pushes the core into a core holder. Prior to the next coring operation, a marker plate is dropped into the core holder to separate the different cores and avoid any mix-ups if the core is broken.

The surface monitoring system allows the operator to take preventive action in case the drilling does not proceed as planned. For example, if the drill bit jams, it can be retracted, the cuttings can be flushed out, the drill bit can be reinserted and a new attempt can be made. Just like in rotary drilling from the surface, the operator has to adjust the drilling parameters to the conditions, and change them if necessary. Two drill bits are available for the MSCT. A diamond drill bit generally used for hard rocks such as limestones, low-porosity sandstones etc., while a polycrystalline diamond compact (PDC) bit is available for softer, less compacted rocks such as shales and porous sandstones. The RSCT/RCOR tool has a core capacity of 30, while the MSCT is usually run with a capacity of 20 cores but an option to upgrade to 50. Cores are generally 1 inch (2.54 cm) in diameter and 1.75 (RSCT/RCOR) to 2.0 inches (MSCT) in length (4.5 to 5.1 cm respectively). These dimensions correspond to standard core plugs and, if a sufficient length has been retrieved, a full petrophysical analysis can be performed on them.

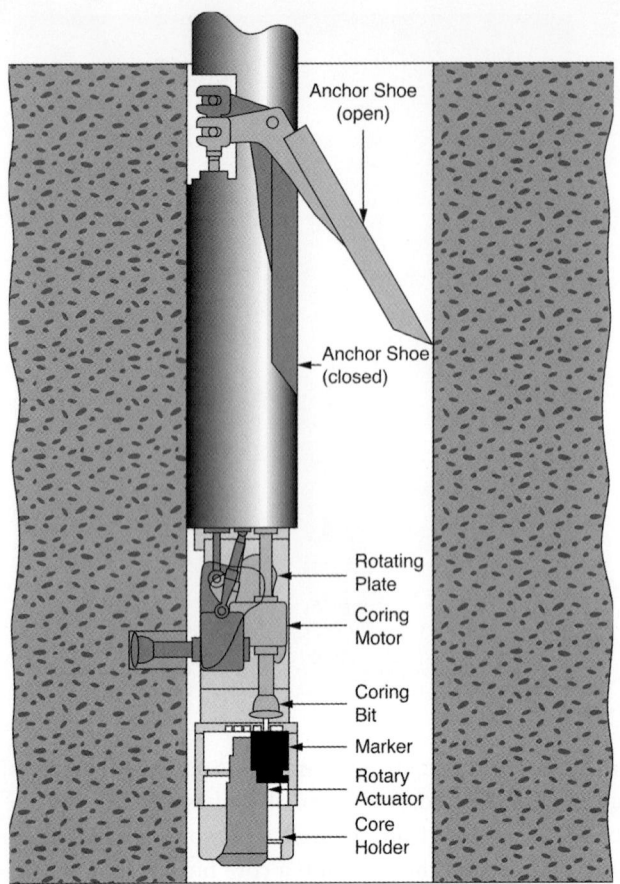

Figure 2.9.7 Sketch of the Mechanical Sidewall Coring Tool (MSCT) anchored and in the process of coring. Once a sufficient core length has been drilled, the bit is retracted and tilted back into a vertical position. The core is pushed into a core barrel and a marker is placed to separate it from the next core.

Figure 2.9.8 shows a FMI borehole image in a fine-grained mixed carbonate/clastic rock sequence. The dark round hole marked with A indicates low resistivity and is the place where a rotary wireline core has been retrieved. The dark vertical groove marked with B is the site where the mechanical core slicer removed a triangular slice of about 40 cm length from the borehole wall. These images are from Schlumberger's test well in Blanco, Texas, where numerous prototype tools had been tested over the years. Careful records of these tests confirmed the origin of these two anomalies on the borehole wall. Similar anomalies on electrical and acoustic borehole images have been reported from percussion sidewall coring operations. Even wireline formation testing can leave similar imprints because the formation fluid retrieved by the probe can change the local resistivity of the rock.

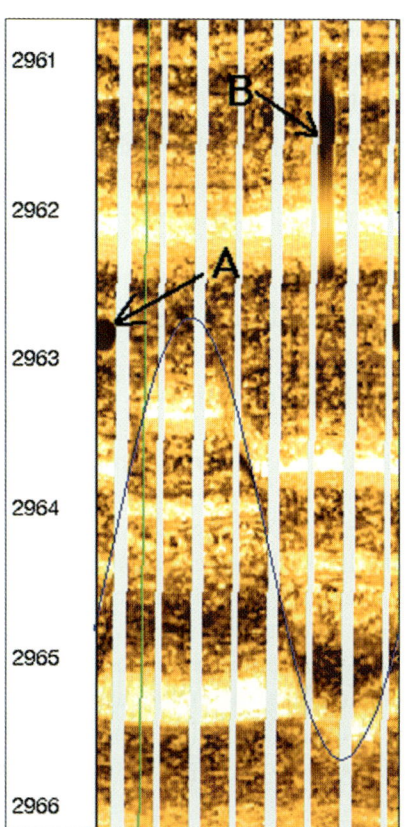

Figure 2.9.8 FMI borehole images showing the location where a wireline rotary core (marked with A) and a core slice (marked with B) had been retrieved. The sinusoid marks a small fault. Laminated calcareous and glauconitic siltstones, Riley Formation (Cambrian), Blanco test well, Texas. Depth in feet.

Figure 2.9.9 shows a core sample retrieved with a rotary sidewall coring tool, together with a triangular rock slice from the test well in Blanco.

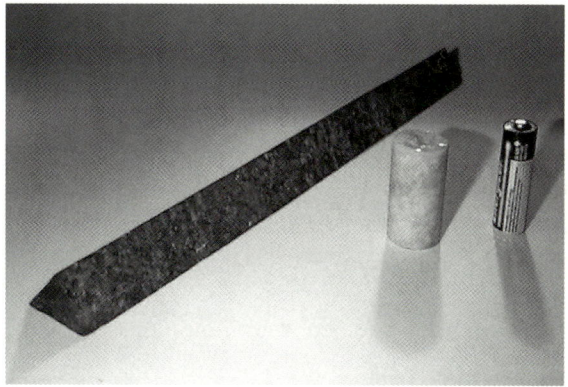

Figure 2.9.9. A core retrieved with a rotary sidewall coring tool, in this case a dense marble. Next to it a triangular slice retrieved in the Blanco test well with the outphased core slicer tool, similar to the one that produced the mark in figure 2.9.8.

Applications. The wireline rotary coring tools usually retrieve mechanically intact cores similar to the core plugs drilled out of conventional core. They can be used for the same applications, which include:

- Porosity, permeability and grain density measurement
- Mineralogical analysis
- Biostratigraphic studies
- Rock mechanical tests
- Source rock analysis
- Residual fluid measurements

2.9.4 A Brief Comparison

Unlike percussion coring tools, rotary coring devices cannot be used in slim holes, but they are applicable in holes up to 12 3/4 inches (RSCT/RCOR) and 17 inches (MSCT). They also retrieve a smaller number of cores, but the sample quality is usually superior, particularly in hard rocks where they are often the only alternative to conventional coring. Rotary cores can be used for full petrophysical laboratory analyses, and the results have been found very useful as inputs into formation evaluation (Hashmy et al., 1985). Percussion cores, on the other hand, have their major application in lithological and biostratigraphic analysis. Both can be used to enhance formation evaluation from logs, particularly when their precise origin is known as in figure 2.9.8.

References

Desbrandes, R (1985) Encyclopedia of well logging. Gulf Publishing Co.
Hashmy. KH, Robinson KA, Rojas JM, Skopec RA (1985) Wireline coring device aids in evaluations of complex geologic areas. 60th Ann Conf Soc Petr Eng: Paper 14299.

Part 3: Applications and Case Studies

Part 2: Applications and Case Studies

3.1 Structural Modeling

3.1.1 General Comments

The "structures" to be discussed in this chapter range from small-scale sedimentary structures to very large-scale tectonic structures.

When data from multiple wells are available, the large-scale structural interpretation is usually quite well constrained, and *contour maps, correlation panels, and cross-sections* are established with the help of markers and datums. Dipmeters and imaging tools serve an important purpose here because they provide additional constraints in the tasks. They allow to establish the local dip for a particular horizon and thus to determine the local contour direction and spacing on isobath maps. On cross-sections, they can help identifying faults in between wells. For example, if dipmeters show the same layer in two wells to be horizontal, but their are encountered at different depths, then a fault in between the wells is possible.

If only one well, or a small number of wells are available, such as in an appraisal situation, the essential problem is that the volume or interval sampled is usually small compared to the total volume of the structure. This conundrum is often only solved through geological conjectures or constraints.

Seismic lines can be of substantial if not crucial help, at least for larger-scale structures that are well resolved by the seismic technique. For structures below about ten meters in size, some of the near-wellbore imaging techniques might be of help, for example sonic imaging or occasionally vertical seismic profiles. There is, however, nothing available at this scale that compares in quality to the contribution of seismic at larger scales. At even smaller scales, structures can be observed directly and almost in their entirety on cores and borehole images, and there is often little doubt about what they correspond to.

A large number of sedimentary and tectonic structures fall into the critical range with scales between 10 meters and 10 centimeters, a scale at which few borehole methods operate. The tools of choice here are the dipmeter and imaging tools (chapters 2.1–2.5). They have the distinct advantage over cores that they are azimuthally oriented and continuously recorded. However, since they observe only a relatively small volume of the structure, their data have to be interpreted. For this, two approaches are in common use. The first one is qualitative and descriptive and usually consists of two steps:

1. An image feature or a dip pattern is detected and interpreted as a particular structure. For example, an abrupt change in dip is attributed to the presence of an unconformity.
2. The structure is then analyzed for its quantifiable attributes. For example, the dip and azimuth of the unconformity surface are determined.

This approach has been illustrated with several examples in chapters 2.1-2.4. Oftentimes, the user sees little need to go further in his analysis because he may have gathered sufficient information at this stage.

In many cases where these data need to be integrated into a reservoir model, however, a more quantitative modeling approach is needed. In the single-tool applications, several techniques were discussed that can be used to do this. These include

- Geometric reconstructions based on dip sequences on a large or small scale (Etchecopar & Bonnetain, 1992; Anxionnaz & Delhomme, 1998).
- Statistical modeling based on dip patterns on a large or small scale (Bengtson, 1981; Luthi et al., 1990).

The geometric reconstructions in figures 2.1.20 to 2.1.23 serve as illustrations for these two basic structural modeling methods. In the following, a few more examples are presented and their implications on reservoir modeling are discussed in more detail.

3.1.2 Combining Seismic and Dip Data

The three-dimensional nature of geological structure is reflected in the dip patterns (chapter 2.1), but if complex structures combine with strong layer thickness changes and faulting, the task of reconstructing the layer geometry can become daunting. In those cases, all data have to be taken into account and combined into one coherent model. Figure 3.1.1 is from a case study discussed in Schlumberger (1997), showing two side-tracked wells on a seismic backdrop in a field in Southern Europe. The section is already interpreted and thus the problem may not appear in its full complexity. For constructing the section, the following steps were taken

- Logs of the two wells were correlated using several prominent markers (chapter 3.4);
- Stick-plots (chapter 2.1) of apparent dips[1] were made along the section and projected onto the seismic line;
- A geometric reconstruction was made for each well using the method proposed by Etchecopar & Bonnetain (1992);
- All information was integrated into a coherent structural model, whereby ambiguous intervals were filled in using geological "best guesses".

Correlation lines and numbers near the tracks of the two wells display results of the log correlation. Some of the layers exhibit thickness changes from one well

---

[1] Apparent dips in the sense used by structural geologists, i.e. the layer dip observed along a given cross section. Dipmeter interpreters use the term to designate a dip relative to a plane orthogonal to the borehole axis.

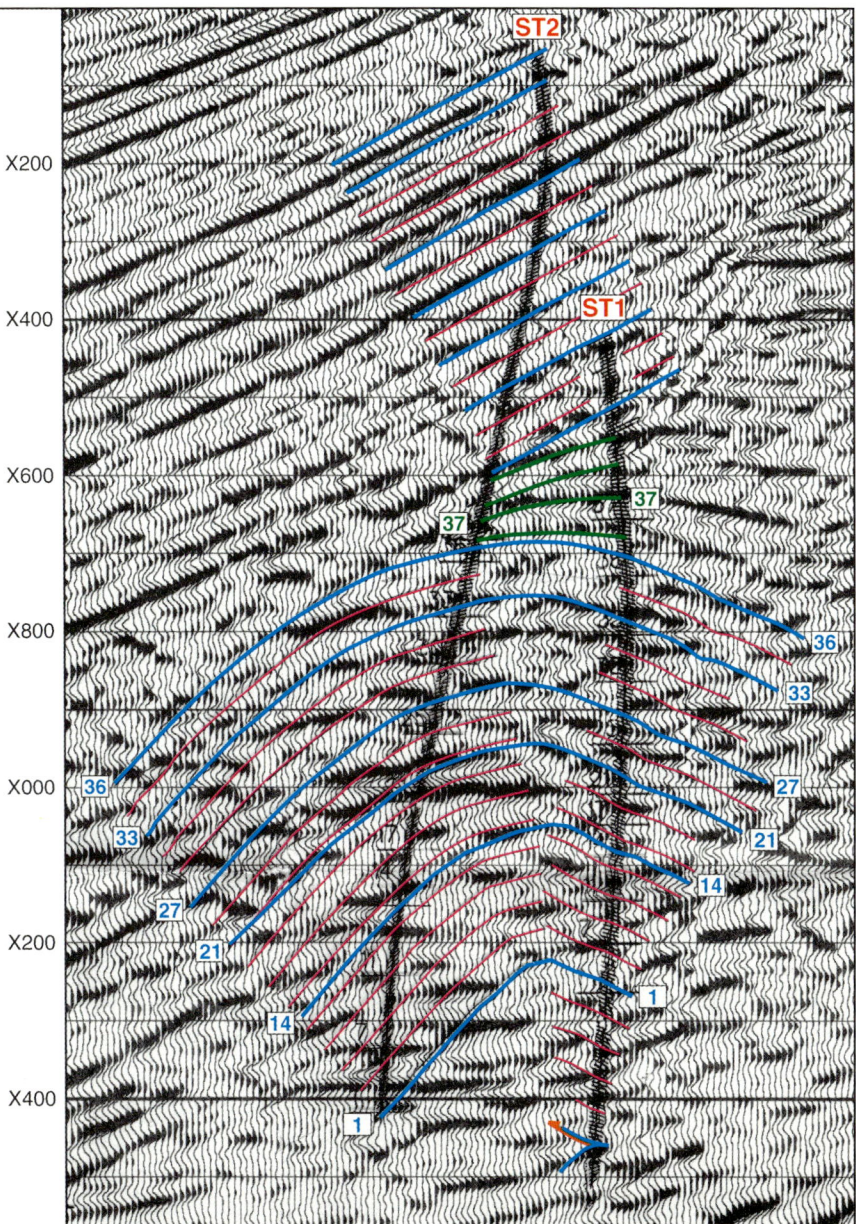

Figure 3.1.1. Correlation of two side-tracked wells using dipmeter results and seismic. The dipmeter stick plots are shown with short black lines along the well tracks. Purple lines are geometric constructions of the near-wellbore structure. Numbers indicate correlative lithostratigraphic markers. The blue and green lines are the well correlations, taking into account all these data. The structure is a ramp anticline with a backthrusted fault at the bottom, and a syntectonic sedimentary draping at the top.

to the other, particularly around marker 37. The dipmeter results in the two wells are displayed as stick plots along the well trajectories with thin black lines, while the 3–D geological reconstructions from the StrucView² program are shown in thicker lines.

The combination and integration of these data resulted in the cross-section shown in figure 3.1.1 and allowed unraveling the layer geometries fairly well. In the lower two thirds of the section, a fold is present with layers thinning from well ST-2 towards well ST-1, as indicated by the blue correlation lines. A sediment wedge drapes over this fold, with a rapid change in layer dips, until in the uppermost part the layers assume a constant dip towards the left (a monocline). The entire section covers over 1000 meters vertically and is interpreted as a ramp anticline that developed during sedimentation. As tectonic activity ceased, the draping took place and finally yielded to constant deposition rates in both well locations. Upon closer inspection of the seismic line as well as the dipmeter survey, a back thrust is identified at the bottom of well ST-1 (ST-2 is not deep enough to see this), and perhaps the décollement can be seen in the lowermost part of the seismic line.

The resulting definition of the anticlinal structure has a great degree of confidence, since all pieces agree with each other. Implications for further field development can be deduced, as it is seen that both sidetracks are located on the flanks while the highest part of the structure is not tapped by either well.

3.1.3 Modeling Cross-Beds

Cross-beds are sedimentary structures formed by migrating dunes and related bedforms. They occur in siliclastic and calcareous sands deposited under high flow energies in practically every depositional environment, and they often form the best reservoir facies type. Cross-beds are important in reservoir modeling because of their large permeability anisotropy, and their complex and variable three-dimensional geometry (Weber, 1986; Stalkup, 1986; Weber & van Geuns, 1990). In the hierarchical nature of cross-bedded deposits, four distinct types of heterogeneities can be distinguished in decreasing order of importance for the reservoir³. These are the interdune layers or first-order bounding surfaces that separate stacks of cross-beds from each other, followed by the set boundaries or second-order bounding surfaces, the reaction surfaces or third-order bounding surfaces, and at the smallest scale the cross-bed laminae which compose the foresets and bottomsets.

---

[2] Mark of Schlumberger, based on the work by Etchecopar & Bonnetain (1992)

[3] The classification based on "bounding surfaces" was established by Brookfield (1997) for eolian sediments and I took the freedom to broaden it here to include all types of cross-bedded deposits.

Figure 3.1.2. Photograph of fluvial cross-bedded sequence in Walton formation, Overlook Mountain, Woodstock (New York). The numbers correspond to the bedding surfaces discussed in the text.
1: contact floodplain shales/channel deposit (bottom) and contact channel/channel; 2: set boundary; 3: possible reactivation surfaces; 4: cross-bed laminae.

Figure 3.1.2 illustrates these four hierarchical bedding elements in a fluvial channel deposit. It has to be noted that a generalization has been made here that may not do justice to all types of deposits. Additionally, not all of these elements occur in all cross-bedded deposits found in the various depositional environments. In general, the reservoir heterogeneities decrease in importance from the first to the fourth type because the permeability contrasts become smaller. Equally, the lateral extents decrease from the first to the fourth type because the events become more and more local.

Cross-beds can be easily identified on dipmeters and borehole images, as had been illustrated in figures 2.1.19 and 2.2.20. Unlike in outcrops, the dips of cross-beds and bounding surfaces can rapidly and accurately be measured. However, their lateral continuity is difficult to assess, and so is their internal geometry and shape. Therefore, modeling is needed, and in the following we will discuss two approaches to do this: A deterministic method similar to the one for large-scale tectonic structures used in figure 3.1.1; and a statistical approach that relies on the distribution of cross-bed directions of a number of beds.

Deterministic Approach

In this method, described by Anxionnaz & Delhomme (1998), the goal is to reconstruct the cross-bedded sequence to a distance about five borehole diameters away from the wellbore. In order to do this, the dips and azimuths of the set boundaries and the cross-beds are first computed, either in interactive or automated (and regularly sampled) mode. Applying the method of Etchecopar & Bonnetain (1992), the translation plane has then to be determined. In a fold, this is the axial fold plane (figure 2.1.21), but in cross-beds, the dips can be translated along the direction of the set boundary and in the direction of the dip azimuth of the cross-beds. This is in essence follows a paleo-horizontal direction, and the approach assumes that dune migration is unidirectional, with a cross-bed geometry invariant perpendicular to migration.

Anxionnaz & Delhomme (1998) found the results from this approach too limited and instead developed an inversion method based on the forward models by Rubin (1987). These models consist of propagating a bedform that is defined by a number of shape parameters through space, at a direction and rate prescribed by more parameters, such that a cross-bedded deposit is formed. From each bedding surface, Anxionnaz & Delhomme (1998) write two equations, one for the geometry of the shape of the bedform (involving the large number of parameters used in the forward model), and one for an approximation of the bedding surface found in the borehole by a plane. For N surfaces they thus obtain $2*N$ equations, a number that may differ from the number of unknowns. Therefore, they use a minimization technique, the conjugate gradient method, to find the best possible

Figure 3.1.3. Block diagram of reconstructed eolian cross-bed sequence, with the conductivity distribution within layers obtained from stochastic conditional simulation (from Anxionnaz & Delhomme, 1998).

solution, or the best fit. They also estimate the conductivities within the reconstructed beds with a geostatistical method, whereby the variogram, or the directional dependence of the conductivity, is obtained from the FMI borehole image. It is then used to extrapolate the conductivities away from the borehole in a stochastic way within each of the reconstructed layers.

An example from eolian cross-beds is shown in figure 3.1.3 in the form of a block diagram. The cross-beds are seen to differ in dip azimuths, suggesting either shifting transport directions, or more likely, curvilinear bedforms. The conductivity distribution shows a thick, resistive layer between the top and the middle cross-bedded sets. No clear bottom set seems to be present, and it is therefore possible that this high-resistivity layer is in fact a first-order bounding surface. This interpretation would have been difficult from the original image alone, but it obviously needs further confirmation, for example through comparison with cores or correlation with other wells.

Statistical Approach

What borehole observations lack in lateral continuity is often compensated for by the vertical succession of layers. In cross-bedded sequences, the information gleaned for each set can consist of thickness as well as dip and azimuth of the foresets. The changes of these parameters with depth can be interpreted in terms of the evolution of the sedimentary system, but another approach is to analyze it from a purely statistical point of view. Specifically, if foreset dip azimuths are measured for each cross-bedded set, they can be interpreted as representing flow directions that evolve with time or depth. Alternatively, they can be seen as a purely statistical scatter originating from the random sampling of geometrically similar sedimentary bodies stacked on top of each other. The azimuth scatter in this scenario thus reflects the three-dimensional geometry of the foresets and, therefore, of the bedforms that deposited them. This working hypothesis had first been advocated by Wurster (1958) and was then elaborated into a computational model by Luthi et al. (1990). Thus a broad foreset azimuth scatter reflects a strong three-dimensionality of the cross-beds, such as in trough cross-bedding. Conversely, a narrow dip azimuth distribution can be attributed to planar cross-bedding, thought to be formed by essentially two-dimensional bedforms.

In the computational procedure an elliptical bedform migrates in a certain direction and leaves behind a deposit. The parameters involved are the curvature of the bedform, its alignment, and its migration direction (which is not always in the direction of the alignment, although in practice the two usually differ little). The deposit is then sampled with a Monte Carlo method, whereby the dip azimuths are obtained for randomly chosen locations. The result is a frequency distribution of the foreset azimuth that would be obtained if a large number of cross-bedded sets with the same properties were stacked on top of each other at random locations and was penetrated by a well.

In the inverse procedure, a data set is compared to a given model, and the parameters of the model as successively changed until a best fit is obtained. The small number of parameters allows for fast convergence and avoids getting trapped in local minima. Luthi et al. (1990) found good fits of the model with data sets from outcrops and wells in eolian, fluvial and shallow marine settings. They all suggest that the bedform shapes are best described by semi-ellipses, probably of transverse bedforms. Some have a higher ellipticity, while others are low and thus more straight-crested, corresponding to the planar cross-bedding type. In a case study from a well into the eolian Rotliegendes in the Southern North Sea, 53 cross-bedded sets were found over a total interval of about 200 meters (Luthi & Banavar, 1988). Their dips and azimuths were obtained from dipmeter results and acoustic borehole televiewer images (chapters 2.1 and 2.3). The narrow azimuth distribution suggests low-curvature bedforms, and the subtle but significant asymmetry in the distribution can be explained by oblique bedform migration (figure 3.1.4). The bedform ellipticity is 0.5, meaning that the long axis (the dune crest) is twice the length of the short axis (the distance to the dune horns). The angle of migration is 45°, which effectively means that the crestlines were oriented (dipping) towards west (270°), but migrated towards southwest (225°), probably because of seasonally varying wind directions. Notice that the azimuth

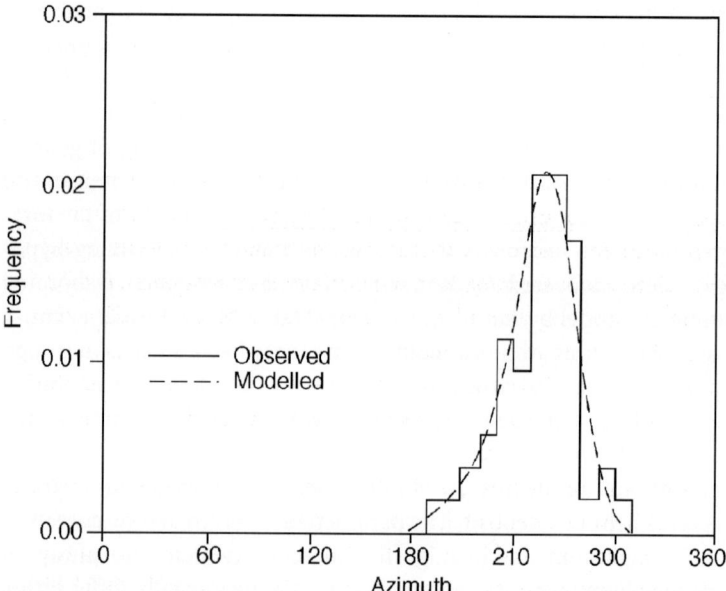

Figure 3.1.4. Cross-bed azimuth distribution from the eolian Rotliegendes Formation, North Sea. The observed azimuths (solid curve) are best matched by a model with a low ellipticity and oblique migration (dashed curve). Redrawn from Luthi et al. (1990).

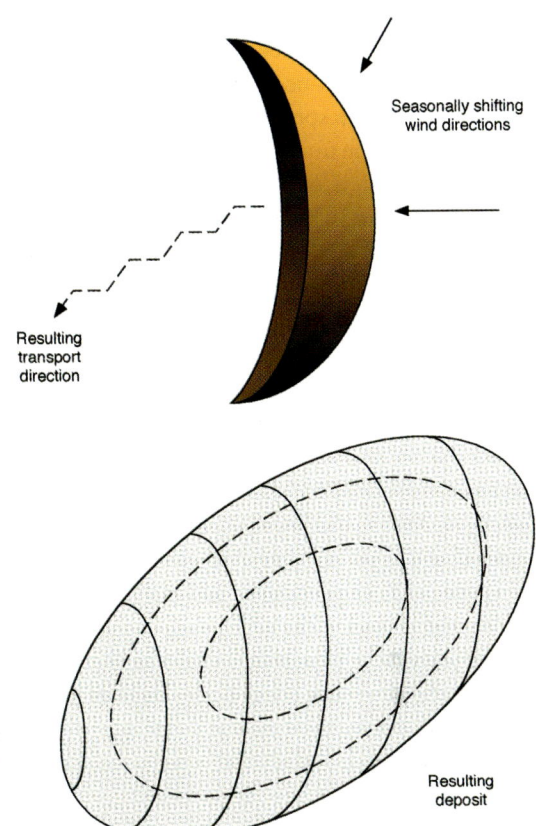

Figure 3.1.5. Transverse dune migration under seasonally shifting wind directions (top). Plan view of the resulting cross-bed deposit with foresets (solid) and isopachs contours (dashed lines) indicated schematically (bottom). North is up.

mean of the foresets is 254° and is different from both the dune orientation and the migration direction. Figure 3.1.5 illustrates the shape of such a transverse dune, its migration under seasonally shifting wind directions, and the resulting deposit in a map view. From the latter, it can be inferred that horizontal wells in a direction parallel to the long axis of the deposit might be an effective way of draining these highly heterogeneous reservoirs.

Modeling the thickness distribution together with the azimuth can further constrain the geometry. For this, the shapes described by Weber (1987) can be used. Each deposit has the form of a spoon, with a maximum thickness in the middle and a monotonic decrease towards the margins (see isopachs in figure 3.1.5 at the bottom). Discrete central thicknesses of the sets have to be used, with increments of 5 feet. For an observed set thickness and azimuth, the spoon is constructed with the next highest central thickness; for example, if the observed thickness is 4.6 feet, a central thickness of 5 feet is taken. By going around the spoon along an elliptical isopach line corresponding to the observed thickness (in the example of 4.6 feet), the program checks if there is a location with the appropriate foreset azimuth. If none is found, the spoon thickness is increased by one increment and the procedure is repeated until a solution has been found. This is done for each cross-bedded set, and it results in a relative position of the well within each set or spoon.

There are some interesting implications following from this model. In the central part of a set, according to figure 3.1.5, one expects to find foreset azimuths close to the dune alignment, or 270°. At the margins, where thicknesses are smaller, the azimuth variation is expected to be much larger. A cross-plot of azimuth versus thickness reveals that the thickest five beds all cluster very closely around 270°, while with thinner beds the scatter becomes increasingly large (figure 3.1.6). This observation implies that if a cross-bedded set shows a foreset azimuth around 270°, the well has probably (but not certainly) penetrated it in the center. If the azimuth deviates significantly from 270°, the well is likely to have penetrated the set at a very marginal location, i.e. at locations corresponding to the horns of the dune. By way of illustration, one set with an azimuth of 196° was found to have a thickness of only 1.6 feet. The statistical calculation gave the result that this set has a probable central thickness of 30 feet, a situation sketched in figure 3.1.7. Interestingly and perhaps not coincidentally this set lies immediately above a set with thickness 26.6 feet with a forest azimuth of 272°, implying that is was likely penetrated close to the center of the deposit. Using Weber's (1987) data for eolian cross-beds, Luthi et al. (1990) estimated that these thicker cross-bedded sets have lateral extents of up to 6000 feet (1.8 km). With these numbers, one can estimate the required well spacing that drains all compartments, or at least all the larger ones.

For practical and more qualitative use, figure 3.1.8 shows foreset azimuth distributions expected in cross-bedded sequences, modeled with the same approach as outlined above. Three curves are shown, representing three crestline curvatures, ranging from relatively straight-crested (0.2) to semi-circular (1.0) and

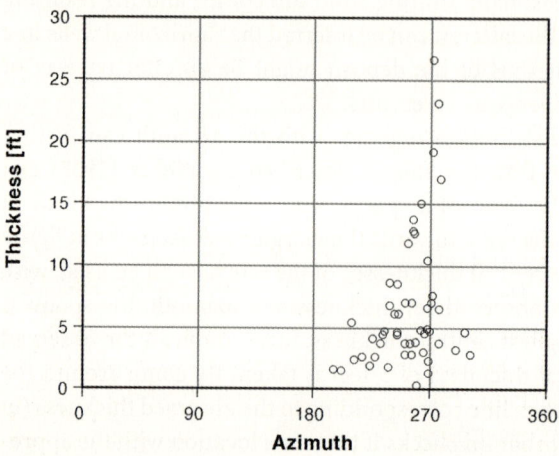

Figure 3.1.6. Set thickness versus foreset azimuth from a well in the eolian Rotliegendes, Southern North Sea. Notice how the scatter decreases with increasing thickness. The thickest sets are the most reliable paleocurrent indicators (see figure 3.1.7).

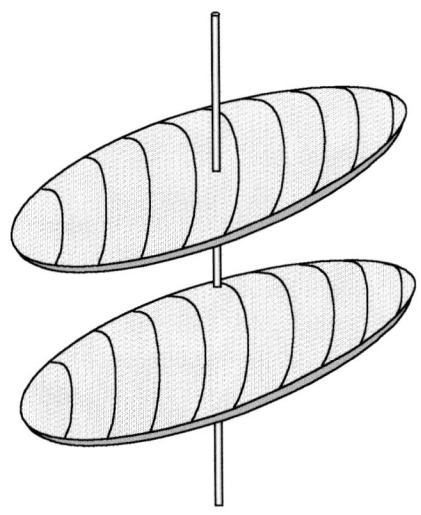

Figure 3.1.7. A conceptual 3–D view of large cross-bedded sets penetrated by a well. The set penetrated in its center (top) has a cross-bed azimuth corresponding to the principal crestline orientation and can be used as paleocurrent indicator. It is also likely to be thicker than the set penetrated at its margin (bottom), whose azimuth can strongly deviate from the mean. It is, therefore, an unreliable paleocurrent indicator.

highly curved (2.0). This range is thought to span a continuum of deposits from planar to trough cross bedding. The figure on the left is for a migration perpendicular to the crestline, giving rise to perfectly symmetrical deposits. At curvatures higher than 1.0, the distribution becomes bimodal with this model. Such bimodalities have sometimes been attributed to changing tidal flows, but as this example illustrates, this need not be the case. The figure on the right is for an angle of migration of 15°. It can be seen that there is increasing asymmetry in the distribution with higher curvatures. The bimodality present in the highly curved

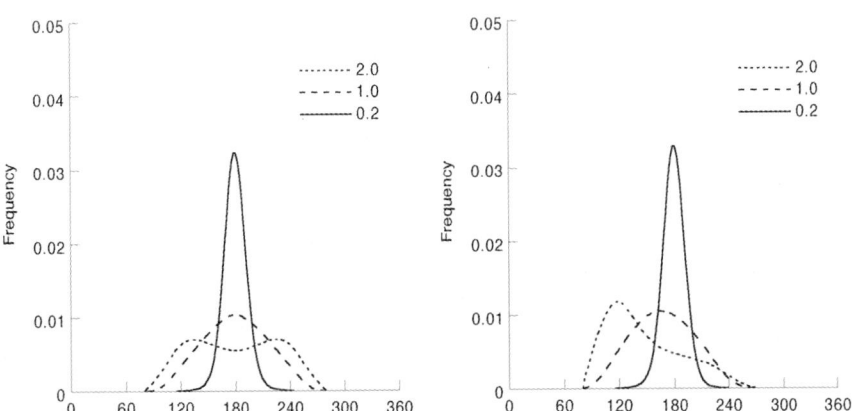

Figure 3.1.8. Expected foreset azimuth distribution in a cross-bedded sequence of similar deposits. Left: Dune migration perpendicular to crestline orientation. Right: Oblique migration at 15°. Three beform curvatures are shown, ranging from relatively straight crested (0.2) to semi-circular (1.0) to highly curved elliptical (2.0). Bedforms arbitrarily oriented E-W and migrating in a southern direction.

bedforms with zero migration angle has disappeared in the oblique migration case. These curves can be used as references for a rapid estimation of bedform geometry in cross-bedded sequences.

3.1.4. Remarks

Quantitative structural modeling as discussed here is not yet commonly used, although software for it is on the market, and continues to appear. One of the reasons for this is perhaps the hope that near-wellbore imaging techniques might soon be developed and will eliminate some of the ambiguities inherent in these computational methods. An active line of research at several institutions is to quantify the geometries of typical sedimentary bodies, either through field measurements on outcrop analogues, or through physical scale models in the laboratory. Drawing from such databases, the user is then able to fit suitable geological templates to the partial view obtained from boreholes and thus construct an object-based reservoir model.

References

Anxionnaz H, Delhomme JP (1998) Near-wellbore 3D reconstruction of sedimentary bodies from borehole electrical images. Trans 39th Symp Soc Prof Well Log Analysts: Paper N.
Bengtson CA (1981) Statistical curvature analysis techniques for structural interpretation of dipmeter data. Am Assoc Petrol Geol Bulletin 65: 312–332.
Brookfield M (1977) The origin of bounding surfaces in ancient aeolian sandstones. Sedimentology 24: 303–332.
Etchecopar A, Bonnetain JL (1992) Cross sections from dipmeter data. Am Assoc Petrol Geol Bull 76: 621–637.
Luthi SM, Banavar JR (1988) Application of borehole images to three-dimensional geometric modeling of eolian sandstone reservoirs, Permian Rotliegende, North Sea. Am Ass Petrol Geol Bull 72: 1074–1089
Luthi SM, Banavar JR, Bayer U (1990) Models to interpret bedform geometries from cross-bed data. J of Geol 98: 171–187.
Rubin D (1987) Cross-bedding, bedforms, and paleocurrents. Soc Econ Paleo Mineral Concepts in Sedimentology and Paleontology 1.
Schlumberger (1997) Italy 2000–Value-added reservoir characterization. Schlumberger Italiana SpA.
Stalkup FI (1986) Permeability variations observed at the faces of crossbedded sandstone outcrops. In: Lake LW, Carroll HB (eds) Reservoir Characterization. Academic Press, pp. 141–179.
Weber KJ (1986) How heterogeneity affects oil recovery. In: Lake LW, Carroll HB Jr (eds) Reservoir characterization. Academic Press, pp. 487–544.
Weber K (1986) Computations of initial well productivities in aeolian sandstones of the Leman gas field, U.K. In: Tillmann RW, Weber KJ (eds) Reservoir Sedimentology. Soc Econ Paleo Mineral Spec Pub 40: 333–354.
Weber KJ, Geuns LC van (1990) Framework for constructing clastic reservoir simulation models. J Petrol Technol: 1248–1297.
Wurster P (1958) Geometrie und Geologie von Kreuzschichtungskörpern. Geologische Rundschau 47: 322–358.

3.2 Bedding and Reservoir Zonation

3.2.1 Introduction

One of the principal tasks of the petroleum geologist is the zoning of well logs or core data into geological units. These units are correlated and inter- or extrapolated across the field in order to obtain genetic units that serve as building blocks of the geological model of a field. Once representative petrophysical properties for these layers have been obtained, a three-dimensional property model can be constructed that is then upscaled, or lumped into larger units that become the building blocks for the reservoir model used in reservoir simulation (Weber & van Geuns, 1990).

Geological zonation would be considerably easier if each depositional process led to a well-defined bed with characteristic geological properties, or lithofacies. However, geological processes are time-variant, source material differs from one place to another, and climate as well as other relevant factors have varied considerably over geological time. Therefore, a given depositional process does not always produce a specific bed – or a sequence of beds – with well-defined characteristics that distinguish it from other sedimentary rocks. Conversely, considerable similarities may exist between beds of completely different depositional origins. Beds may also exhibit strong intrinsic vertical variations in their internal composition or structure, such as in the well-known example of the "Bouma sequence" found in turbidite deposits (Bouma, 1962). Thus, the geologist is often only able to unravel the nature of a deposit through painstaking detective work.

The geological zonation and classification task is therefore much more complex than in biology, where a species may show small variations within its kind but is distinctly separate from others. The geologist often has to place layer boundaries where gradual or subtle changes occur, simply because he knows by experience that this will allow him to separate two genetically different layers, well aware that another geologist might place the boundary somewhere else or use a different classification altogether. Such different ways on how to zone a geological sequence is one of the reasons why there are no formal rules for sequence stratigraphy.

In theory, however, classifying well logs into zones is straightforward since they are digital in nature and constitute reproducible physical measurements of the rock sequences traversed by the wellbore. Doveton (1994) describes the main methods in current use, all of which are based on computational techniques developed in other fields. Two main approaches are distinguished:

- *Unsupervised methods,* in which the data points are analyzed for groups or clusters occurring within them, and the interpreter then assigns each cluster to a lithology; and

- *Supervised methods*, in which a database exists that the measurements are related to, i.e. the classes correspond to lithologies and no intervention by the interpreter is needed.

These two groups of classifications will be briefly reviewed in the following.

Unsupervised Classifications

In unsupervised classifications using logs no *a priori* knowledge is available in the computational procedure. The underlying philosophy, as stated by Serra & Abbot (1980), is that

"...the set of log measurements taken at any level in a well ... constitutes a description of the rock at that level ... in terms of resistivity, density, hydrogen index, spontaneous potential, levels of natural gamma-ray radiation etc."

Serra & Abbott (1980) go on

"... to define electrofacies as the set of log responses that characterizes a sediment and permits the sediment to be distinguished from others."

The term electrofacies is thus analogous to lithofacies, which is defined as the sum of lithological characteristics of a rock that distinguishes it from others. Serra & Abbott (1980) propose methods of determining electrofacies using spider plots, and, in a more automated way, clustering techniques. Well aware that the lithology signal is rather weak in many logs when compared to the porosity and fluid signal, they discuss ways of enhancing the geological information by giving greater weight to logs with a relatively stronger response to lithology, such as natural gamma ray logs (chapter 2.7.4). They also propose using information from the dipmeter logs (chapter 2.1) by extracting certain curve characteristics, for example the sharpness of the peaks, or the number of peaks and troughs per unit interval. These properties are represented in "synthetic logs" that have the same sampling density – typically two samples per foot – as the normal logs. Such logs may contain information on the bedding characteristics that cannot be obtained from other logs. Underlying this effort is the assumption that if appropriate well logs are available, electrofacies can be equated with facies obtained from core descriptions.

Wolff & Pelissier-Combescure (1982) describe the commercial implementation of the electrofacies concept. The product, FACIOLOG*, is a zonation of well logs into electrofacies after a series of processing steps that include

- principal component analysis, creating logs with a more direct relationship to rock properties than the original logs;
- clustering, first into all local maxima, then into the desired number of classes, or modes; and a

---

* Mark of Schlumberger

- graphic display in a manner similar to geological columns from outcrops or from core descriptions.

Principal component analysis is a standard method in multivariate data analysis (Davis, 1986) that aims at detecting a simple underlying structure in a set of multivariate observations. Such structures are expressed in patterns of variances and covariances between the variables. These can be combined into a symmetrical matrix whose eigenvectors are aligned with the direction of the variation of the input data. The number of eigenvectors is equal to the number of variables, with the first one – the principal axis, or principal component – representing the direction of maximum variance, and its length (the eigenvalue) the amount of the variance.

In practice, principal component analysis can thus be used to reduce the dimensionality of the multivariate space. Logs often are highly correlated with each other and through principal component analysis the essential features can be extracted. This is conceptually illustrated in figure 3.2.1, where a crossplot of the density and the neutron porosity exhibits a great degree of correlation. One can readily distinguish points roughly aligned on the line corresponding to limestone, and another group of points in the vicinity of the dolomite line. Dolomite has a higher grain density than calcite, and therefore shows a higher bulk density than limestone for a given neutron density. By determining the main variance in this data set, the two principal component axes PC1 and PC2 are detemined (figure 3.2.2). Switching now to a coordinate system with these two axes, we observe that

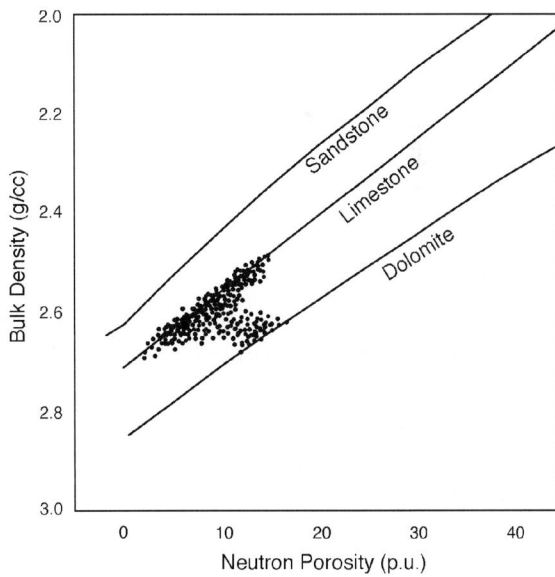

Figure 3.2.1. A neutron/density cross plot of data from a conceptual limestone/dolomite sequence.

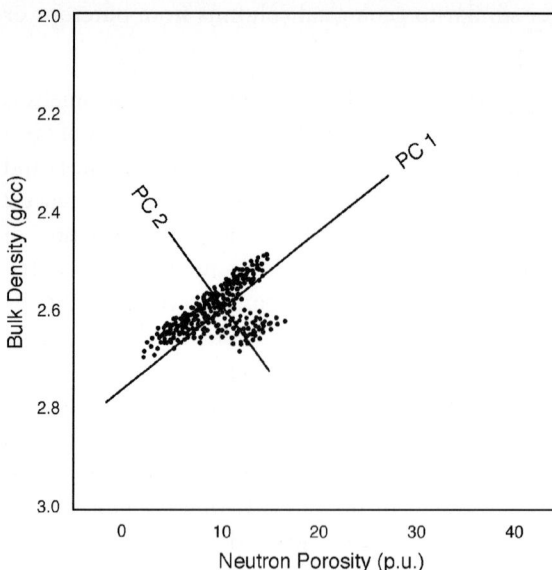

Figure 3.2.2. The same data as in figure 3.2.1, projected onto two principal component (PC) axes.

- PC1 has a larger variance and PC2 a smaller variance than either of the original variables; and
- PC1 aligns with porosity while PC2 reflects the grain density of the rock.

Projection of the data points onto these axes gives PC logs. If a third variable, for example the sonic log, is introduced, three PC logs are obtained, but it is likely that well over 90 % of the total variance of the data is contained on the first two PC axes. In the clustering procedure, therefore, one can opt for using only two PC logs rather than the three original logs without losing significant information but with a potentially important reduction in computing time. If more than three logs are used in the analysis these considerations become even more important. One can also opt for using only selected PC logs as input to the clustering procedure. For example, if only PC2 is used of the example in figure 3.2.1, a very simple discriminator of lithology is obtained (similar to the discriminant function). If the user's objective is a lithological zonation, he can improve the result by including PC logs that respond primarily to lithology (and perhaps bedding through dipmeter-derived synthetic logs), and by omitting PC logs that contain primarily information on porosity and type of fluids. In practice, when using multiple logs, distinguishing the information carried by each PC log is more difficult than in the simple case discussed here.

The clustering process in FACIOLOG includes two steps. The first one consists of hierarchical clustering (Davis, 1986), in which points are connected to each other in PC space. For this, similarity measures between data points have to be computed, for example as the inverse of the Euclidean distance be-

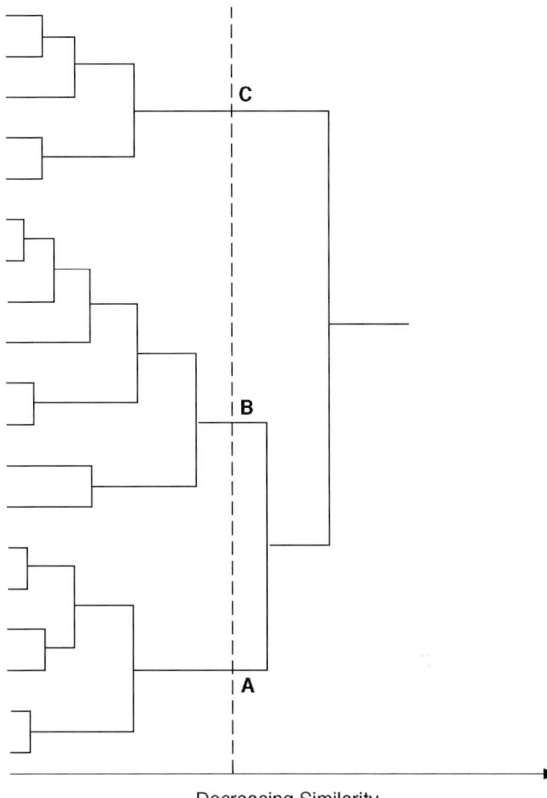

Figure 3.2.3. A dendrogram obtained from clustering the data points of the previous figures, At a given level of similarity, the number of resulting clusters can be read off as the number of intersected links. For the level indicated by the dashed line, three clusters are obtained.

tween two points in multidimensional space. Initial clusters – sometimes referred to as *local modes* – can then be determined as those points whose similarities with the five nearest neighbors (for example) are above a certain threshold value. These points are local maxima if a density map of the data distribution in space was computed. Clustering then proceeds in a hierarchical manner by merging first the closest points with each other, and successively more remote points until all data points merge.

The levels of similarity at which points merge are represented in a dendrogram (figure 3.2.3). At any given similarity, the number of clusters can be read from a dendrogram. By choosing a suitable similarity threshold, the number of clusters is determined as well as all the data points belonging to it, which are all the points connected to one particular cluster in the direction of higher similarity. In figure 3.2.3, the threshold indicated by the dashed line results in three clusters. The data points belonging to each cluster are shown in figure 3.2.4, on a PC coordinate system, although a coordinate system featuring the original log axes could also be used. The three clusters are seen to represent different facies, two limestones (differing in their porosity) and one calcareous dolomite. This

263

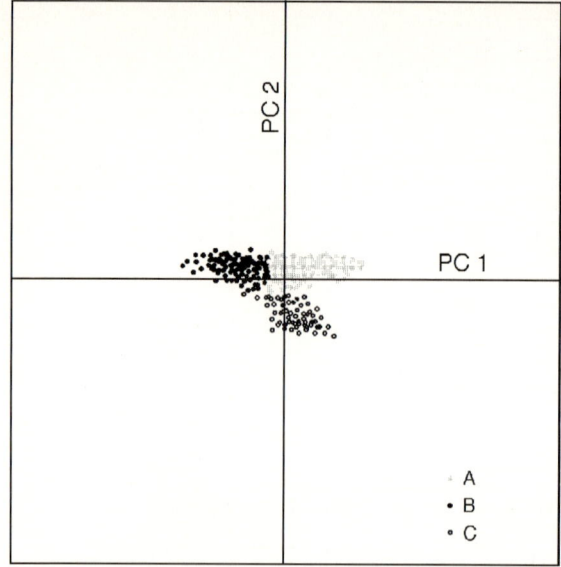

Figure 3.2.4. The three clusters obtained from the dendrogram, plotted in a PC coordinate system.

interpretation is obtained from inspection of the data points on the original cross plot. In general, though, prior to assigning such lithofacies terms to the clusters it is advisable to carefully analyze all information, particularly if cores are available.

For each final cluster, or electrofacies, there is an average log response that is obtained by projecting the PC values of these cluster modes back on the original log axes. Thus, "squared logs" are obtained with constant values for a given electrofacies, which can be particularly useful if the resulting zonation is used as input into a reservoir model. The final result of the zonation can be plotted as shown in figure 3.2.5, with the original logs next to the electrofacies zonation and a description of the proposed lithofacies.

Figure 3.2.6 shows an example of an electrofacies zonation using ROCKCELL[*] (Schlumberger, 1997), a successor to FACIOLOG. ROCKCELL is specifically designed for multiwell facies analysis from logs and contains several classification and estimation techniques including multidimensional regression and artificial neural networks. Its classification approach is based on the assumption that in one or more "key wells" the relationship of log response to rock facies can be established. The cross-plot and clustering methods described above are important tools in this phase. Core data is used to validate the grouping, zonation, and facies assignment. In the next step, the relationship of log responses to facies is defined in a model, for example a multidimensional histogram or an artificial neural network. These mathematical relationships between sets of numerical values (the log readings) and symbolic values (the facies) are used in the last step to

---

[*] Mark of Schlumberger and AGIP

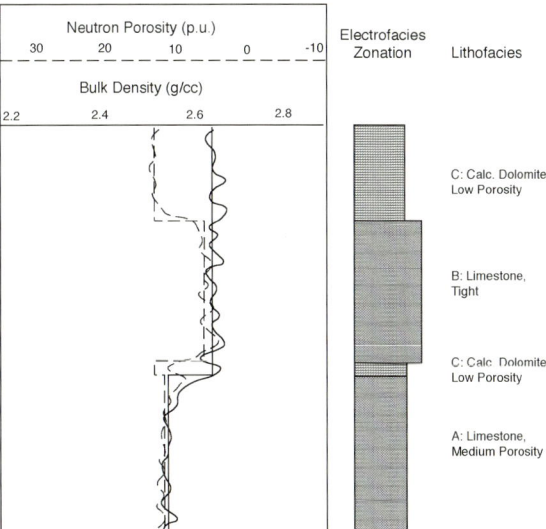

Figure 3.2.5. The three clusters are assigned a lithology and plotted versus depth (right). Their average log values are plotted as squared logs (left), together with the original logs.

assign facies in all other wells based on log readings. Although ROCKCELL is similar to FACIOLOG, it is more interactive and contains a larger number of tools to perform the zonation. It is specifically designed for an interactive workstation environment.

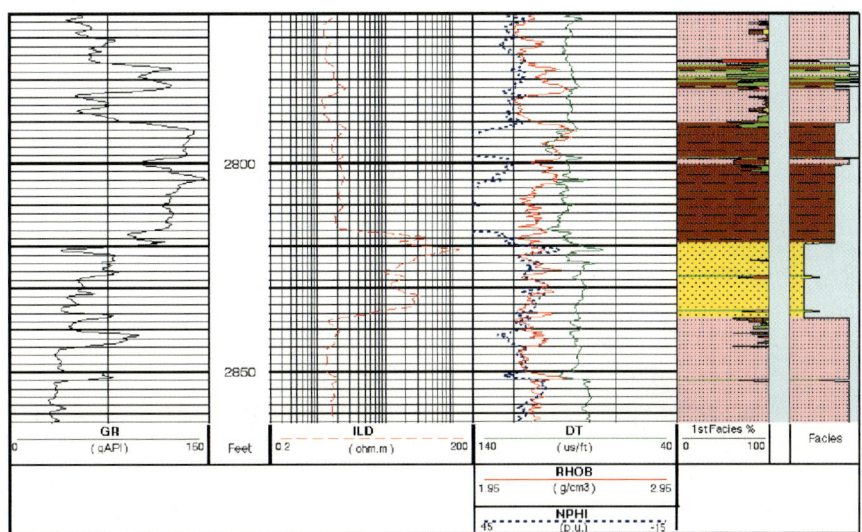

Figure 3.2.6. An example of log zonation using ROCKCELL. The logs in the three tracks on the left include gamma ray (GR), deep induction resistivity (ILD), sonic transit time (DT), bulk density (RHOB) and neutron porosity (NPHI). Lithofacies probabilities (1st Facies %) and final zonation are shown on the right.

There are numerous other unsupervised classification schemes for well logs, some of which are described in Doveton (1994, chapter 4). They all have the advantage of being applicable regardless of lithology and types of input logs and they typically furnish a result for every data point, or every level of the well logs. On the other hand, the significance of the resulting zonation has to be figured out by the interpreter, sometimes in a rather tedious procedure, particularly when no "ground truth" in the form of core data is available.

Supervised Classifications

In these methods, an *a priori* knowledge of the classes exists in the form of a database, and therefore, the resulting zones have a predetermined significance. For example, the response of a limestone with 10 % porosity is well known for the common well logging tools, and consequently any given measurement can be compared to this response and its probability of belonging to this class can be evaluated. Delfiner et al. (1987) describe a program called LITHO*, which uses an extensive database of log responses for common sedimentary and some igneous rocks. Each lithology is represented as an ellipsoid in the multidimensional log space, a simple example of which is shown in figure 3.2.7. The outline of the ellipsoid corresponds to a certain probability, with larger ellipse circumference corresponding to lower probabilities. Ellipsoids of different lithologies overlap partially, and a probabilistic procedure is used to obtain the most likely lithology at a

---

* Mark of Schlumberger

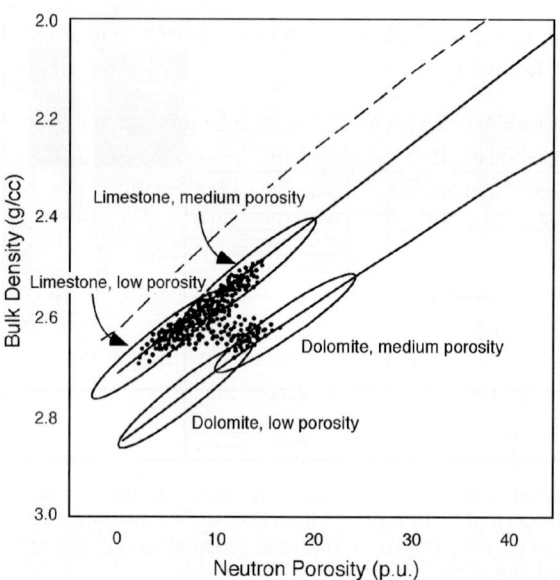

Figure 3.2.7. Example of a lithological data bank used in supervised classifications such as LITHO. The ellipses define areas where the indicated lithofacies fall with a high probability. The data points shown are the same as in figure 3.2.1.

given depth. This "posterior probability" is obtained by combining the probabilities in all applicable cross-plots for a given lithofacies. Usually, the lithofacies with the highest posterior probability wins, but if the lithofacies with the second- or third-highest probability occurs immediately above and below this data point, it might be given preference. This procedure is called the "vertical continuity logic". In LITHO, the database can be updated with new lithofacies, for example if they are locally different in composition or simply not available in the default database. Stowe & Hock (1988) have developed such a customized database using LITHO for the Zechstein carbonates of northern Germany. They obtained good results by distinguishing 48 carbonates and 24 evaporites, a much higher number than present in the original database but needed for their facies analysis.

Figure 3.2.8 shows an example of a zonation obtained from LITHO, together with annotated lithologies. While unsupervised classifications are data-driven, purely statistical methods whose results have to be diligently interpreted, supervised classifications have the advantage of an immediate association of each data point with a lithological descriptor. Experience has shown, however, that it is difficult to obtain a consistent quality of results. Reasons for this can be that

- Different logging suites exist in the wells;
- Different contractors logged the wells;
- Some wells were drilled with oil-based, others with water-based mud;
- Variable borehole conditions exist due to different drilling techniques.

Reservoir Zonation Using Geological Logs

Reservoir zonation has greatly benefited from the development of new geological well logs, particularly in the following areas:

- *Analysis of thin beds.* Dipmeters (chapter 2.1) and borehole imaging logs (chapters 2.2 – 2.5) are capable of distinguishing beds one to two orders of magnitude thinner than with conventional well logs. This becomes particularly important in reservoirs where thin sand layers constitute a significant part of the reservoir.
- *Bedding information.* Borehole images provide two-dimensional information on bedding that no other well logs contain (chapter 2.2.5). This becomes important in reservoirs where small- to medium-scale heterogeneities influence the production characteristics, for example in different sandstone types found in deltaic sequences, or in certain platform carbonates.
- *Mineralogical composition.* Nuclear spectroscopy tools (chapter 2.7) provide geochemical information that can result in more accurate mineralogical compositions. This may lead, for example, to distinctions between arkosic and argillaceous sandstones, or between dolomites and dolomitic limestones.

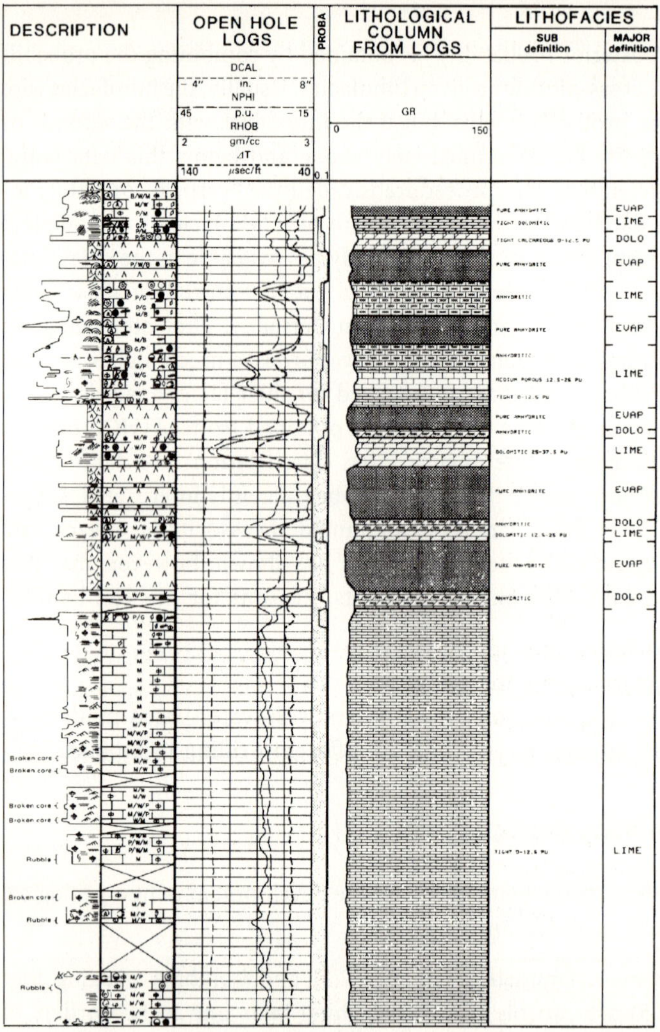

Figure 3.2.8.
An example of LITHO results (right) together with the logs (middle) and the core description (left) over a carbonate/evaporite sequence from the Middle East (from Delfiner et al., 1987)

3.2.2 "Thin Beds"

Bed Thickness and Tool Resolution

Well logs – like all measurements with physical sensors – have an intrinsic limitation in their resolution, i.e. in their ability to distinguish two adjacent events. The smallest event for which a true reading is obtained is referred to as intrinsic resolution[*]. It is often related to the sensor configuration, such as the coil spacing

---

[*] Intrinsic resolution can also be defined as the distance on the tool response function between the two points at half the maximum response value (Theys, 1991).

in induction tools, or the source-detector spacing in nuclear tools. Table 3.2.1 lists ranges of intrinsic resolutions for the common logging tool groups available on the market. The table also lists optimal log resolution ranges, which is the resolution obtained when considering not only the sensor characteristics, but also the sampling rate and data averaging on the well site. Some of the lower resolution values are obtained with special processing techniques aimed at enhancing the vertical resolution. Table 3.2.1 shows that from highest to lowest resolution there is a ratio of 500, and that no tool has a resolution better than 0.5 centimeters. In reality, resolutions are often not as high as those listed in table 3.2.1 because borehole conditions are not ideal, for example if borehole rugosity or mudcake is present, or if the tool movement is irregular. These "effective" vertical resolutions determine in practice whether the log analyst can see and evaluate a particular bed.

The inability to properly evaluate reservoirs with beds thinner than the effective resolution of a log is referred to as the "thin bed problem". The term "thin" is obviously relative and depends on the types of logs, or rather, on the log with the lowest resolution used in the log analysis. For example, if a logging suite comprises a gamma ray, a density/neutron and an induction measurement, then the petrophysical evaluation will not be able to compute fluid saturations with a resolution better than that of the induction log. Well aware of this limitation, engineers and petrophysicists have put considerable efforts into improving tool resolutions as much as possible without sacrificing the measurement accuracy. Progress has been made on two fronts. First, on the sensor technology side,

Table 3.2.1 Ranges of intrinsic vertical resolution and optimal effective resolution for the principal log measurement types[a].

Measurement	Range of Intrinsic Vertical Resolution		Optimal Log Resolution	
	inches	cm	inches	cm
Microelectrical Imaging	0.2 – 0.3	0.5 – 0.75	0.2 – 0.3	0.5 – 0.75
Dipmeters	0.4	1.0	0.4	1.0
Electromagnetic Propagation	2.0	5.0	2 – 4	5 – 10
Microspherically-Focussed Log	2 – 3	5 – 7.5	3 – 4	7.5 – 10
Laterolog	24	60	24	60
Induction Log	36 – ~96	90 – ~250	36 – ~96	90 – ~250
Density	2 – 15	5 – 40	4 – 18	10 – 45
Neutron Porosity	12 – 24	30 – 60	12 – 30	30 – 75
Gamma Ray Logs	8 – 12	20 – 30	18	45
Ultrasonic Imaging	0.2 – 0.4	0.5 – 1.0	0.4 – 1.0	1.0 – 2.5
Array Sonic Tool	6 – 48	15 – 120	12 – 48	30 – 120

[a] Compiled from oil service company catalogs, Theys (1991), and other sources

higher-resolution sensors were developed and sampling rates were increased to as high as 1.2 inches (3 cm) from the previously common 6 inches (15 cm). Additionally, considerable work was undertaken to reduce the sensitivity to the borehole environment – or at least to understand it so it can be corrected – and to obtain accurate depths of each measurement by adding accelerometers as auxiliary sensors to the tools. Finally, data processing algorithms were developed that improve measurement resolution by taking into account the tool response to thin beds. The recent logging literature contains an abundance of papers on these subjects, some of which will be reviewed later in this chapter.

For reservoir zonation, thin beds are also a problem, although perhaps to a lesser degree than for petrophysical analysis. We have already seen in the conceptual example in figure 3.2.5 that there was a "thin" dolomite bed present shortly below the middle of the interval shown, even though no vertical scale is given. Upon closer inspection we see that the log response, although different from the layers above and below, is not exactly the same as the much thicker dolomite bed at the top of the interval. Particularly the neutron log, which has a lower resolution than the density log (table 3.2.1), does not reach the values expected in the dolomite. These data points were attributed to the dolomite cluster C simply because the other clusters differed even more from them. The thick bed at the top primarily determines the squared logs for the dolomite since it represents many more data points than the thin bed. But the squared logs distinguish the thin bed better from its neighboring beds than the original logs, a result that is similar to somewhat deconvolution and median filters. However, we do not know if it was correct to assign these few data points to the dolomite facies, because there are inherently two possibilities:

1. The bed belongs indeed to the same lithofacies, but the neutron log does not read the true value because the bed is too thin, or
2. The bed is a different dolomite, somewhat more calcareous and less porous than the thick bed at the top.

We cannot rule out any of these possibilities from the information available. If, however, we have a high-resolution measurement that tells us exactly *where the bed boundaries* of this dolomite are, we can make a forward model of the neutron tool response assuming that it has the same properties as the thick dolomite bed. If the resulting neutron log from this computation looks like the one that was measured, we know that hypothesis 1 is true. If not, we have to change the properties of the bed until we find a good match and we know that hypothesis 2 is true.

We could, of course, assume arbitrary locations of the bed boundaries and perform repeated forward modeling of the tool response until we obtain a satisfactory match. In this case, we will likely find several combinations of bed boundaries/bed properties with a good match, illustrating that the problem is underdetermined. Knowledge of the correct position of bed boundaries, therefore, is a necessary ingredient. High-resolution logs such as dipmeters and borehole imaging logs are the tools of choice for this because they have intrinsic vertical resolutions of at least one order of magnitude better than any other log.

Thresholding for Sand Count

Obtaining correct positions of bed boundaries is not only important for the zonation problem outlined above. A commonly used technique for obtaining sand count is to select a suitable threshold on the gamma ray log, typically half-way between a known shale reading and a known sandstone reading. Every data point below this threshold is considered sandstone, while everything else is attributed to the shales. This method works fine as long as the beds are thicker than about half the intrinsic resolution of the tool. Assuming for example a 30 cm-thick sandstone bed surrounded by shale, and a tool resolution of 50 cm, the gamma ray reading in front of it will be influenced to more than half by the sandstone and the tool reading will be closer to the sandstone gamma ray value. If, on the other hand, the sandstone bed is only 10 cm thick, the surrounding shale dominates the readings, the measurement will be closer to the shale value, and no sandstone bed will be detected by the thresholding procedure (figure 3.2.9). Naturally, for thin shale beds within a thick sandstone bed the same applies. This may result in an erroneous sand count, particularly when the beds are in general thin, i.e. at or below the resolution of the tool used for thresholding (Hackbarth & Tepper, 1988; McGann et al., 1988; Sullivan & Schepel, 1995; Reid & Enderlin, 1998). The resulting zonation also does not properly reflect the bedding sequence. In figure 3.2.9, the thinning-upward of the sandstones in the upper part of the sequence has been lost on the zonation obtained from thresholding.

A Turbidite Case Study

Bed Thickness Distributions. Figure 3.2.10 shows a photograph of a 30 – cm core interval, taken under ultraviolet light, from a turbidite sequence in the Gulf of Mexico. The sandstone beds are bright yellow because of the fluorescence of the oil they contain, and beds well below one centimeter thick can be seen. A digital scan of the brightness down the centerline of the photograph is also shown on the figure. This curve has 2048 data points and has a sharp response, reflecting the oil-filled sandstone beds. For comparison, the 27 curves from one pad of the Formation MicroScanner* (FMS, chapter 2.2) are shown next to this curve, at the probable depth match of the two data sets. Such a match is difficult to obtain since the data comes from a depth of over 3 kilometers and the number of sandstone beds per meter is typically between ten and one hundred, resulting in a large number of possible matches. One meter corresponds to a depth uncertainty

---

* Mark of Schlumberger

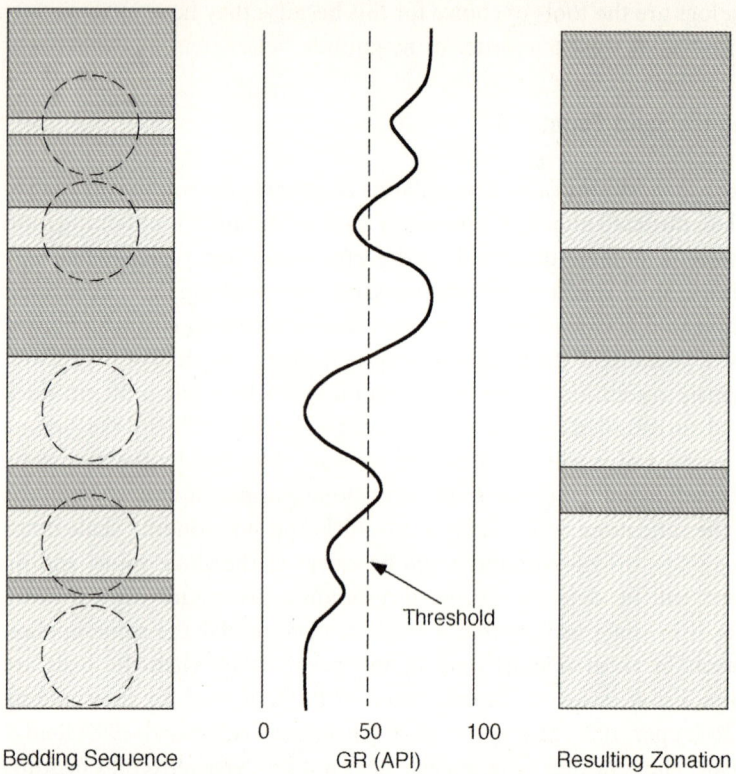

Figure 3.2.9. Conceptual sketch of sand/shale zonation by thresholding the gamma ray log to obtain a "sand count". Beds can be found with this method as long as they are thicker than about half the intrinsic vertical resolution of the tool, indicated by the "spheres of influence" positioned over the center of the sandstone beds in the left column. The resulting zonation misses the thin sandstone and shale beds (right column), showing massive beds instead, and can give a wrong net sand volume.

of 0.033 %, but fortunately some of the thicker beds in the well provide reliable tie-ins for both core and borehole images. The match is further complicated because the two data sets come from two physically different locations, one from the slabbed core surface originally inside the core barrel, the other from the wellbore surface on the outside of the drill bit. The discontinuous nature of several beds in figure 3.2.10 illustrates this. The thickest bed in this interval, however, shows a good match and is well resolved by the curves on the borehole image. It corresponds to an increase in resistivity caused by the presence of oil. The two thinner beds below are merged into one. They are each just below one centimeter in thickness, and so is the shale layer in between. The FMS has an intrinsic vertical resolution of 7 millimeters, but in the presence of borehole rugosity and with conductive mud in the borehole, the current focussing is not optimal and tool resolution deteriorates. In this case, the "effective" resolution of the tool seems to

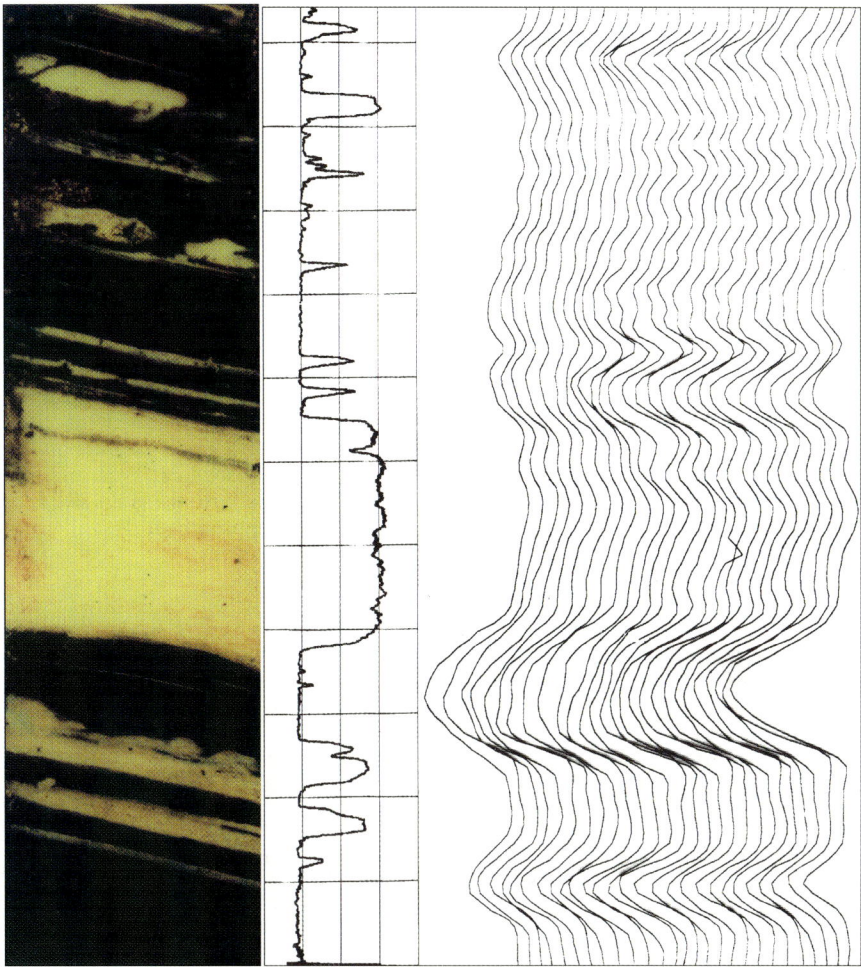

Figure 3.2.10. Photograph of a core from the Gulf of Mexico showing oil-filled turbidite sands (yellow bright) and shales (left). Length of core is 30 cm, photograph taken under UV light. Digitized scan of the brightness down the middle of the core photograph is shown in middle, and the 27 FMS curves of one pad recorded in the same well and depth-matched to the core on the right.

be around twice the intrinsic vertical resolution, or about 15 millimeters. Most of the other sandstone beds in this short section are considerably thinner and cannot be resolved by the FMS. Although the resolving power of the FMS in this case is not optimal, it is still at least one order of magnitude better than other logging tools, which typically provide only two data points over the interval shown. If the simple thresholding technique discussed above was applied to the FMS, the main beds would be captured in the resulting binary "sand count" log, but all thinner beds would be left out.

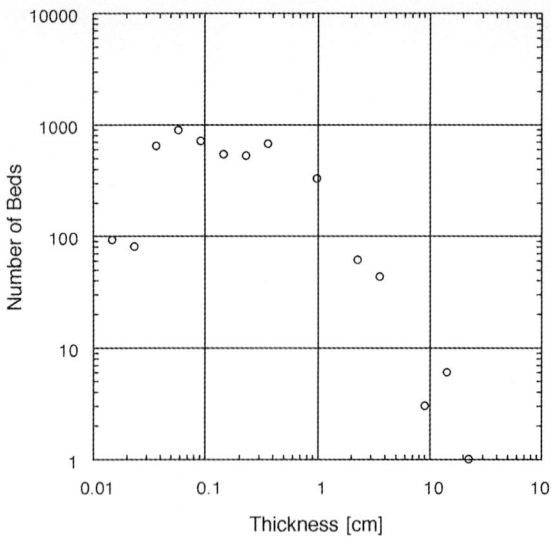

Figure 3.2.11. The number of sandstone beds as a function of thickness, determined from the brightness scans on cores, shown in the previous figure. Number of beds decreases at very low thicknesses because of sensor limitations and because the grain size is approached.

In order to assess the importance of these thinner beds, a complete digitization of the core photographs over almost 60 meters (200 ft) was undertaken. The resulting digital scan line comprised over 600,000 data points and was used to determine the bed boundaries. Thresholding, for reasons discussed above, was deemed insufficient and a procedure based on the first derivative of the brightness curves was used instead, where negative and positive peaks indicated transitions from shale to sandstone and vice versa. A total of 4590 sandstone beds and an equal number of shale beds were found. The sandstone thickness distribution in figure 3.2.11 shows the number of beds to increase with decreasing thickness: the thickest bed is around 22 centimeters thick, but almost 1000 beds are found with thicknesses of around one millimeter. Below 0.5 millimeters the number of beds decreases because the sampling rate of 0.15 millimeters is approached, and because no bed can be expected below the grain size, which is between 0.2 to 0.5 millimeters.

Interestingly, the shale thickness distribution (not shown) almost exactly matches the sandstone distribution except for a few beds that are up to 2 meters thick. Multiplying the number of beds for each class with their median thickness results in a net sand count for that class, shown in figure 3.2.12. Over 50 % of the total net sand thickness of about 25 meters (c. 83 feet) is accounted for by layers between one millimeter and one centimeter thick, i.e. by layers that cannot be resolved with the FMS or any other log.

Volumetric Distributions from Power Laws. If all beds had the same lateral extent, the volumetric contribution of each bed class would be proportional to its net sand thickness shown in figure 3.2.12. However, such a depositional process is difficult too imagine. A much more plausible geological scenario is that the

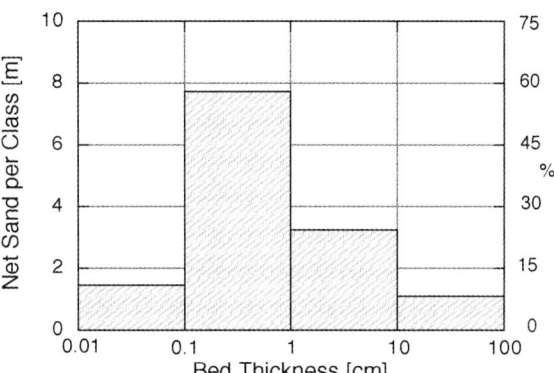

Figure 3.2.12. Net sand thickness per bed thickness class, in meters (scale on left) and percentage (scale on right).

thinner beds have, on average, smaller lateral extents and thus smaller volumes. These scaling properties of turbidites are important if one wants to predict the reservoir volumes and construct three-dimensional reservoir models. Rothman et al., (1994) and Malinverno (1997) have related bed thickness distributions to the geometry and volumetric distribution of the turbidites beds. They focus in particular on the power-law behavior of turbidite thickness distributions measured in a vertical section. If N is the number of beds whose measured thickness is greater than or equal t, then a power-law distribution is expressed as

$$N(t) \propto t^{-\beta} \tag{3.2.1}$$

The scaling exponent β is also referred to as the fractal dimension. Several authors have reported a power-law behavior of turbidite bed thicknesses over more than two orders of magnitude (Hiscott et al., 1992; Rothman et al., 1994), similar to what has been observed on other naturally occurring events such as earthquake magnitudes. Malinverno (1997) investigated the question how the thickness distribution of turbidites relates to their volumetric distribution assuming that this follows also a power law of the form

$$N(v) \propto v^{-c} \tag{3.2.2}$$

where v denotes the volume, and c the scaling exponent of the volume distribution. The smaller c becomes, the more the total volume is accounted for by the largest members. Malinverno (1997) studied how the relationship between β and c depends on the following factors:

1. the scaling between bed length and bed thickness at the bed depocenter, expressed by an exponent γ;
2. the rate of decrease in bed thickness away from the depocenter, expressed by a; and

3. the spatial distribution of the bed depocenters, expressed by *d*.

The scaling factor γ is an exponent relating the thickness *t* to the width *w* in the form of

$$w \propto t^{-\gamma} \tag{3.2.3}$$

For $\gamma = 0$, the bed width is the same irrespective of the bed thickness, while for $\gamma = 1$ the width increases linearly with increasing thickness (figure 3.2.13). The significance of the rate of decrease in thickness away from the depocenter *a* is illustrated in figure 3.2.14. Malinverno (1997) found that *a* does not enter into the

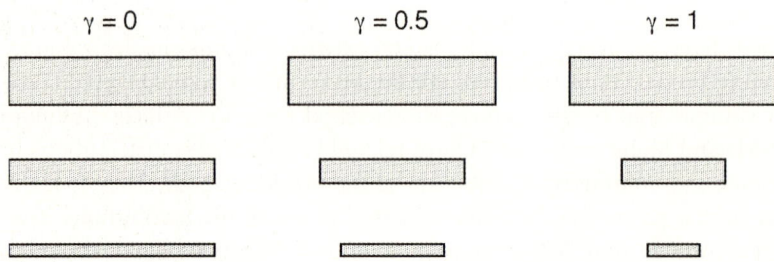

Figure 3.2.13. The parameter γ controls the width/thickness relationship of turbidite beds (after Malinverno, 1997).

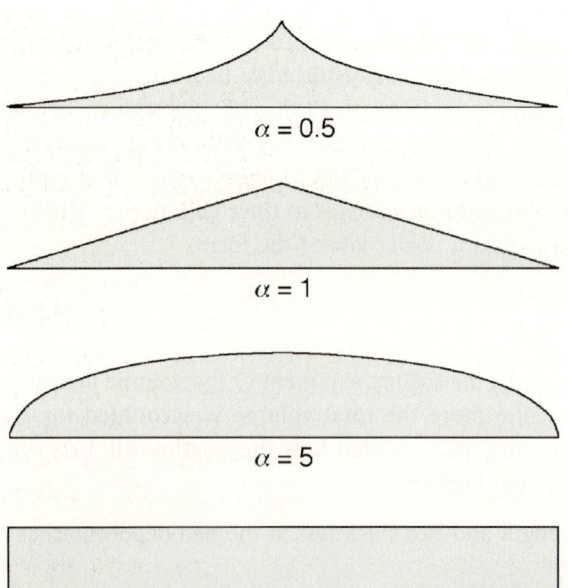

Figure 3.2.14. The parameter *a* controls bed shape, or the rate of thickness decrease away from the axis (after Malinverno, 1997).

relationship between the exponents of the thickness and volume power law distributions. However, it contains information that is needed for the three-dimensional reconstruction of the bed volumes as will be discussed later on. Finally, the exponent d describing the spatial distribution of the depocenters can be visualized as the number of dimensions at which the beds scattered in space: If their depocenters are all exactly above each other and thus on a point on a map, $d = 0$. If they follow a straight line, $d = 1$, and if they are randomly scattered on the map, $d = 2$. Malinverno's (1997) main result is that

$$\beta = c + \gamma(2c - d) \quad (3.2.4)$$

For a basin with finite size, the larger turbidite beds cannot be randomly distributed for lack of space. As a result, there will be a thickness distribution that shows one exponent for the smaller beds (β_{small}) and one for larger beds (β_{large}). Malinverno (1997) found that in this case

$$\gamma = \frac{\beta_{large} - \beta_{small}}{2} \quad (3.2.5)$$

and

$$c = \frac{\beta_{large}}{1 + \beta_{large} - \beta_{small}} \quad (3.2.6)$$

The cumulative thickness distribution plot of our turbidite example, shown in figure 3.2.15, also reveals such a dual power law behavior, with $\beta_{large} = 1.7$, and $\beta_{small} = 0.6$. We assume now that the simple model presented here applies and that the turbidites were deposited in a confined setting such as an intrabasin, with a resulting distribution of the turbidite bodies indicated schematically in figure 3.2.15. From equation 3.2.5 we then obtain $\gamma = 0.55$, meaning that beds with larger thicknesses are also possess greater widths. The scaling is not linear, however, since by doubling the thickness the width increases only by a factor of about 1.6 (that is, the thickness/width ratio is a function of the thickness, approximately as shown in the middle of figure 3.2.13). Also, from equation 3.2.6 we obtain $c = 0.77$, a result that indicates a strong volumetric contribution of the few thick beds occurring in this sequence. In fact, using Malinverno's (1997) figure 3 we estimate that 20 % of the total volume is accounted for by the thickest bed alone. The volume distribution versus cumulative bed thickness for an exponent $c = 0.77$ is plotted in figure 3.2.16. The total number of beds, or volumes, is kept the same (4590) and the volumes are relative with $v = 1$ for the smallest bed. While the bed thickness distribution (figure 3.2.15) covers approximately three orders of magnitude from thinnest to thickest bed, the volume distribution spreads over almost five orders of magnitude from smallest to largest. The most voluminous bed is found to be about 60,000 times larger than the smallest one. An approximate volumetric distribution derived from figure 3.2.16

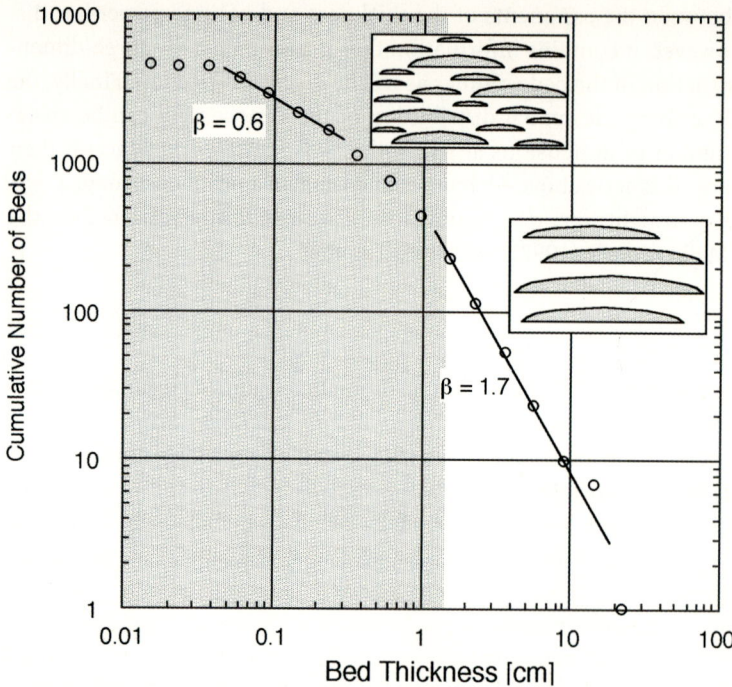

Figure 3.2.15. The cumulative number of beds versus bed thickness, both plotted on a logarithmic scale. Two straight lines can be fitted, one for the thinner and one for the thicker beds. This suggests a different arrangement of the beds with respect to each other, as sketched in the insets. Grey area indicates thicknesses below resolution of FMS.

illustrates the importance of the largest beds (figure 3.2.17): The nine beds with thicknesses over 10 centimeters represent about 55 % of the total volume. This is to be contrasted with the "layer-cake" model shown in figure 3.2.12, where this

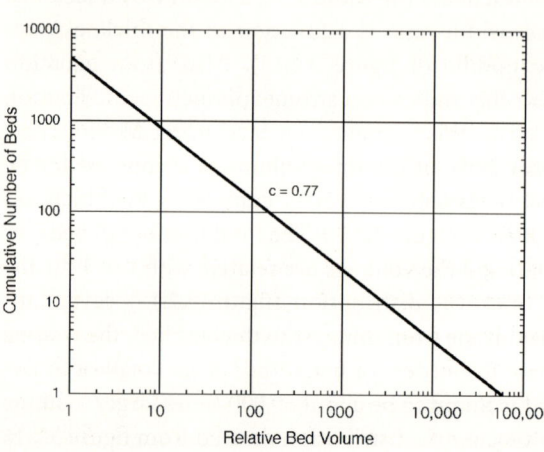

Figure 3.2.16. The cumulative number of beds versus the relative bed volume for $c = 0.77$. Notice that there is a difference of almost six orders of magnitude in relative volumetric importance between the thinnest and the thickest bed.

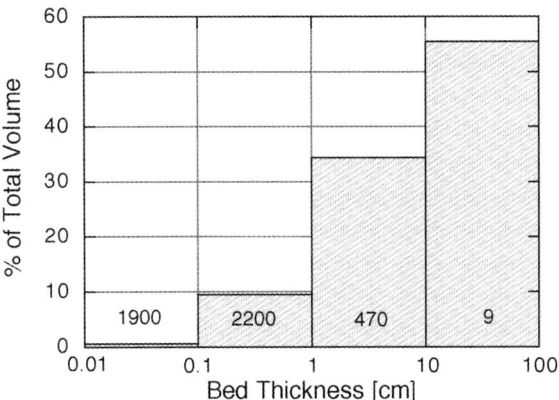

Figure 3.2.17. Relative volumetric distribution of well turbidites, estimated from the power law relationship of figure 3.2.16.

bed thickness class accounts for only about 10 % of the volume. The three-dimensional shape of the turbidite bodies is undoubtedly more realistic than the layer-cake geometry, leading us to conclude that the thicker beds have an exponentially higher volumetric importance than the thinner beds. However, this exceeding volumetric importance of thicker beds is not as high as it would be in the case of complete geometric similarity (with $\gamma = 1.0$), where a bed three orders of magnitude thicker would have a volume nine order of magnitude larger. It is important to realize that the term "volume" here refers to the turbidite sand volume *connected to the wellbore*. In the case of the three-dimensional model, there are numerous small (and thin) sand layers, possibly equal in volume to ones in the layer-cake case, that may be close to the wellbore but not hydraulically connected to it. This is exacerbated by the observed dual-power behavior of the bed thickness that implies a random geographic distribution of the smaller layers and a clustering of the thicker layers around the depocenter. Whether the oil in place in the smaller layers will be recovered or not depends on the well spacing, the hydraulic interconnectivity between the sand layers, and the permeability of the shales.

While in the layer-cake model the limited resolution of the FMS could lead to a serious underestimation of the net sand volume in the reservoir, the three-dimensional model accounts much better for the sand volumes connected to the wellbore. Specifically, if only beds above one centimeter are detected, comparison of figures 3.2.12 and 3.2.17 reveals that only 33 % are accounted for in the layer-cake model, but close to 90 % in the three-dimensional case.

Volumetric Distributions Using Physical Scale Models. The previous discussion has used the rather vague term "three-dimensional" model. In fact, from the thickness distributions observed in a one-dimensional scan, no information can be gleaned on the actual three-dimensional shape of the turbidite bodies (whence Malinverno's conclusion that the shape exponent a does not enter into the relationship between β and c as long as it is the same in all beds). Like in all but the

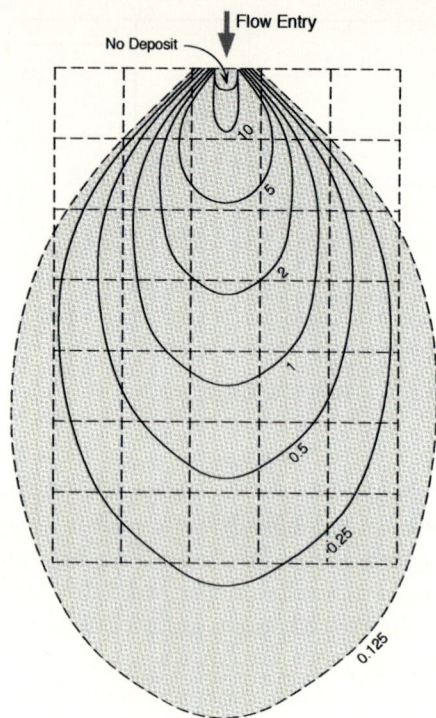

Figure 3.2.18. Isopachs of an experimental turbidite deposit, in millimeters (after Luthi, 1980). The dashed grid has a spacing of one meter.

simplest stereological models, such higher-dimensional reconstructions require some *a priori* knowledge on the shape of the object. While much outcrop work has been done on turbidite geometries, few observations are truly three-dimensional. By contrast, physical scale models, although using a simplified topography, can provide a full three-dimensional picture of the depositional thickness. The experiments of Luthi (1980) in a large tank using fine-grained sand, silt, and chalk, can provide a template for the three-dimensional reconstruction. A deposit from one such experiment is depicted in figure 3.2.18. It shows isopach contours somewhat resembling beaver tails that are elongated in the direction of the flow (determined by gravity) and decrease monotonically away from the source. The thickness distribution of this deposit can be obtained by sampling it at regular or random locations, whereby the minimum bed thickness is given by the grain size of 32 microns. Although some extrapolation is necessary because of the limited size of the experimental deposit (dashed lines in figure 3.2.18) it seems to be clear that there is a relatively large number of locations with small thicknesses, and only very few with thicknesses close to the maximum of about 10 millimeters. The resulting distribution is shown with a solid line in figure 3.2.19. If we imagine now a stack of identical deposits but distributed at random locations, and if we drill an imaginary well through this stack, then the resulting thickness distribution would be equal to the one we obtained from the experi-

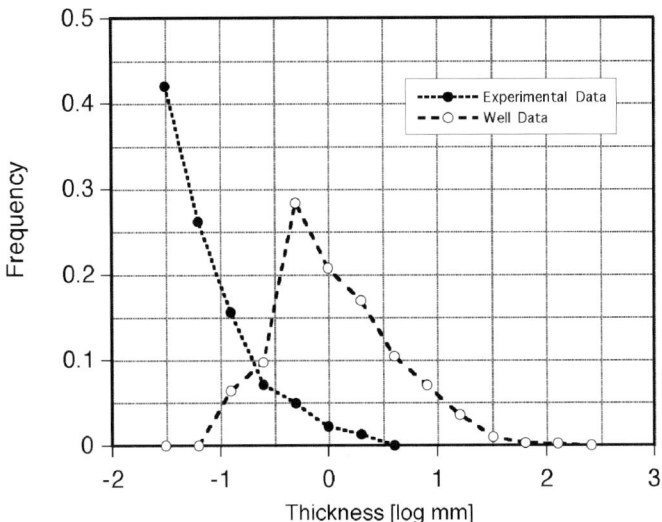

Figure 3.2.19. Bed thickness frequencies for experimental turbidite (figure 3.2.19) and well turbidites (figure 3.2.11).

ment. The thickness distribution of the actual well turbidites, however, differs from this because

- they are much larger than the laboratory model, and
- they have variable sizes.

The dashed line in figure 3.2.19 shows the well data, confirming that they are thicker and have a broader shape. Their distribution is limited at the lower end by the grain size, as discussed above[2]. Our goal is now to find the scaling factor that relates the experimental thickness distribution to the well thickness distribution. Assuming that the well turbidites have a shape that is geometrically similar to the experiment, the experimental distribution $E(t)$ can be related to the well distribution $F(t)$ by a scaling factor distribution S in the form

$$F(t) = S(n) * E(t) = \sum_{n=1}^{N} E(t-n)S(n) \qquad (3.2.7)$$

where * designates a convolution operation and n a multiplier. Both t and n are expressed as logarithms with base 2. The scaling distribution $S(n)$ can be obtained by transforming both E and F into the frequency domain with a Fast Fourier transform and dividing them with each other (e.g. Press et al., 1992). This yields S in the frequency domain, and, after applying an inverse Fourier transform, one obtains S in the original length domain, or rather, in multiples of a unit length.

---

[2] In this discussion we omit considerations on the dynamic similarity of the processes.

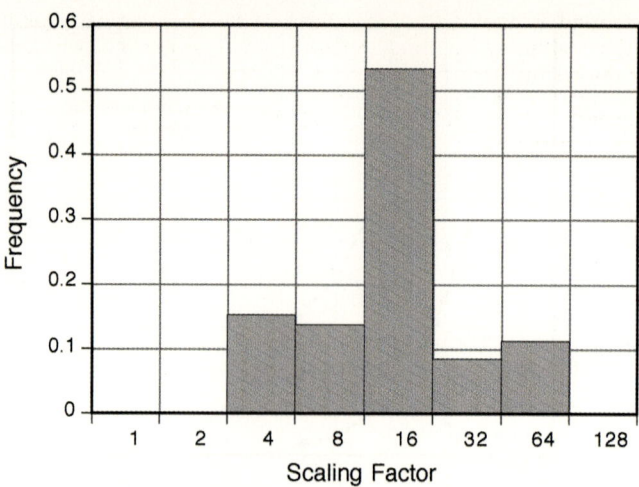

Figure 3.2.20. Scaling factor distribution between well and experimental turbidites.

In practice, the data have to be prepared in a prescribed way that explains the different graphic representation in figure 3.2.19 compared to those in the previous paragraph. The resulting scaling factor – expressed in powers of 2 – is shown in figure 3.2.20. There is a sharp maximum at 16, indicating that there may have been a typical process at work that resulted in deposits of roughly equal size and geometrically similar to the experimental deposit, thereby supporting the analogy. The scaling factors 4, 8, 32 and 64 all show similar frequencies of around 10 %. With the maximum width of the experimental deposit estimated at 12 meters, most beds therefore have a width of about $12 \times 16 \approx 200$ meters and a length of approximately $18 \times 16 \approx 300$ meters. Those 12 % of the total number of beds with a scaling factor of 64, however, will have typical extents of about 750×1200 meters. The volumetric distribution can be calculated from the scaling factors, raised to the cube, weighted by their frequencies, and multiplied by the volume of the experimental deposit of $0.04 m^3$ (table 3.2.2). The relative volumes are shown in figure 3.2.21. As in the previous estimations from power laws, an overwhelming importance of the thicker beds is observed: Beds with a scaling factor of 64 contribute 12 % to the total number of beds, but 87 % to the volume.

The total volume connected to the wellbore, according to table 3.2.2, shows that the probable volume connected to the wellbore amounts to 6.7 million m^3. At an average porosity of 37 % and an oil saturation of 90 % this amounts to a volume of stock tank oil initially in place (STOIIP) of about 2.2 million m^3, or roughly 16 million barrels. Assuming a recovery factor of 45 %, the well can thus be expected to produce a little under 7 million barrels over its lifetime from this interval if no other well interferes with draining these layers. Obviously, these numbers are dependent on the model we used and should merely illustrate

Table 3.2.2. Scaling factor and associated frequency obtained from comparing the well turbidites with the experimental turbidite. The resulting volumes illustrate the overwhelming importance of the thicker (or larger) beds.

Scaling Factor S	Frequency f	Volume V in m^3 (^{a})
4	0.15	1,836
8	0.13	12,301
16	0.52	391,071
32	0.08	481,216
64	0.12	5.775,052
Sum	1.00	6.661,926

^{a} $V = n_o \cdot f \cdot S^3 \cdot V_o$ ($n_o = 4590$, total number of beds; $V_o = 0.04$ m^3, volume of experimental deposit)

that, if the deposits are known to have a typical geometric shape, certain estimates on reservoir volumes can be made.

Comparison of the two approaches. It is interesting to compare these results to the ones obtained from the power law considerations, where we observed that beds thicker than 1 millimeter account for about 10 % of the total number of beds but contribute 90 % to the connected volume. Here, beds with a scaling factor of 64 constitute 12 % of the total number of beds and 87 % of the volume. This is an interesting but not a very accurate correspondence, since thickness distributions and scaling factors cannot be equated. Both approaches are statistical in nature and do not tell us precisely what extent a particular layer has. Each bed has a certain probability of belonging to a given bed size class with a maximum thickness equal or larger to the its thickness. The only major difference is that the physical scale model provides a template that can be used to construct a semi-deterministic shared-earth model around the observations in the wellbore.

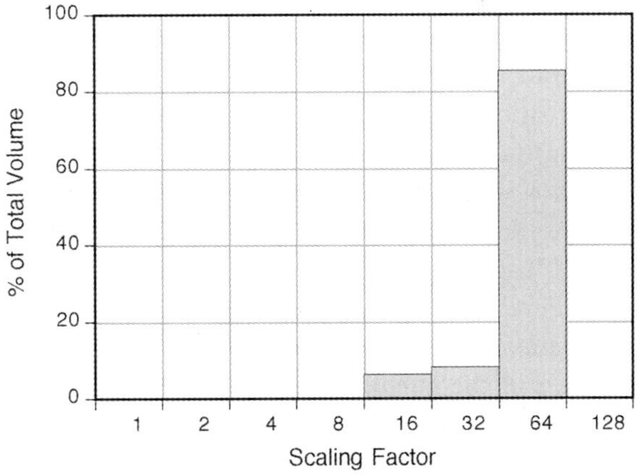

Figure 3.2.21. Relative volumetric distribution of the well turbidites, estimated from the scaling with the experimental turbidite.

This is achieved by inserting turbidite bodies with the shape of the template while honoring the statistical distribution given by the scaling factors and the observed thicknesses in the wellbore.

Sedimentological aspects. Throughout the foregoing discussion we have omitted the sedimentological aspects and focussed entirely on geometric relationships. The Tertiary turbidites in the Gulf of Mexico are often described as non-conventional turbidites because of the numerous intraslope basins that control depositional patterns. In this example, the geologist performing the core analysis interpreted the thinner beds as levee deposits or contourites, the thicker beds as channel deposits, and the mudstones as interchannel deposits. In the geometric modeling we assumed the deposits to be continuous, with a relatively thick deposit in the central, upstream part as shown by figure 3.2.18, gradually thinning away in lobe-like shapes. A transverse cross-section of these deposits would look similar to the topmost sketch in figure 3.2.14, except that the cusp in the center is more rounded. Such profiles have been reported from several recent submarine fans (e.g. Normark, 1978) although for a complete fan system, not individual layers. These beds seem to correspond best to deposits found in the middle to outer fan, where typically facies C, D, and E are found (Howell & Normark, 1982). In deeper parts of this well turbidites with many characteristics of channel deposits occur, indicating a depocenter shift with time. The essential result, however, is that the few thicker layers are likely to have a volumetric contribution far exceeding the one of the much more numerous thinner layers. Although these thinner layers cannot be resolved by common logging tools, borehole images are capable of detecting layers above 1 – 2 centimeters in thickness and, therefore, account for the bulk of the reservoir volume.

Sharpening the Tool Response

The quest for better tool resolution has taken two different roads, one to improve the intrinsic sensor resolution, and the other to develop appropriate algorithms to improve the sharpness of the response. Sometimes the two approaches are combined.

Increasing sensor resolution is usually limited by the physics of the measurement and constrained by economics. Nuclear tools, for example, can be built with smaller detectors, which can result in an improved vertical resolution. Smaller detectors, however, mean smaller count rates and therefore a poorer signal/noise ratio for a given logging speed (or the time available to collect data from a given interval in the borehole). This can be corrected with a slower logging speed, but economic considerations might not allow spending more time on the logging operation.

Much of the attention has focussed on the induction tool (Barber, 1988), a fundamental measurement in many areas but with a poorer vertical resolution than other tools (table 3.2.1). Halliburton's High-Resolution Induction Tool

(HRI**) offers a resolution of one foot with six measurements at depths of investigation of 24, 30, 40, 50, 60 and 90 inches (Strickland et al., 1987). Schlumberger's Array Induction Imager Tool (AIT*) provides induction resistivities with a vertical resolution of 1, 2 and 4 feet, each with depths of investigation of 10, 20, 30, 60 and 90 inches (Barber & Rosthal, 1991; Barber et al., 1995). Both tools combine new coil arrangements with resolution-enhancing processing techniques. A comparison of the deepest AIT-curve with a FMI image is shown in figure 3.2.22 over an interval with numerous resistive beds of variable thickness. The induction curve is seen to have high readings in the two or three beds with thicknesses over one foot (30 cm), but for thinner beds the response degrades as previously illustrated in figure 3.2.9. Also shown are other measurements, notably a shallow-resistivity (R_{x0}) measurement with two resolutions, one at 8 inches (20 cm), the other one – the micro-cylindrically-focussed log (Eisenmann et al., 1994) – at one inch (2.5 cm). The latter is seen to have an excellent resolution almost as high as the FMI, resolving all but the thinnest beds shown by the FMI. On the right-hand side is a new high-resolution density measurement with a resolution of one inch (2.5 cm), compared to two lower-resolution density measurements. Again, the response is much better in the thinner beds than with the conventional log, although the difference is not as great as in the induction/microresistivity comparison.

These high resolutions are obtained with new sensor designs combined with resolution-enhanced processing techniques. In order to achieve higher resolution, the sensors have to be made less sensitive to borehole irregularities such as mudcakes (or "stand-off" in general), and washouts (Goligher at al., 1996). These improvements are often achieved through computer-simulations of the sensor responses. The resolution-enhancement techniques are generally of a type termed "alpha-processing" in which a higher-resolution measurement is used to increase the resolution of a measurement with lower resolution. In order to achieve this, the high-resolution measurement is degraded such that it matches the resolution of the log whose resolution needs to be enhanced (Galford et al., 1986; Flaum et al., 1989; Nelson & Mitchell, 1990). The difference between the original and the degraded log is then used to estimate the correction for the low-resolution log as illustrated schematically in figure 3.2.23. Other enhancement techniques include deconvolution of the original log using the sensor response (the "vertical response" or "point-spread function"). The log is considered a result of the formation property convolved with the sensor response in the form of equation 3.2.7, where E now corresponds to the formation property, S the sensor response function, and F the resulting log. Deconvolution proceeds in the frequency domain under the constraint that there is no "blind" frequency in the response function, otherwise the procedure fails (no division of F by S is possible). For this reason, deconvolution is not always applicable and

---

** Mark of Halliburton
* Mark of Schlumberger

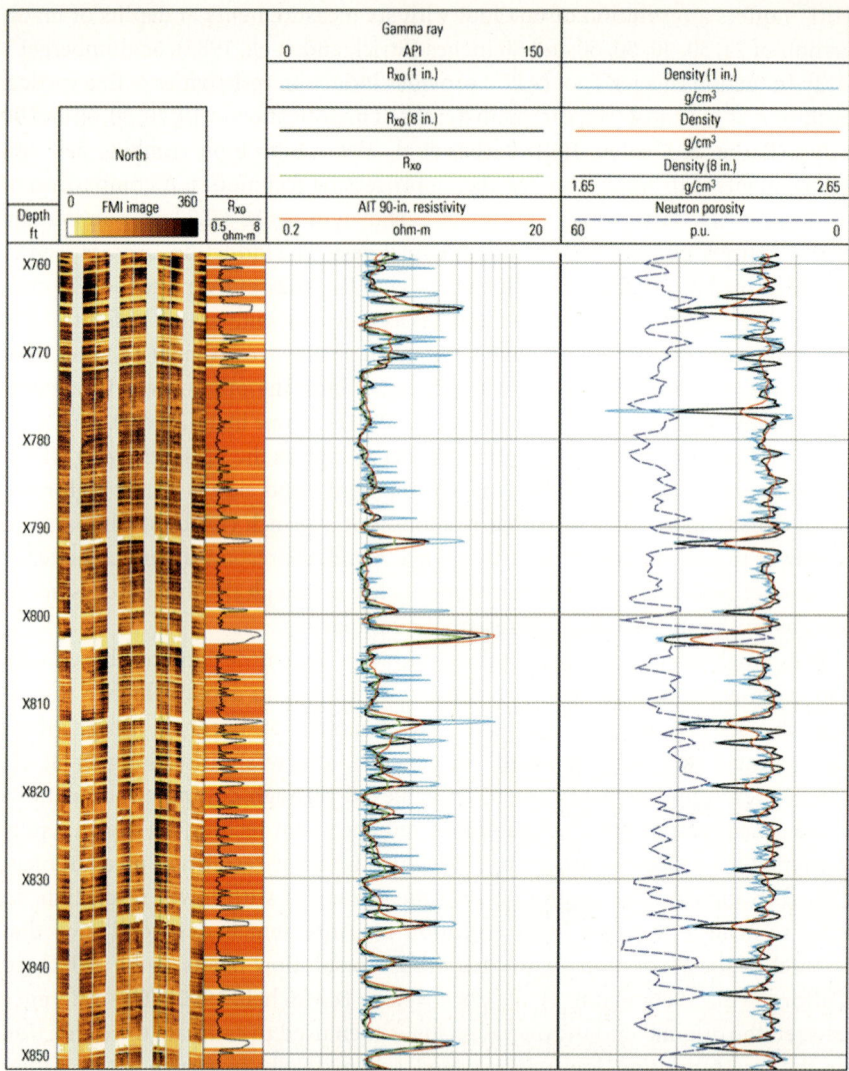

Figure 3.2.22. A thinly bedded sequence with complete suite of logs, including FMI borehole images (left). See text for discussion.

other processing techniques such as alpha-processing may be preferred (Looyestijn, 1982). However, alpha-processing itself has its own pitfalls, particularly if washouts are present in the borehole because they might affect the high-resolution reading but not the lower resolution log (Minette, 1990).

The *need for enhancing log resolution* can usually be assessed by comparing logs with different resolution. If the higher-resolution logs are much more active than those with lower resolution, particularly over potential reservoir intervals, there is a need for resolution enhancement. Figure 3.2.22, for example, shows that

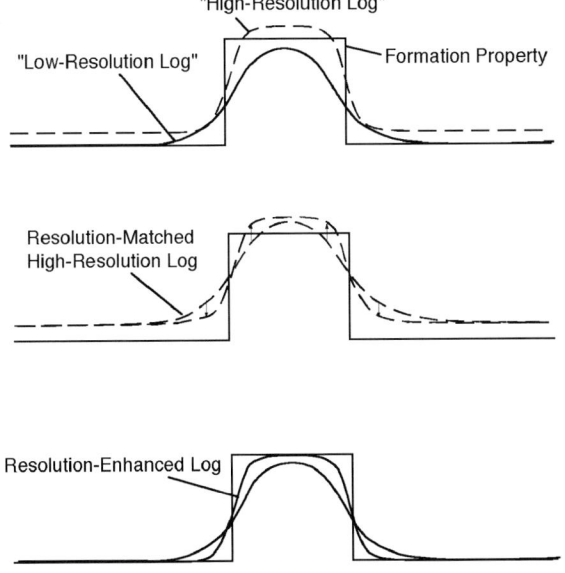

Figure 3.2.23. The principle of alpha processing. The high-resolution log gets matched to the low-resolution log (middle). The difference is used to sharpen the low-resolution log (bottom).

the neutron log is much more sluggish than the high-resolution resistivity logs, and that, therefore, it can benefit from resolution enhancement.

Enhanced-resolution logs provide better results when thresholding because they detect bed boundaries more accurately. They can also be used to improve petrophysical results, particularly for pay thickness and saturation estimation. Allen (1984) developed such a method, which he named "Laminated Sand Analysis". The neutron and density logs are first enhanced using their near-detector response as the high-resolution log, and are then used to compute the formation porosity. From the electromagnetic propagation tool (EPT*) the bound water saturation is computed (see chapter 1.3). The free-water saturations in these layers are then adjusted until the modeled induction tool response matches the measured induction log (which has a lower resolution than the other tools, see table 3.2.1). The approach provides a petrophysical evaluation with a resolution approaching that of the highest-resolution tool included in the computation. Allen (1984) found this method to work well in Tertiary clastics of the Gulf of Mexico. Dipmeters have been tried as high-resolution logs, but with limited success, as they cannot reliably distinguish water-bearing sand layers from shale streaks. Geochemical logs benefit from resolution enhancement, because their ability to estimate the cation exchange capacity can resolut in improved bound-water saturations.

Another method of resolution enhancement consists of determining the bed boundaries from a high-resolution log such as the FMI or FMS. These boundaries

---

* Mark of Schlumberger

can be determined interactively or with an automated boundary-detection algorithm (Trouiller et al., 1989). Figure 3.2.22, for example, shows sharp bed boundaries on the FMI and most other logs. The beds are then used to define a layered model for which properties are determined that explain the actual log reading. This is an inverse problem with a multitude of solutions, and therefore the process has to be user-guided, particularly regarding the choice of bed boundaries. Within each bed, a resistivity is assumed, for example the one corresponding to the deep resistivity log reading in the middle of the bed. The tool responses are then computed for this layered model and compared to the actual readings. If there are differences, the layer properties have to be adjusted until an acceptable fit is found for all logs. The resulting squared logs can be input into a petrophysical evaluation where porosities and saturations are computed. This in turn can be used for reservoir assessment.

An example of squared logs obtained with this procedure is shown in figure 3.2.24. There is a high apparent dip due to a significant borehole deviation, and all depths, starting with the bed boundaries, are referenced to the bottom of the

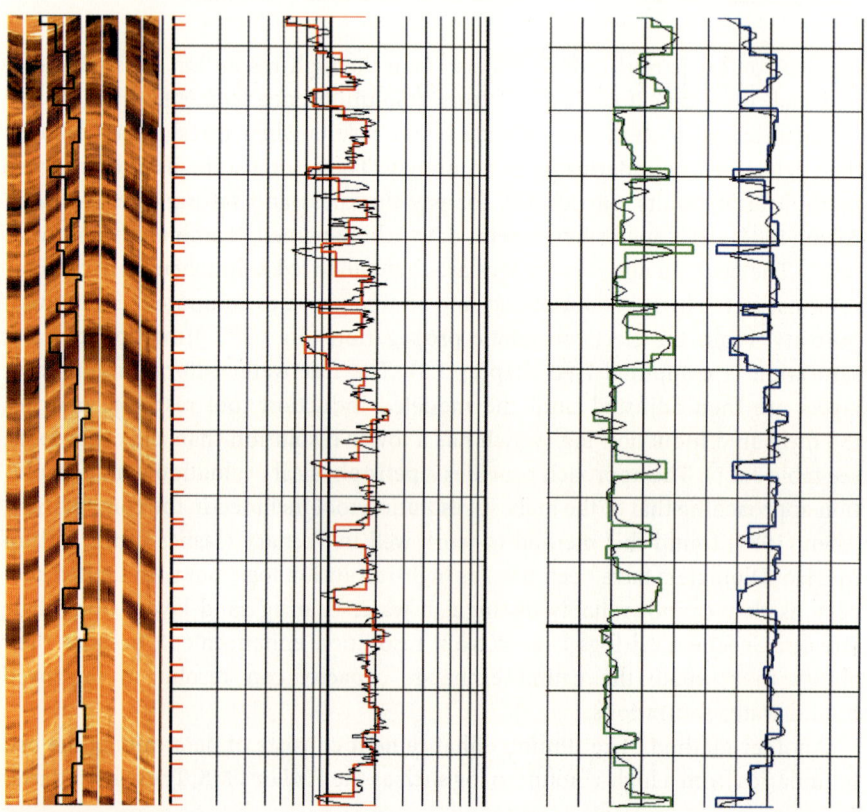

Figure 3.2.24. Original and high-resolution squared logs (middle-resistivity; right-density/neutron) obtained using FMI borehole images (left). Explanations in text.

hole. This is where the nuclear tool sensors are most likely to read because the tool follows the bottom of the hole, and the spring, located at the rear side of the sensors, causes the sensor to be oriented downwards. The bottom is where the sinusoids on the FMI image are highest, i.e. about in the middle of the image (or south). The layer boundaries are first detected on the FMI and are displayed with red barbs. The resistivity, density, and neutron logs are then given initial values (guided by the actual reading) for each layer. Subsequent adjustments of the values are made until a statisfactory match is obtained between modeled and measured log. The resulting squared logs are shown superposed onto the measured logs. They illustrate that significant differences occur in some thinner layers.

The example shown in figure 3.2.25 also uses a curve extracted from the FMI tool (left on the figure) to guide the evaluation. This curve has been thresholded into three classes, corresponding to shale, shaly sand, and sand respectively and indicated by different shades. This simple zonation can be used for a sand count where cores are not available. The petrophysical interpretation, first made with the actual measured logs, was then guided by this detailed, squared curve similar to the example in figure 3.2.24. These two petrophysical interpretations are shown next to each other. The results without resolution enhancement (middle track) suggest a steady albeit variable amount of clay over the entire zone. The

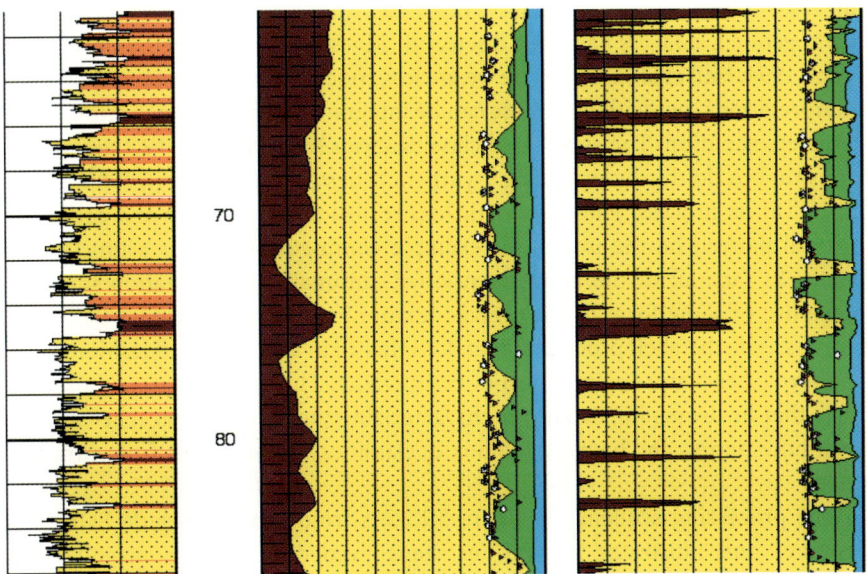

Figure 3.2.25. An example how resolution enhancement improves petrophysical results. The log on the left is obtained from the FMI; colors represent three thresholded classes. The conventional petrophysical results show a shaly sandstone throughout (middle), while the resolution-enhanced interpretation shows alternating thin sand and shale layers (right). Vertical scale in feet.

core porosity data, marked by circles and triangles, does not match well and is in general underestimated by the logs. The results using resolution-enhancement (right track) show a much sharper response, with well-defined shale layers separating clean sands. The porosities now match the core data much better, although in the upper part, where the sand beds become thinner, there is still a sizeable discrepancy. The net result of this evaluation is that from thresholding the sand count becomes more accurate, and from the petrophysical evaluation the porosities and the saturations are more accurate. In this case the enhanced-resolution analysis shows higher porosities and higher oil saturations. Similar results are reported by Ramamoorthy et al. (1995). If the additional sand count is taken into account, the resulting net oil estimation is considerably higher than with the conventional analysis. In "laminated" sands, the result is often – but not always – in the direction of an upward revision of the oil-in-place estimation (Reid & Enderlin, 1998).

3.2.3. Bedding from Image Texture

As discussed in chapter 2.2.5, microelectrical borehole images contain unique information on bedding characteristics *within* geological layers because of their high vertical resolution, and their two-dimensional nature. The patterns produced by various bedding types are reflected in the image texture on the borehole images, and several automated ways of segmenting them into zones corresponding to geological layers had been discussed (Rivest et al., 1992; Delhomme, 1992; Luthi, 1994; Hall et al., 1996). Here we are less concerned with theses tool-specific interpretations, but with how these can be integrated with other logs into a more complete zonation. The supervised classifications mentioned in 3.2.1 are an obvious tool to do this by including "synthetic" logs (Serra & Abbott, 1980) in the classification process. These synthetic logs record the occurrence of a particular textural pattern on the image, for example conductive spots, as a function of depth. In fact, all of the above-mentioned automated segmentation methods make use of some form of such image-texture logs. However, rather than providing an off-the-shelf procedure that does this all in an automated fashion, newer approaches feature a great deal of interactivity. The user can select those features that he deems important for characterizing the rock sequence, he can edit layer boundaries, and he has a toolbox of classification programs from which to select.

The processing sequence in figure 3.2.26 is from Schlumberger's BORTEX[*] program and illustrates the various steps needed to extract image-texture logs. The program can work with curves from dipmeters or from microelectrical images. After dip compensation, the components common to the various curves are retained and stacked into one single channel. This curve is then squared in

---
[*] Mark of Schlumberger

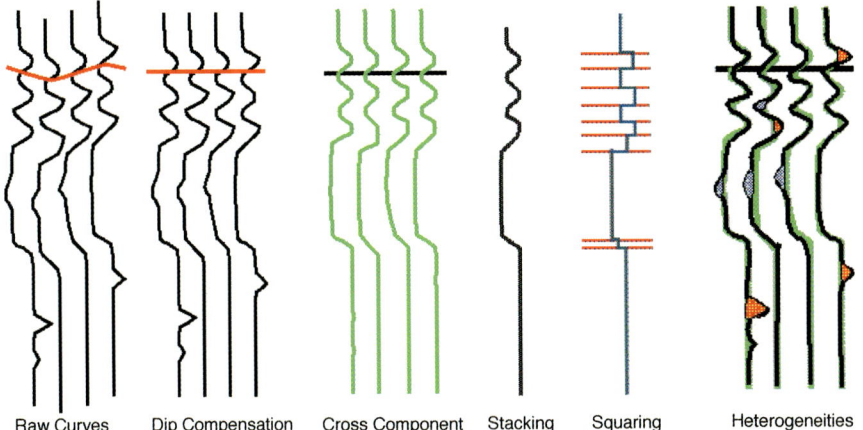

Figure 3.2.26. Extracting heterogeneities from dipmeter or borehole image curves. Curve stacking of dipmeter or borehole image data, and extracting the cross-component, can be used to obtain high-resolution squared logs such as in figure 3.2.24. The difference between the cross-component and the actual curves is a measure of the heterogeneity of the formation (right).

order to obtain bed boundaries and average "conductivities" for each layer. The result can be edited manually or changed by adjusting the parameters in the squaring algorithm. The residual between the cross-component signal and the raw curves can be used to quantify the number, size and shape of heterogeneities, which can be either resistive or conductive. More textural information can be extracted from the stacked conductivity curve, for example its variance, or "activity", over a sliding window. Figure 3.2.27 shows the segmentation obtained from a FMI after squaring the stacked average conductivity curve (middle track). The resulting layer zonation is shown together with some textural curves in the right track. The user may prefer grouping all thin layers in the highly resistive lower half of the interval into one or two layers, or he may leave it because he is looking for subtle textural differences in this potential reservoir zone. For each layer now, the open-hole logs such as the gamma ray, the density, or the neutron log, can be analyzed to obtain an electrofacies, in a supervised or unsupervised classification scheme. The stacked conductivity curve is used to assign thin beds, for example shale streaks in a sandstone that may be too thin to be resolved by the open-hole logs.

Further processing of the residual image or dipmeter curves yields the textural curves that can be used, in a supervised classification, to obtain "morphofacies", i.e. a facies zonation based on the image morphologies. Two types of curves can be distinguished:

- Curves containing information on the previously defined beds, such as bed thickness, or average bed conductivity; and

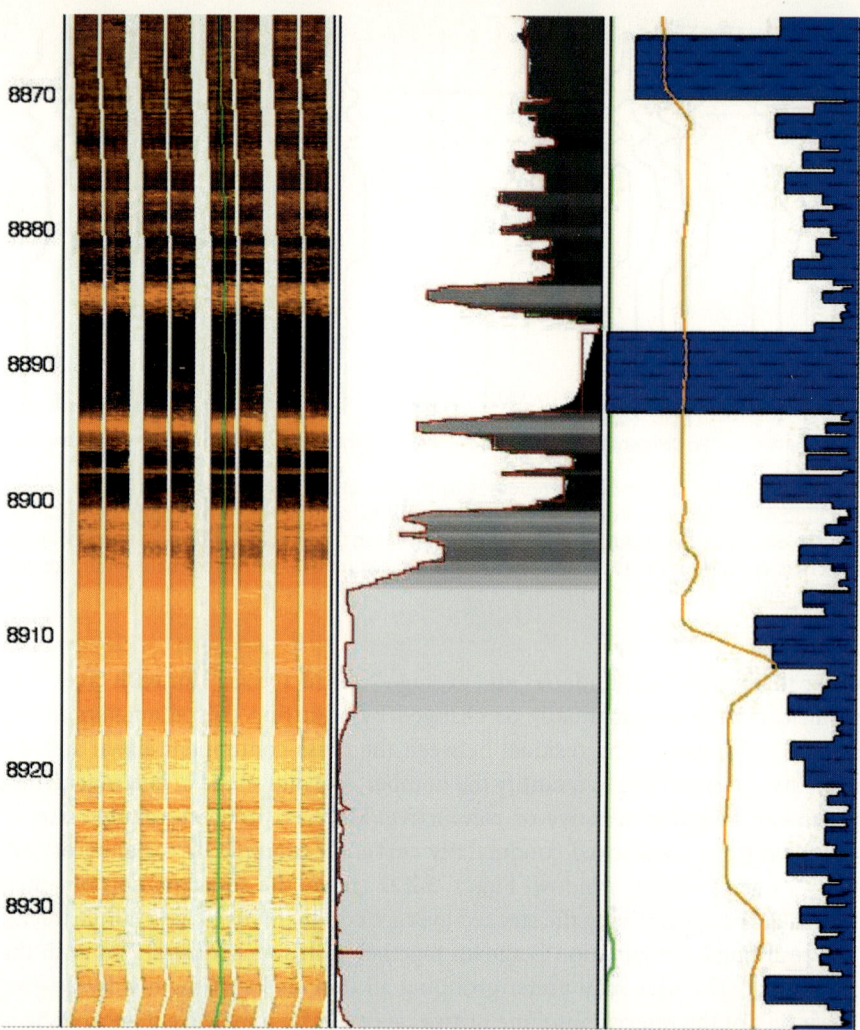

Figure 3.2.27. Results obtained from the processing sequence in figure 3.2.26. The original FMI (left) is subdivided into thin layers of constant conductivities which are then grouped into "squared" logs limited by the bed boundaries (middle). The squared log is used to display the bed thickness (right). Other logs reflect heterogeneities within beds, and the number of conductive events.

- Curves containing image-textural information irrespective of the bed boundary locations, for example the size and connectivity of conductive or resistive spots on the image.

Both types of curves can be computed as a sliding average and then be clustered into a user-defined number of such morphofacies. In the example shown in figure 3.2.28, the left track contains bed-pertinent curves, while the two tracks in the

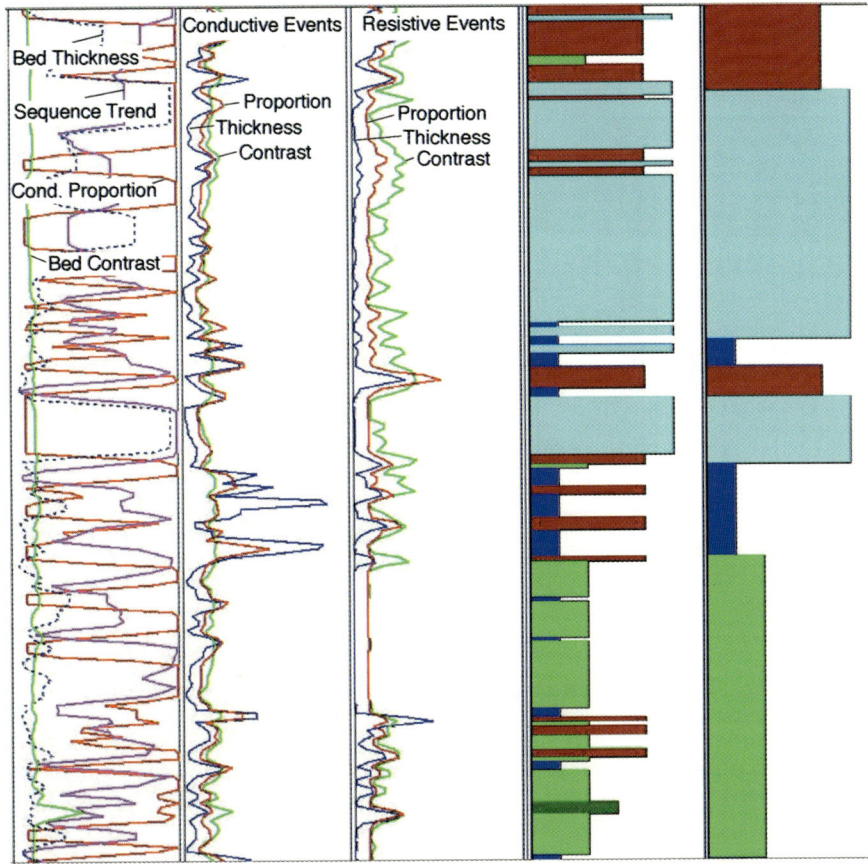

Figure 3.2.28. "Textural logs" (left) and resulting zonation, without and with vertical continuity logic (right).

middle contain textural information on conductive and resistive events (spots, thin layers etc.) found on the images. Some of these metrics are related to each other, but in complicated ways depending on the algorithms used to compute them. The two tracks on the right show the resulting zonation into five morphofacies after clustering, with and without vertical continuity logic.

The interactive nature of the approach is illustrated by the screen snapshot in figure 3.2.29. The raw data (in this case dipmeter curves) are shown in one window, while in other windows the interpreter has cross-plots of the data points that help him decide in the clustering phase, results of the zoning, and typical intervals illustrating each morphofacies.

The approach discussed here may be somewhat conceptual, largely because the methodology as well as the logs are quite recent developments. However, an increasing number of interesting field examples is being reported. Delhomme et

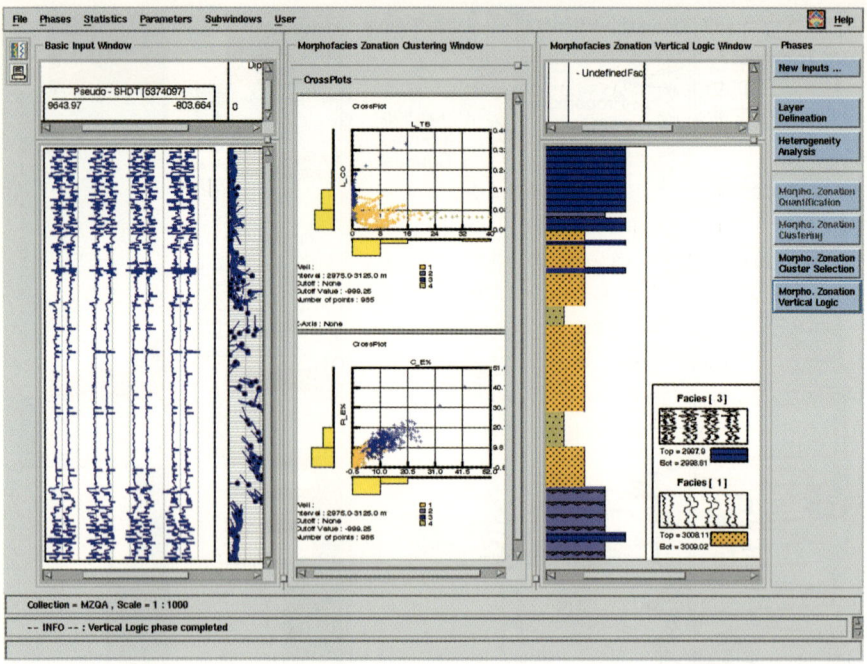

Figure 3.2.29. Screen shot illustrating the interactive nature of textural zonation of dipmeter curves on a workstation. The dipmeter data are in the left window, while cross-plots of textural curves properties are in the middle window. The right window shows a textural zonation together with sample intervals of the typical curve characters for the different texture classes.

al. (1996), for example, have looked at the anisotropy as derived from the analysis of FMI images in a gas reservoir. The estimated the horizontal and vertical conductivities from the images using an averaging procedure over an area of 6×6 cm. They then computed the electrical anisotropy as the ratio of the two conductivities in the form of a log sampled every 6 centimeters. Comparing the results on several sands with flow data from production logs and nuclear magnetic resonance data (chapter 2.6), they found that the laminated sands had much better productivities than the sands with a mottled aspect on the images, which was attributed to slumping or bioturbation.

3.2.4 Conclusions

An important development is that there has been a significant move away from automated zonation programs towards a more interactive approach. Rather than relying on one single algorithm, a toolbox of several computational methods is put at at the user's disposal. Modern geological well logs contribute significantly to a better, more robust geological zonation that takes into account a wider range of rock properties. In chapter 3.4 we discuss ways of performing field-wide zonation and correlation across a field to obtain a geological model. When combined with petrophysical data and dynamic reservoir data, these geological models provide the basis for the upscaling process done prior to reservoir simulation.

The discussion on tool resolution and thin bed analysis illustrates that much progress has been made in an area that may become ever more important, particularly with an increasing number of new fields located in turbidites on the continental slopes. Three-dimensional reconstruction of such sedimentary systems is important in order to assess well productivities and reserves in place. Layer-cake models may look appropriate if only a single well is considered, but are inadequate on a field-wide basis. High-resolution logs can provide one of the keys to build better geological models in these reservoirs.

References

Allen DF (1984) Laminated sand analysis. Trans 25th Symp Soc Prof Well Log Analysts: Paper XX.
Barber TD (1988) Induction tool vertical resolution enhancement – physics and limitations. Trans 29th Symp Soc Prof Well Log Analysts: Paper O.
Barber TD, Rosthal RA (1991) Using a multiarray induction tool to achieve high-resolution logs with minimum environmental effects. 66th Ann Conf Soc Petr Eng: Paper 22725.
Barber TD, Orban A, Hazen G, Long T, Schlien R, Alderman S, Seydoux J (1995) A multi-array induction tool optimized for efficient wellsite operation. 70th Ann Conf Soc Petr Eng: Paper 30583.
Bouma AH (1962) Sedimentology of some flysch deposits. Elsevier, Amsterdam.
Davis JH (1986) Statistics and data analysis in geology. John Wiley Sons, New York.
Delfiner PC, Peyret O, Serra O (1987) Automatic determination of lithology from well logs. Soc Petrol Eng Formation Evaluation 3: 303–310.
Delhomme, JP (1992) A quantitative characterization of formation heterogeneities based on borehole image analysis. Trans 33rd Symp Soc Prof Well Log Analysts: Paper T.
Delhomme JP, Bedford J, Colley NM, Kennedy MC (1996) Permeability and porosity upscaling in the near-wellbore domain: the contribution of borehole electrical images. 71st Ann Conf Soc Petr Eng: Paper 36822.
Doveton JH (1994) Geologic log analysis using computer methods. Am Assoc Petrol Geol Computer Applications in Geology 2.
Eisenmann P, Gounot MT, Juchereau B, Trouiller JC, Whittaker S (1994) Improving Rxo measurements through semi-active focusing. 69th Ann Conf Soc Petr Eng: Paper 28437.
Flaum C, Galford JE, Hastings A (1989) Enhanced vertical resolution processing of dual detector gamma-gamma density logs. The Log Analyst 30: 139–149.
Galford JE, Flaum C, Gilchrist WA, Duckett SW (1986) Enhanced resolution processing of compensated neutron logs. 61st Ann Conf Soc Petr Eng: Paper 15541.
Goligher A, Scanlan B, Standen E, Wylie AS (1996) A first look at Platform Express measurements. Oilfield Review, Summer issue: 5–15.
Hackbarth CJ, Tepper BJ (1988) Examination of BHTV, FMS, and SHDT images in very thinly bedded sands and shales. 63rd Ann Conf Soc Petr Eng: Paper 18118.
Hall J, Ponzi M, Gonfalini M, Maletti G (1996) Automatic extraction and characterization of geological features and textures from borehole images and core photographs. Trans 37th Symp Soc Prof Well Log Analysts: Paper CCC.
Howell DG, Normark WR (1982) Sedimentology of submarine fans. In: Scholle PA, Spearing D (eds) Sandstone deposition environments. Am Ass Petrol Geol Memoir 31, 365–404.

Looyestijn WJ (1982) Deconvolution of petrophysical logs: Applications and limitations. Trans 23rd Symp Soc Prof Well Log Analysts: Paper W.

Luthi S (1981) Experiments on non-channelized turbidity currents and their deposits. Marine Geology 40: M59–M68.

Luthi SM (1994) Textural segmentation of digital rock images into bedding units using texture energy and cluster labels. Math Geol 26: 181–196.

Malinverno A (1997) On the power law size distribution of turbidite beds. Basin Research 9: 263–274.

McGann GJ, Riches HA, Renoult DC (1988) Formation evaluation in a thinly bedded reservoir. A case history: Scapa field, North Sea. Trans 29th Symp Soc Prof Well Log Analysts: Paper V.

Minette D (1990) Thin bed resolution enhancement: potential and pitfalls. Trans 31st Symp Soc Prof Well Log Analysts: Paper GG.

Nelson RJ, Mitchell WK (1990) Improved vertical resolution of well logs by resolution matching. Trans 25th Symp Soc Prof Well Log Analysts: Paper JJ.

Normark WR (1978) Fan valleys, channels, and depositional lobes on modern submarine fans: characters for recognition of sandy turbidite environments. Bull Am Assoc Petrol Geol 62: 912–931.

Press WH, Teukolsky SA, Vetterling WT, Flannery BP (1992) Numerical recipes. Cambridge Press.

Ramamoorthy R, Flaum C, Coll C (1995) Geologically consistent resolution enhancement of petrophysical analysis using image log data. 70th Ann Conf Soc Petr Eng: Paper 30697.

Reid R, Enderlin R (1998) True pay thickness determination of laminated sand and shale sequences using borehole resistivity image logs. Trans 39th Symp Soc Prof Well Log Analysts: Paper GGG.

Rivest JF, Beucher S, Delhomme JP (1992) Marker-controlled segmentation: an application to electrical borehole imaging. J Electron Imaging 1(2): 136–142.

Rothman DH, Grotzinger JP, Flemings P (1994) Scaling in turbidite deposition. J Sedim Res A64: 59–67.

Schlumberger (1997) ITALY 2000 – Value-added reservoir characterization. Schlumberger Italiana SpA.

Serra O, Abbott HT (1980) The contribution of logging data to sedimentology and stratigraphy. Soc Petrol Eng Ann Conf: Paper 9270.

Stow I, Hock M (1988) Facies analysis and diagenesis from well logs in hte Zechstein carbonates of northern Germany. Trans 29th Symp Soc Prof Well Log Analysts: Paper HH.

Strickland R, Sinclair P, Harber J, DeBrecht J (1987) Introduction to the high resolution induction tool. Trans 28th Symp Soc Prof Well Log Analysts: Paper E.

Sullivan KB, Schepel KJ (1995) Borehole image logS: Applicati9ons in fracutred and thinly bedded reservoirs. Trans 36th Symp Soc Prof Well Log Analysts: Paper T.

Theys P (1999) Log data acquisition and quality control. Editions Technip.

Trouiller JC, Delhomme JP, Carlin S, Anxionnaz H (1989) thin-bed reservoir analysis from borehole electrical images. 64th Ann Conf Soc Petr Eng: Paper 19578.

Weber KJ, Geuns LC van (1990) Framework for constructing clastic reservoir simulation models. J Petrol Technol: 1248–1297.

Wolff M, Pelissier-Combescure J (1982) FACIOLOG – automatic electrofacies determination. Trans 23rd Symp Soc Prof Well Log Analysts: paper FF.

3.3 Fractured Reservoir Analysis

3.3.1 Introduction

Fractured reservoirs are among the most prolific reservoirs in the world because of the fractures' ability to sustain very high flow rates. Examples where fracturing provides the primary production mechanism include the Tertiary Asmari formation in the fields of Southwest Iran (Lees, 1933; also Levorsen, 1967), the Maastrichtian/Danian chalks of the Ekofisk and other fields in the North Sea (Van den Bark & Thomas, 1980, Fritsen & Corrigan, 1990), the Permian Spraberry formation in West Texas (Wilkinson, 1953), the granitic basement in Southeast Asia (Areshev et al., 1992; Tandom et al., 1999), the Austin Chalk in Texas, the Monterey formation in California, and many more. In all cases, a fracture network has been established through shear or tensile rupture of the rocks by tectonic forces. Often, the actual hydrocarbon reservoir is provided by matrix porosity, not the fracture system, but the fractures are crucial in providing a permeable conduit to drain the fluids – a mechanism equally used in hydraulic fracturing. The matrix porosity may be very high as in the case of the North Sea chalk, where it reaches over 40 %, or very low as in the basement rocks where it typically is no more than a few percent. Reservoir engineers call these "dual porosity systems" and have found various ways of simulating the flow in such reservoirs (Van Golf-Racht, 1982). Since fractures, and particularly large-scale shear zones, can extend very far, water breakthrough can occur very rapidly. It is, therefore, important to drain the reservoir carefully despite the temptation of having a high production from only a few wells. Completions are often bare-foot, i.e. without casing in the reservoir, because the fractures communicate well enough with each other and the wellbore. In cases where a dominant fracture orientation is established and the fractures are more or less vertical, horizontal wells oriented at a right angle to the fracture strike have proven successful as they intersect a maximum number of fractures.

Nelson (1985) gives an account of the geological analysis of naturally fractured reservoirs. The traditional analytical techniques of such reservoirs in wellbores are well tests and cores. Classical open-hole logs such as porosity, resistivity, and gamma ray measurements have at best a subtle response to fractures. Only resistivity logs, particularly the laterologs, have been used with moderate success, because the different current paths between the deep and the shallow measurement are affected in a different way by fractures (Sibbit and Faivre, 1985). In recent years, however, a number of well log techniques have become available which can provide important parameters to characterize fractured reservoirs. These include

- Borehole images (electrical and acoustic) can provide *fracture morphology, dip, azimuth and aperture* (Cheung & Heliot, 1990; also chapters 2.2 and 2.3 in this book)

- Stoneley wave reflections from array sonic logging can provide *fracture location and aperture* (Hornby et al, 1989). This technique is particularly powerful when combined with borehole images (Hornby and Luthi, 1990).
- Wireline testing with a dual-packer system allow selective testing of the *fracture productivity* once identified on borehole images. This technique is schematically sketched in figure 3.3.1.
- Production logging with spinners also allows determining *fracture productivity*, particularly when processed with techniques that accurately determine the inflow from each flowing interval (often point entries). Comparison of the individual fracture flow rates with their geometry can lead to important insights into the fracture pattern.

In the following two case studies of fractured reservoirs are discussed, one from carbonates and one from basement rocks, with the goal of demonstrating the use of new well logging techniques in fractured reservoir analysis.

3.3.2 A Fractured Carbonate Case Study

This example is from a field in the Mediterranean region, which contains thick sequences of Mesozoic carbonates that form prolific oil reservoirs in many countries. Despite low porosities of the reservoir rocks, the wells are highly productive, because of abundant fracturing that occurred during the Tertiary compressional phase and that may have been enhanced during subsequent episodes of uplifting and partial subaerial exposure.

The scope of study is to classify the various types of fractures and fracture-related features based on their morphology as seen on electrical borehole images. Additionally, the relationship of fracture types with production data obtained from well tests is analyzed in order to develop a reservoir model, which will be helpful in designing an appropriate production strategy.

Data Used. Aside from a standard suite of open-hole logs, electrical borehole images recorded with the FMI[*], and flow rates from well tests over numerous intervals were available. Borehole images were acquired over a total interval of 1800 meters. They are generally of good quality, although some sections are washed out by several inches. The images were processed with the standard processing chain (chapter 2.2.2) and subsequently displayed on an workstation for interactive analysis, notably for fracture typing, measurements of fracture dips and azimuths, and calculating fracture apertures. A few cores were taken with good recovery; they were used to relate the logs and images to lithofacies types. The well is vertical.

---

[*] Mark of Schlumberger

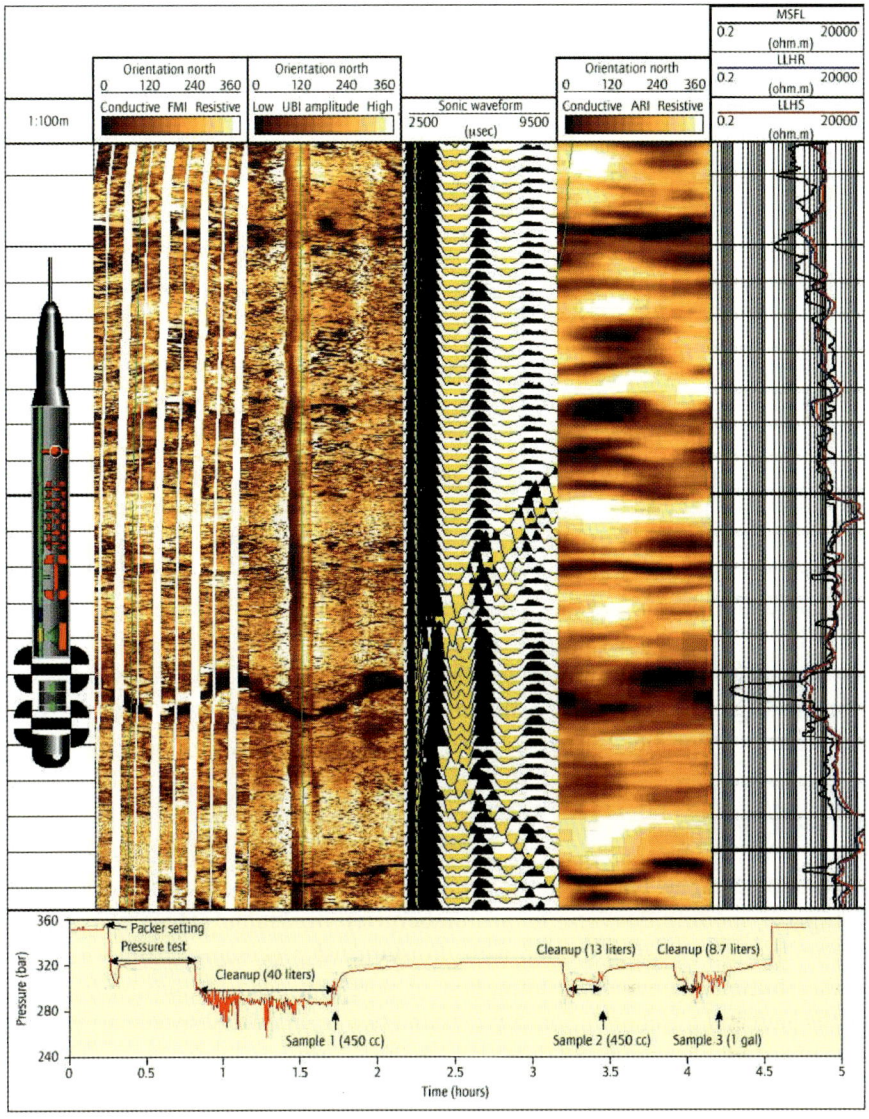

Figure 3.3.1. Probing a fracture with the dual packer MDT wireline testing tool. The measurement cycles consists of several pressure tests followed by cleanups at the end of which fluid samples are taken (bottom). The logs are from left to right: FMI, UBI, waveforms from the DSI array sonic tool, ARI, and HALS laterolog resistivities[*]. Depth grid line spacing is one meter.

---

[*] Marks of Schlumberger

Table 3.3.1. The six producing intervals identified from well tests. A is about 1500 meters above F.

Interval	Thickness (m)	Production (bbls/d)
A	5	10,000
B	3	320
C	3	480
D	8	⎫
E	5	⎬ 750
F	9	⎭

Geological Background. The reservoir section covers 1800 meters, ranging in age from Barremian to Senonian. It contains primarily platform carbonates, mostly limestones but occasional dolomite sections occur as well. Tertiary clastics provide an unconformable cover acting as a seal. The anticlinal structure is contained within a strike-slip zone, forming a flower-type structure with dips of up to 40 degrees towards the bounding faults on each side.

Well Production. The well was tested over many sections and the results, indicated the producing intervals shown in table 3.3.1, ranked from top to bottom. These intervals were given particular attention in this study. As can be seen, this well has a very high flow rate, which is predominantly from the uppermost zone of the well. A major goal of this study is to understand why this zone is so prolific.

Fracture Analysis

Methodology. The FMI images were interactively analyzed on a workstation using the module BorView in Schlumbergerís GeoFrame software. For this, the images were first processed with the standard processing chain, which includes button alignments, elimination of potential dead buttons, corrections of the gain and offset of button currents and two separate tool speed corrections, one using the accelerometer data from the tool, and one using the images. The images were then loaded into the BorView module, which allowed interactive inspection, classification, and quantification of the various fracture types (Cheung & Heliot, 1990).

Fracture Classification. Since much of this analysis relied on borehole image data, the fracture classification is essentially *morphological*. It contains not just fractures, but fracture-related features such as breccias and faults. Six such "fracture" classes were distinguished and are schematically illustrated in table 3.3.2. Some of the fracture types have a well-defined relationship to bedding, notably the bedding-confined fractures and some of the solution-enhanced fractures, which follow the bedding surfaces. Therefore, bedding planes were also measured at more or less regular intervals, such that their relationship to fracturing could be established.

Table 3.3.2. Fracture classes, their morphological expression and the measurements performed on each class. Colors refer to subsequent figures.

Fracture Class	Morphology	Measurements	Color
Planar Fractures		Dip/Azimuth/Aperture	Blue
Solution-Enhanced Fractures		Dip/Azimuth/Aperture	Purple
Bedding-Confined Fractures		Dip/Azimuth (if possible)	Orange
Wide Conductive Zones		Dip/Azimuth	Red
Brecciated Zones		Dip/Azimuth (if possible)	Yellow
Induced Fractures		Azimuth (Dip is ~90°)	Black

Because the fracture classification adopted here is morphological and based on visual inspection, it is subjective and different interpreters might obtain different classifications. Not all fractures visible on the FMI images were marked and measured. In heavily fractured zones, notably with natural, solution-enhanced and bedding-confined types, only the major fractures were picked. In the brecciated zones, hardly any planar features exist and thus no reliable measurement of either bedding or fracturing is possible. The emphasis in these zones was placed on picking a few features such that the zone shows up on a composite plot.

Although the classification is essentially morphological, attributing a genetic origin of the various features is still possible, albeit with variable degrees of certainty. Table 3.3.3 contains this genetic classification for each fracture type.

The following contains a brief description of the six fracture classes identified in this well. Figures 3.3.2 and 3.3.3 illustrate these features from various intervals of the well. Most of these intervals show a predominance of one type of feature, but it is common for several types of fractures to coexist at a given interval.

Planar Fractures (fig. 3.3.2 left): These are almost certainly caused by tectonic forces, notably shearing and tension. Aperture calculations can quite reliably be performed on them since they correspond best to the modeled case. However, they are relatively rare.

Table 3.3.3. Presumed genetic origins of the six fracture classes.

Fracture Class	Possible Origins
Planar Fractures	Natural fractures caused by tectonic forces (shearing, tension)
Solution-Enhanced Fractures	Dissolution-enlargement of natural fractures and of bedding planes
Bedding-Confined Fractures	Tectonic forces (tension); fractures terminate at bedding planes
Wide Conductive Zones	Unclear; can be shattered shear zones, leached fractures etc.
Brecciated Zones	Brecciation in fault zones; collapse breccias in karstic caves
Induced Fractures	Tension fractures caused by rock stresses and mud pressure in borehole

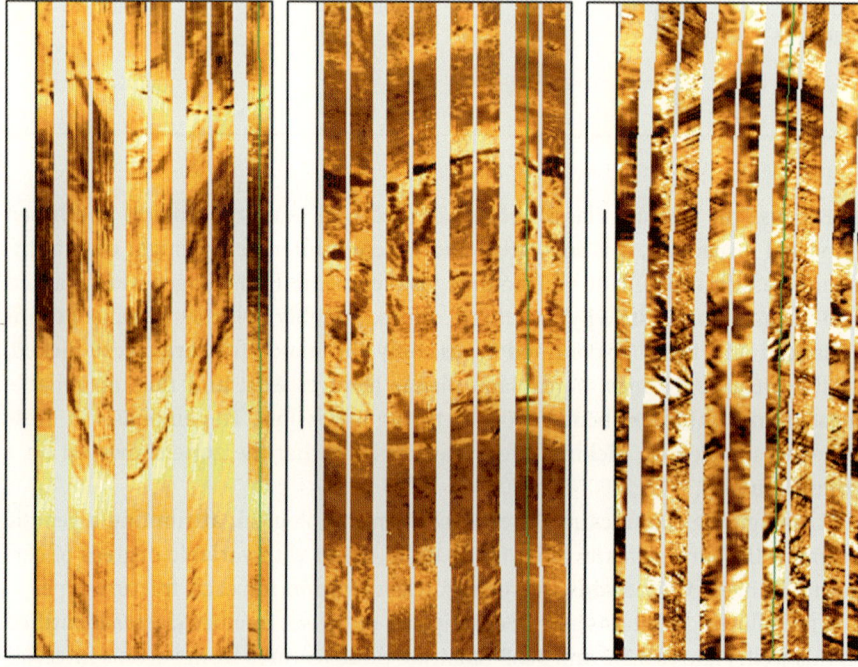

Figure 3.3.2. Left: Planar and some solution-enhanced fractures. Middle: Solution-enhancements following bedding planes. Right: Bedding-confined fractures. Notice how they terminate at the bedding boundaries. Scale bars are one meter.

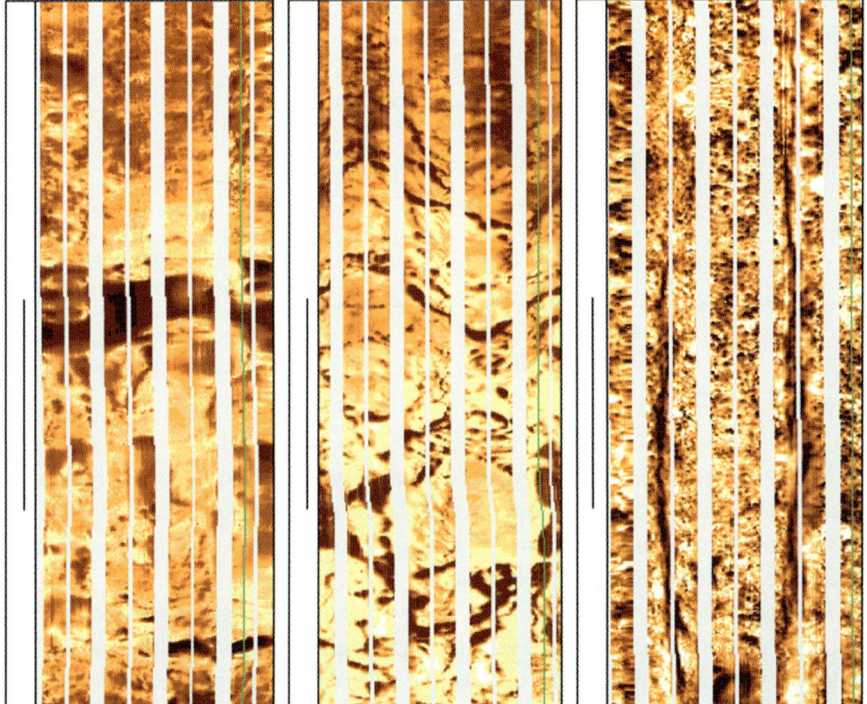

Figure 3.3.3. Left: Wide conductive feature. Middle: A brecciated zone with solution-enhanced fractures. Right: Induced vertical fractures striking approx. E-W. Scale bars are one meter.

Solution-Enhanced Fractures (fig. 3.3.2 middle). Most of these were formerly planar, natural fractures, but subsequent water circulation leached and cemented them in variable degrees such that they show a large variation in openings along their trace. Although apertures vary, they can be calculated. Some seem to follow bedding planes.

Bedding-Confined Fractures (fig. 3.3.2 right). These relatively rare fractures occur where distinct and frequent bedding planes occur. They are perpendicular to the bedding planes and, as observed in outcrops, usually denser in thinner beds. They form an intersecting fracture network.

Wide Conductive Zones (fig. 3.3.3 left). These purely morphological features may be anything from "superfractures", i.e. widely open fractures, or fracture zones, to porous, conductive layers of uncertain origin.

Brecciated Zones (fig 3.3.3 middle). These zones are quite characteristic on the images because of the large clasts, generally surrounded by conductive material. Often there seem to be abundant solution channels, fissure, fractures and caves, and sometimes there is evidence of faulting. Two different types occur, tectonic breccias and collapse breccias.

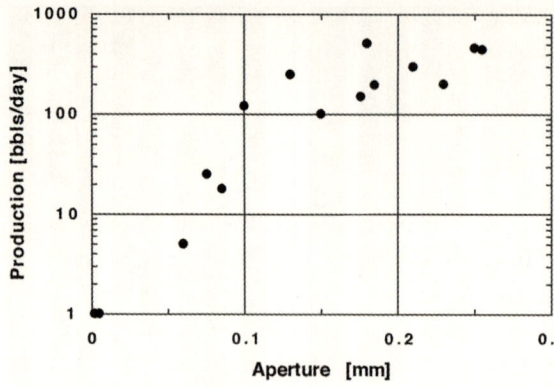

Figure 3.3.4. Fracture aperture versus oil production from wells in Texas, USA. The data points follow roughly the cubic law suggested by theory (Courtery M. Grace and B. Newberry).

Induced Fractures (fig. 3.3.3 right). These features are very characteristic and therefore a genetic name has been given. They are vertical to subvertical, aligned in the direction of maximum stress, and are caused by tensional failure due to excessive mud pressure in the borehole.

Fracture Aperture Calculation. Fracture apertures are important because the permeability of a fracture scales with the square and the flow with the cube of the fracture aperture (chapter 2.2.5). Thus, the difference between a fracture 100 micron wide and one only 10 micron wide can amount to a factor of about one thousand in terms of production. The crossplot in figure 3.3.4 shows a comparison of fracture aperture data against fracture production compiled from several wells in Texas. This data set has been very carefully analyzed regarding both the fracture aperture calculations as well as the individual productions. The flow rates follow roughly a cubic law, although downhole pressure varies by about a factor of three, illustrating the paramount importance of fracture aperture. Notice that production seems to become significant only once fractures reach several tens of microns in width.

The fracture apertures in figure 3.3.4 have been calculated using the previously described technique of Luthi & Souhaité (1990). It is well suited to the conditions in this well because it was drilled with water-based mud and the apertures seem to be in the correct range, i.e. between a few microns to about one millimeter. Above these apertures, the Stoneley reflection method described by Hornby et al. (1989) has to be used. The Azimuthal Resistivity Imager (ARI), responds to fractures in a similar fashion as the FMI and can be used for wider fractures as demonstrated by Faivre (1993).

The modeling of fracture aperture responses of the FMI has only been done for planar, "infinite" fractures. Therefore, among the fracture classes shown in table 3.3.2, only planar fractures qualify in the strictest sense. However, since most fractures seem to be solution-enlarged to some degree, the algorithm was also applied to them. The "wide conductive features" can be considered planar and "infinite", perhaps even more so than some planar fractures, but it is impossible to determine a background or matrix current due to the wide nature of the features.

Results

Figure 3.3.5 shows an example of the results of the fracture analysis over zone E on a detailed scale. The type of fracture is indicated in colors as indicated in table 3.3.2. The apertures are shown on a logarithmic scale because of the wide range found in such reservoirs. Both the hydraulic as well as the mean aperture were calculated, with the hydraulic aperture having higher values because it is computed as the cubic mean. Additionally, fracture dips and azimuths are displayed for those fractures where they could be determined. Bedding over the entire interval was found to dip between 20° and 40° towards NE.

Figure 3.3.5. Fracturing in zone E, showing an abundance of SSW-dipping fractures intersected by a major N-dipping feature interpreted to be a fault with associated fracturing and tectonic brecciation. Left: Images with fracturing traced by a semi-automated computer program. The colors correspond to the fracture apertures. Middle: Dip and azimuth of fractures in standard tadpole presentation. The colors correspond to the fracture types (yellow – brecciation; blue – planar fractures; purple – solution-enhanced fractures). Right: Mean fracture apertures on a logarithmic scale from 0.1 (left) to 1000 microns (right). Two averaging methods have been used, a linear mean and a hydraulic mean using a cubic law (higher values). Scale bar on left is one meter.

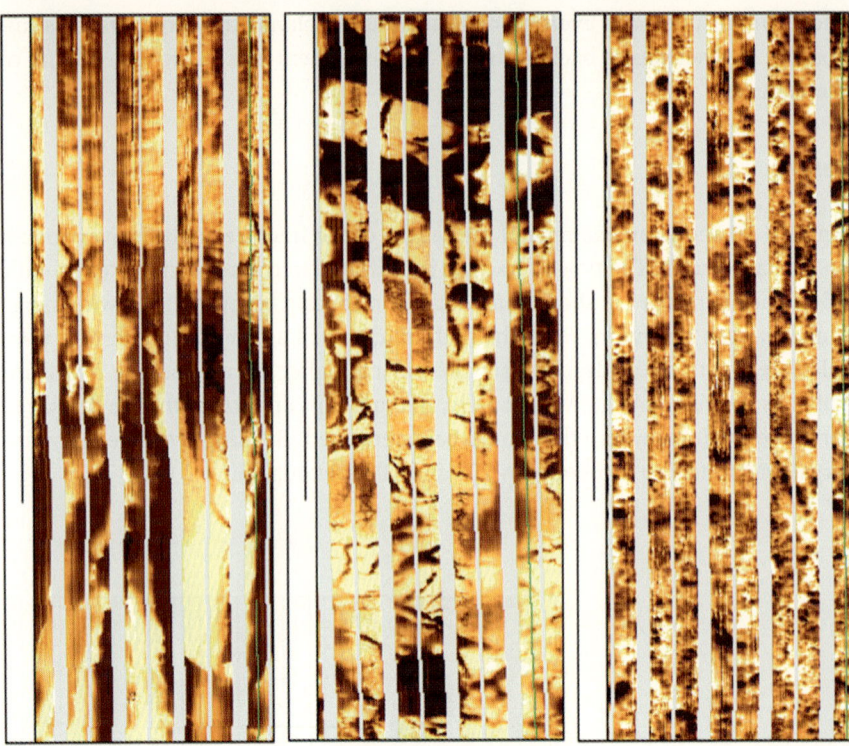

Figure 3.3.6. Images of the main producing intervals. Left: Detail of zone A, which is highest in well and most productive. Steep conductive features are interpreted as faulting, with secondary fracturing evident above and below. Middle: Brecciation with associated fracturing, typical for zones D, E and F. Right: Circular conductive patches in dolomitic, productive intervals B and C, interpreted to be vugs.

Zone A deserved major attention because it produces over 80 % of the total oil. It is located a few tens of meters below the Tertiary cap rock. Inspection at various image enhancements and scales reveal this zone to be very complex, containing abundant brecciation and steep NW-dipping natural fractures (figure 3.3.6 left). It probably is a *partially leached or karstified fault zone*. No fracture aperture calculations are possible within this zone, as it is too wide and irregular. Above this zone there are abundant solution-enhanced, narrowly spaced fractures dipping steeply towards SE as shown in the summary plot in figure 3.3.7. Below this zone, there are abundant solution-enhanced fractures dipping steeply towards W and NW. This change in dip azimuth supports the presence of a fault. Apertures of these fractures are small and range from 1 to 10 microns, but the fault zone itself has certainly openings many times wider than this.

Over the remainder of the well, there are numerous *solution-enhanced fractures* (figure 3.3.2 middle). They often dip at around 60° towards SW, implying that they are perpendicular to bedding. However, other orientations occur as

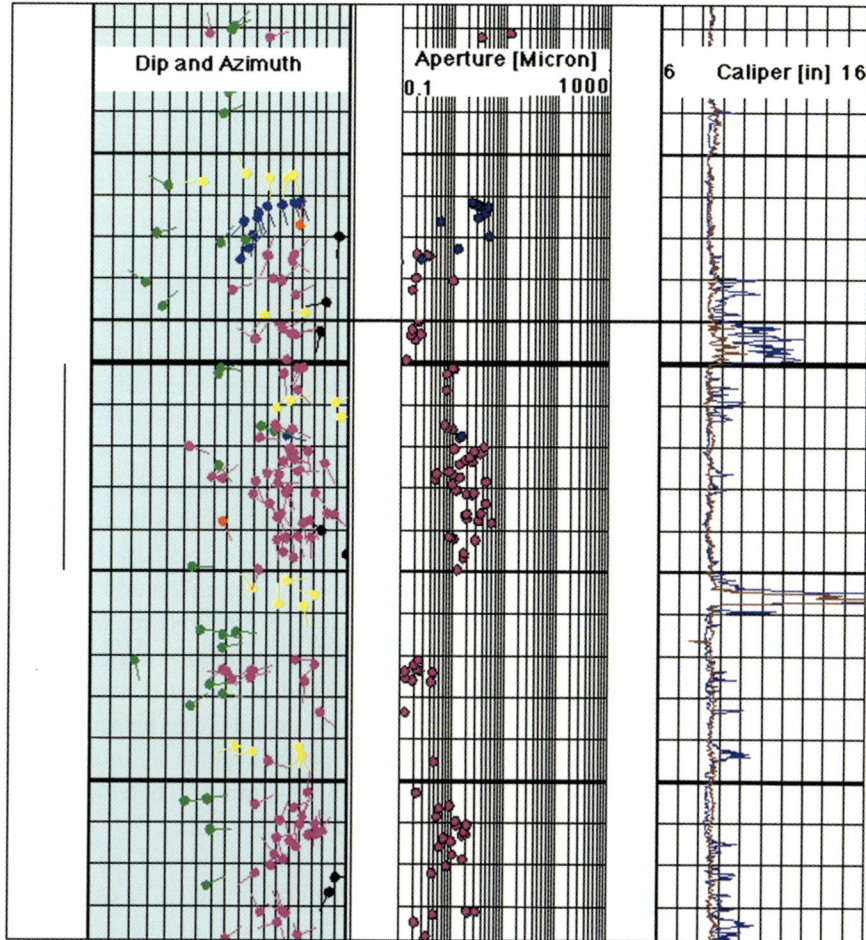

Figure 3.3.7. Result overview over approximately 250 meters in the upper part of the well. Notice the vertical clustering of brecciated zones (yellow), usually associated with washouts as indicated by the calipers. These zones are not productive unless associated with abundant planar and/or solution enlarged fractures. Such a zone is seen in the upper part. It corresponds to zone A and is the most prolific zone in the well. Scale bar on left corresponds to 50 meters.

well, particularly in the lower portion of the well. The fracture apertures of this type are generally small, i.e. below ten microns.

Of particular interest is the occurrence of *brecciated zones*. Two types were distinguished, with the first one showing abundant clasts but no particular association with fracturing (figure 3.3.2 middle). It is often associated with large hole washouts, and it is interpreted to be a karstic collapse breccia, although some zones might be of depositional origin and some might have a tectonic component. None of them seem to produce oil.

By contrast, the second type of brecciation is associated with abundant fracturing (figure 3.3.6 middle), often with a very dense network of either planar or solution-enhanced fractures, all dipping in the same direction. In these zones we find invariably steeply dipping conductive features which we believe to be fault planes (figure 3.3.5). Their strike direction is often parallel to the associated fracturing. No hole washouts occur in these zones. They are likely to *be fault zones with tectonic brecciation and shear fracturing*. Examples are zones D, E, and F. All these zones are productive, and all except for the top zone dip steeply toward south. Their fracture apertures are not necessarily higher than in the surrounding zones, only in zone E they reach their maximum values of a little over 100 microns, which certainly accounts for the production from this zone (figure 3.3.5). In the others, the fault planes – which were not quantified in terms of apertures for reasons described earlier – and the abundance of thin fractures are likely to account for the production.

The local occurrence of bedding-confined fracturing and of the wide conductive zones does not seem to have any relationship to production, although these features might potentially be very permeable. Induced fractures have been observed almost throughout the well. In general they strike east-west and are subvertical, suggesting an east-west maximum horizontal stress direction. However, there are also other directions of induced fracturing which might require a more detailed analysis. Finally, the productive zones B and C are not related to any fracturing at all. Instead, they both occur in dolomites and closer inspection reveals them to contain large, elliptical conductive patches, which we interpret as being vugs (figure 3.3.6 right; no core available in these zones). Therefore, these seem to be the only zones to produce directly from the matrix.

Figure 3.3.8 is a Schmidt stereographic plot showing the poles of the fracture classes in the topmost interval of this well comprising 400 meters. The color-coding corresponds to table 3.3.2. Bedding is seen to dip towards northeast at about $20-30°$. There are solution-enhanced fractures which are subparallel to bedding, and others, forming a relatively well-defined cluster, that seem to be perpendicular to bedding and dipping steeply towards southwest. Still other fractures, notably those classified as natural, seem to have a NE-SW strike with very steep dips.

Discussion

This fracture study is based predominantly on FMI electrical borehole images. The advantage of these data is that they are recorded downhole and that they are very sensitive to the presence of fractures, usually filled with conductive borehole fluid. Based on the image morphology, six classes of highly conductive features are identified, most related to some fracturing process: Planar fractures, solution-enhanced fractures, bedding-confined fractures, wide conductive zones,

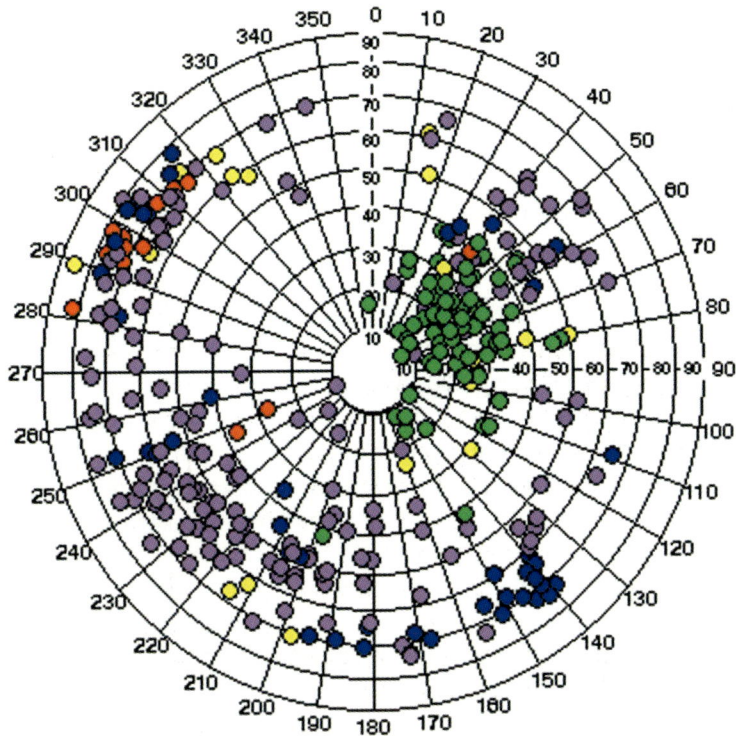

Figure 3.3.8. Schmidt stereo plot (upper hemisphere) of the uppermost interval of the well, covering 400 meters. Bedding (green) dips towards NE at about 20–30°. Fracture clusters are found parallel and orthogonal to bedding, as well as striking NE-SW and dipping at about 70°. The cluster with poles in the NW quadrant comes from the karstic fault zone which constitutes the major producing zone of this well.

brecciated zones and induced fractures. The most abundant type is the solution-enhanced fracture, but there is also a significant number of brecciated zones.

Particular attention has been paid to zones identified as productive during well testing. With two exceptions, these zones all show brecciation and abundant fracturing. We attribute them to faulting, perhaps in association with karstic leaching. Such leaching might be particularly important in the uppermost productive interval, which produces over 80 % of the total flow of the well. Interestingly, these zones all strike E-W or NE-SW, suggesting that they are related to the general northeastward thrusting that formed the field structure. Those fractures oriented at higher angles and perpendicular to the principal stress seem to have been closed as they are found to be non-productive. Therefore, a directional well due S/SW or N/NE could be planned in order to intersect a larger number of fractured zones and thus increase production. It is quite possible that such a well does not need to be drilled very deep into the reservoir, as connectivity is likely to be good throughout the reservoir, perhaps with the exception of the vuggy zones.

Abundant fracturing occurs throughout the well, albeit in various intensities. Solution-enlargement of the fractures is common, but apertures seem to be quite small, in the range between 1 and 10 microns. At such low values, fractures are not very productive, and their contributions to flow might be difficult to distinguish from inflow from the matrix porosity. However, zone E shows quite wide fractures of over 100 microns and is indeed one of the best producing intervals. Some of the larger solution-enhanced features resemble the outcrop analogue depicted in figure 3.3.9 but do not seem to be directly associated with the principal production zones. Two dolomitic intervals, zones B and C, produce from vugs and are therefore the only zones producing directly from the matrix.

Another type of breccia seems to correspond to collapse breccias as described by Kerans (1989) in the Ordovician Ellenburger Formation. He noticed that much of the production comes from intervals with solution-collapse breccias, but at present we cannot establish such a connection in this field. Thus the primary production in this well seems to be from brecciated and/or karstified fault zones. These drain smaller fracture networks that, in turn, drain the matrix porosity. Figure 3.3.9 illustrates several of these features on an outcrop of Cretaceous lines tones.

Table 3.3.4 summarizes the productive zones and their main characteristics based on the present analysis.

Figure 3.3.9. Solution-enlarged fractures in a Cretaceous rudist limestone on the coast of Dalmatia, Croatia. These rocks are thought to represent a an outcrop analogue of the case study discussed in this chapter.

Table 3.3.4. Summary of productive intervals and their characteristics based on the present study.

Zone	Productive Pore Types	Probable Origin	Production (bbls/d)
A	Steep, dense fracture network with brecciation	Faulting (NE-SW) with karstic leaching?	10,000
B	Vugs in dolomite	Leaching	320
C	Vugs in dolomite	Leaching	480
D	Brecciation/fracturing	Faulting (E-W)	
E	Fracturing/brecciation	Faulting (ESE-WNW)	750
F	Brecciation/fracturing	Faulting (E-W)	

3.3.3 A Fractured Basement Case Study

Introduction

Most hydrocarbon-producing basement reservoirs are horst structures that were faulted and fractured during the uplift. Their tops are often weathered and altered, but beginning and end of the weathered zone may be difficult to determine. Draping shales or evaporites provide the cap rock at the top, and the rock is sourced from the flanks of the horst. Debris on the flanks derived from the top of the horst structure might provide some additional reservoir volume. The well discussed here is from such a typical basement-reservoir setting.

The data set contains a complete set of modern well logs and is thus well suited to demonstrate their use for fracture characterization and modeling. These include electrical borehole images (FMI, ARI), ultrasonic borehole images (UBI), array sonic log (DSI), plus a neutron/density log, a laterolog and a natural gamma ray log. Production logging was been done with the PLT in several passes up and down the hole. The total production of this well during the initial flow tests was almost 12,000 barrels per day.

Figure 3.3.10 displays all logs and images acquired in this well over a section covering about 110 meters, together with a petrophysical interpretation. Additionally, one pass of the production log is shown on the leftmost track. The section covers an interval just below the weathered zone and consists of foliated granites and granodiorites.

Data Analysis and Results

The example may best be discussed by starting with the conventional open-hole logs. The gamma ray log shows that the uranium content is low, and that potassium is the main contributor to natural radioactivity, undoubtedly contained in potassium feldspars and perhaps some mica. The caliper log is straight except

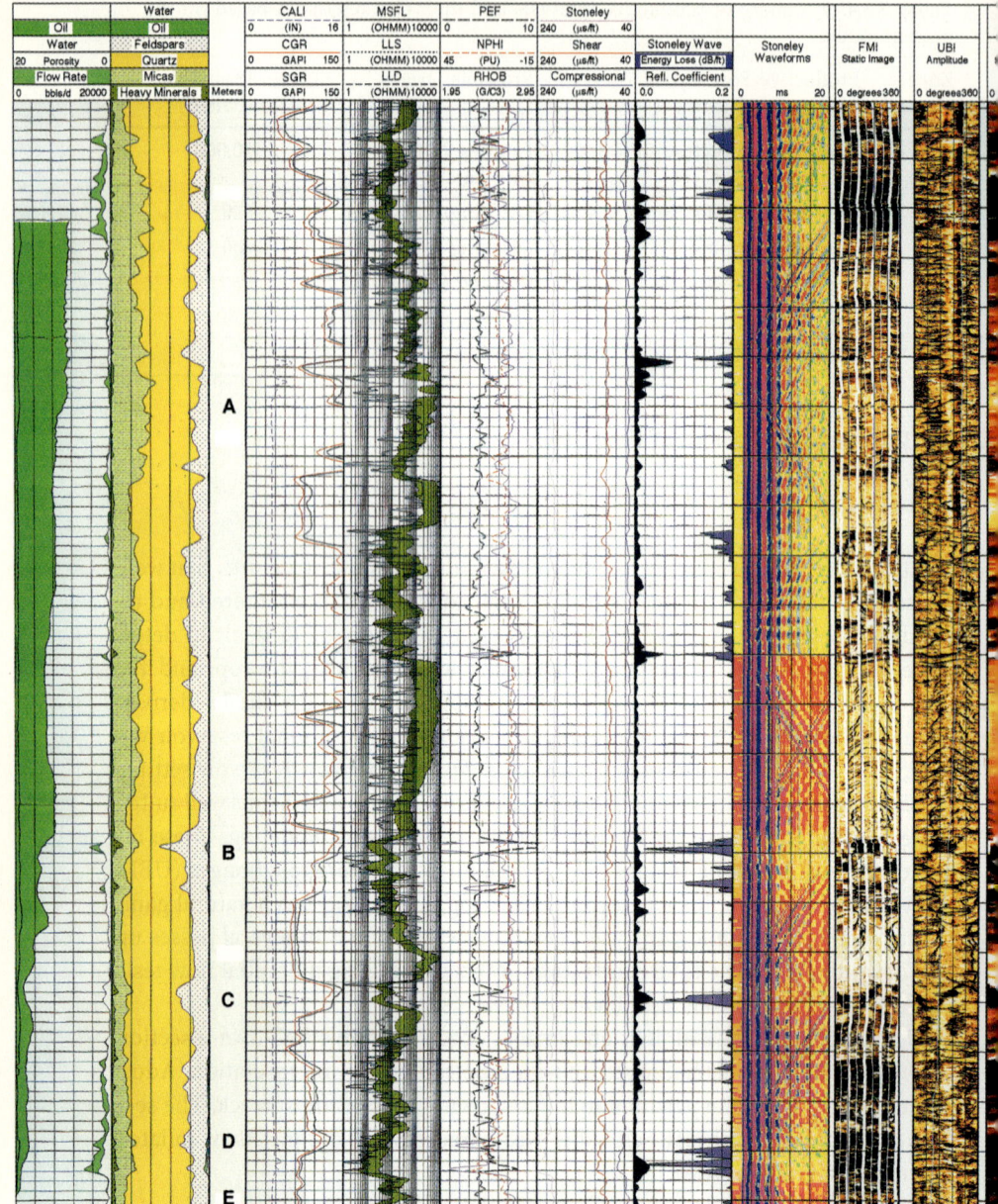

Figure 3.3.10. Data from a well in fractured basement. From left to right: Cumulative flow rates, petrophysical analysis, four tracks with open hole logs, two tracks with Stoneley wave data, three image logs. Vertical interval covers 110 meters.

for well-defined thin zones where it is up to three inches larger than the bit size. These zones can be matched to fractures and breakouts as will be seen later. However, it is important to notice that not all fracture zones show borehole enlargements. The resistivity logs show large and abrupt variations. In the more resistive zones, the deep reading goes up to 10,000 ohm-meters, while it falls to below 10 ohm-meters in the more conductive intervals. The shallow laterolog traces the deep laterolog, but is generally about one order of magnitude lower in resistivity. This is mostly due to the invasion of conductive drilling fluid, which affects the shallow resistivity more than the deep one. The microresistivity log is very active because of the rapid succession of highly resistive basement rocks, and fractures and related features that are filled with conductive borehole mud. An interesting observation can be made on the nuclear logs: All the three measurements (density, neutron porosity, photoelectric factor) seem to have a baseline, from which they sharply depart at discrete intervals. The photoelectric factor and neutron porosity increases while the density decreases, suggesting that this is an effect caused by mud invading fractured and borehole-enlarged zones. The three sonic slownesses (for the compressional, shear, and Stoneley waves) do not exhibit any anomalies and are rather constant. The Stoneley wave trains, on the other hand, show strong reflections as seen on the variable display log. These are caused by extra amounts of fluids contained in washouts and fractures, where the Stoneley pressure pulse is partially reflected. The Stoneley energy obtained from the reflection coefficient is shown by the curve filled in purple, and it corresponds to the amount of excess fluid (over the one contained in the borehole). If the reflection is caused by a fracture alone, this energy is a measure of the fracture width. However, care has to be taken to distinguish it from a local washout, and unfortunately the two often occur simultaneously.

The three tracks on the right are all borehole images. Many features discussed in chapters 2.2 and 2.3 can be identified:

- *Key seating* on the UBI in the upper half of the interval (despite an almost vertical well)
- *Breakouts* on the UBI at various depths, generally aligned NE-SW and thus suggesting a SE-NW maximum horizontal stress direction
- *Induced fracturing* on the FMI and UBI, best seen as short dark wisps connecting between foliations and aligned SE and NW, supporting the maximum horizontal stress direction deduced from breakouts
- *Foliation*, often only faintly visible on both FMI and UBI, and generally dipping north at intermediate angles. The foliation planes seem to have opened up (delaminated) in the direction of maximum stress on the borehole wall.
- *Natural fractures* on all three images (ARI, FMI, and UBI), cross-cutting foliations and extending across the entire borehole. The fact that they are seen on the ARI indicates that they are deep features and therefore of tectonic origin, not drilling-induced.

Among these features, the natural fractures are of greatest interest. Closer inspection shows that there does not seem to be a preferential orientation, though most of them dip at about 60 °C. They occur most of the time as single fractures, but fracture clusters can occasionally also be observed. Most fractures are associated with a Stoneley reflection and a spike on the nuclear logs as described above. Some of them show increases on the caliper. From the production log we can see that the flow comes from these features, although the high flow rates cause turbulence which affects the spinner readings and thief zones may additionally affect the readings.

In order to assess the inflow from the various intervals of interest, the five production logging passes uphole and downhole are first averaged. Then the fractures are determined and, assuming each of them to be a point entry, the inflow is calculated for each of them through a best-fit procedure of the entire flow profile. Five major production zones are identified in this way, while other possible zones were shown to be non-productive or perhaps thief zones. For convenience, these five productive fractures and fracture zones have been labeled from A to E from the top downward. Their individual contribution is plotted on figure 3.3.11, together with a three-dimensional projection of their geometry. This figure is best compared with the borehole images and logs in figure 3.3.10.

Zone A is a single fracture that produces about 1800 barrels per day. Based on the Stoneley reflection results, one might be tempted to locate it about 5 meters higher, but that zone turns out to be washed-out where no inflow takes place. Between zones A and B several fractures zones are seen from the Stoneley reflections as well as the borehole images. They turn out to be non-productive and one of them, located about mid-way between the two zones, could well be a thief zone. Zone B consists of three productive zones, with inflows ranging from 1500 to 3000 barrels per day from each. The dips vary greatly and it is quite possible that the mutual intersection of these fractures led to the high production rates of oil from this zone. Zone D is a single, highly dipping fracture producing almost 2000 barrels per day. It has a typical signature on all logs and images, and the high conductivities on the ARI shows this fracture to be deeply invaded by mud at the time of logging, suggesting it to be highly permeable. Zones D and E are somehow related but dip in different directions. They seem to consist of several parallel fractures, with a combined production of almost 2500 barrels per day.

Thus, the general picture is one of a very high production from a surprisingly small number of fractures with no preferred orientation.

Discussion

Two questions arise from this analysis: What is the nature of these fractures, and where is the fluid storage? Closer inspection of the borehole images reveals that the fractures are in fact rather wide features (this is difficult to see on the com-

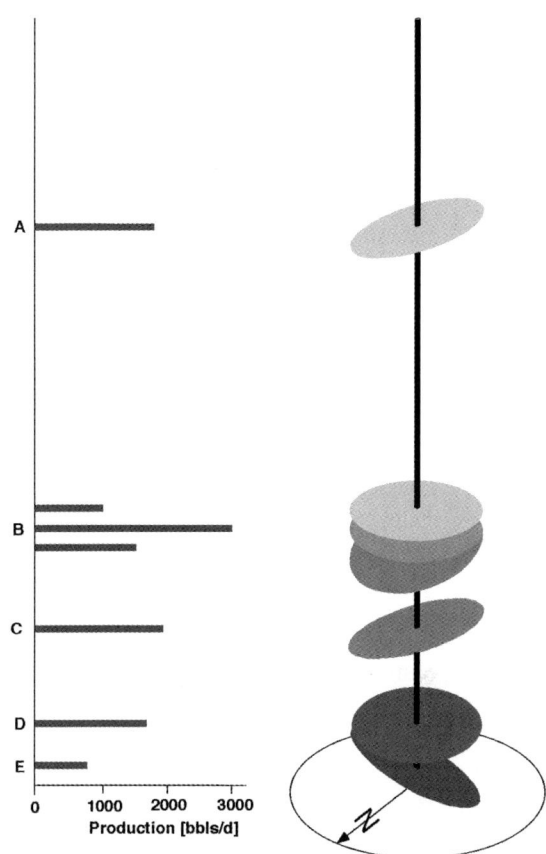

Figure 3.3.11. Inflow rates of oil calculated from the five production log passes, assuming point entries at the fractures locations (left). Five productive zones emerge. Their geometry is shown in the 3-D model at right in a vertically compressed scale that flattens the dips (courtesy J. van Doorn). Vertical interval covers 110 meters.

pressed images shown in figure 3.3.10). These large widths and significant borehole damage did not allow for a combined Stoneley wave/FMI interpretation as proposed by Hornby et al. (1990). Rather, emphasis had to be put on a combination of the production logs with borehole images, supported by the other logs. It is impossible to assess whether any movement has taken place along the fractures, but in view of the large widths this is quite possible and they might correspond to faults or shear zones. However, unlike in the previous example, little to no associated secondary fracturing is observed and thus fluid flow is still essentially through point entries. The reservoir storage question is even more difficult to address, particularly if only data from one well are available. The petrophysical results in figure 3.3.10 suggest certain intervals up to ten meters thick to contain some matrix porosity, generally in the 2–4 % range. These zones are conspicuous on the ARI image as dark intervals and have oil saturations of around 50 %. The pores in these zones could be shrinkage pores formed during contraction of the magma, dissolution pores during circulation of interstitial waters, fracturing, or a combination of these (Areshev et al., 1992). The possible reservoir storage in the interval may be supplemented by additional porosity found in the outer parts of the basement horst, as well as in adjacent sediments.

From the geometric relationships shown in figure 3.3.11 it seems that no preferred orientation of the fractures occur. Such 3-D model of fracture systems can be used as input into the reservoir simulator.

References

Areshev EG, Le Dong T, Thuong San N, Shnip OA (1992) Reservoirs in fractured basement on the continental shelf of southern Vietnam. J. Petrol Geol, 15: 451–464.
Cheung PSY, Heliot D (1990) Workstation-based fracture evaluation using borehole images and wireline logs. Ann Conf Soc Petrol Eng: Paper 20573.
Hornby BE, Johnson DL, Winkler KW, Plumb RA (1989) Fracture evaluation using reflected Stoneley-wave arrivals. Geophysics 54: 1274–1288.
Hornby BE, Luthi SM, Plumb RE (1992) Comparison of fracture apertures computed from electrical borehole scans and reflected Stoneley waves: An integrated interpretation. The Log Analyst, 33, 1: 50–66.
Faivre O (1993) Fracture evaluation from quantitative azimuthal resistivities. 68th Ann Conf Soc Petrol Eng: Paper 26434.
Fritsen A, Corrigan T (1990) Establishment of a geological fracture model for dual porosity simulation on the Ekofisk Field. In: Buller AT, Berg E, Hjelmeland O, Kleppe J, Torsaeter O, Aasen JO (eds) North Sea Oil and Gas Reservoirs – II. Proc 2nd Int Con North Sea Oil and Gas Reservoirs. Norwegian Institute of Technology„ Graham & Trotman Ltd, pp 173–184.
Kerans, C (1988) Karst-controlled reservoir heterogeneity in Ellenburger carbonates, West Texas. Bull Am Assoc Petrol Geol, 72: 1160–1183.
Lees GM (1933) Reservoir rocks of Persian oil fields. Bull Am Assoc Petrol Geol, 17: 229–240.
Levorsen AI (1967) Geology of Petroleum. W.H. Freeman, San Francisco.
Luthi SM, Souhaité P (1990) Fracture apertures from electrical borehole scans. Geophysics, 55: 821–833.
Nelson RA (1985) Geological analysis of naturally fractured reservoirs. Gulf Publishing Co.
Tandom PM, Ngoc NH, Tija HD, Lloyd PM (1999) Identifying and evaluating producing horizons in fractured basement. Asia Pacific Inproved Recovery Conf Soc Petrol Eng: Paper 57324.
Van den Bark E, Thomas OD (1990) Ekofisk: First of the giant oil fields in Western Europe. In: Halbouty MT (ed) Giant Oil and Gas Fields of the Decade 1968–1878. Memoir 30, Am Assoc Petrol Geol, pp 195–224.
Van Golf-Racht TD (1982) Fundamentals of fractured reservoir engineering. Developments in Petroleum Science 12, Elsevier, Amsterdam.
Wilkinson WM (1953) Fracturing in Spraberry reservoir, West Texas. Bull Am Assoc Petrol Geol, 37: 250–265.

3.4 Well Correlation

3.4.1 Introduction

Correlation or *stratigraphic correlation* is a geological term referring to the process by which two or more geological intervals are equated even though they are spatially separated. The International Stratigraphic Guide (Hedberg, 1976) subdivides it as follows:

"There are different kinds of correlation depending on the feature to be emphasized: Lithologic correlation demonstrates correspondence in lithologic character and lithostratigraphic position; a correlation of two fossil-bearing beds demonstrates correspondence in their fossil content and in their biostratigraphic position; and chronocorrelation demonstrates correspondence in age and in chronostratigraphic position."

Correlation is a principal task for many geologists, particularly in the oil and gas industry, but also in mining and hydrogeology. Cuttings, cores and well logs constitute the main data used for correlation, with well logs playing a special role because they are long, continuous recordings. Additionally, they are useful because they are unbiased physical measurements and often available in a large number of wells in a field or a basin. Doveton (1994) finds that

"It is ironic that petrophysicists are only minority users of wireline logs. They are easily outnumbered by geologists, who apply logs routinely as frameworks for lithostratigraphic correlation ... However, even today, the task of correlation is usually a laborious manual process."

He goes on to provide a comprehensive summary of stratigraphic segmentations of well logs, automated correlation methods, and reservoir mapping and modeling methods. In this chapter, we will not try to reproduce his excellent summary, but will focus on how new logging methods, combined with modern geological concepts such as sequence stratigraphy and computer-aided techniques, can improve well correlations.

Most conventional open-hole logs including electrical, nuclear and acoustic logs are primarily sensitive to the properties of the rocks and the fluid they contain. In fact, most of them are more sensitive to the fluids, and the lithological signal is often only secondary. Nevertheless, if these logs are used for well correlation, an essentially lithostratigraphic correlation will be obtained[1]. Many geologists prefer to use just the gamma ray logs, or a combination of gamma ray logs with spontaneous potential (SP) logs, because they are commonly available and they carry a relatively strong lithological signal. The nuclear spectroscopy logs discussed in chapter 2.7 carry an entirely lithological signal, and with them a reliable *lithostratigraphic correlation* can be obtained. If they are com-

---

[1] The term *petrophysical correlation* is not used, although here it might be more appropriate than lithostratigraphic correlation.

bined with logs carrying textural information such as the nuclear magnetic resonance logs, and bedding information from borehole images, a complete *lithofacies correlation* can be obtained.

Borehole images and dipmeters, however, can contribute more than facies information to well correlation. Their most important use might well be the structural information they contain, which can be extrapolated away from the wellbore, tied into seismic lines and into existing correlations and maps.

No logging tool is capable of aiding in *biostratigraphic correlation*. However, the paelomagnetic logging tool discussed in chapter 2.8 can provide a chronological zonation and, if several wells are available in one field, a *chronostratigraphic correlation* can be made. An example of this has been shown in figure 2.8.8 from two nearby wells of the Aquitaine Basin in France. While this in itself is an exciting new development in logging, it becomes particularly powerful when combined with lithostratigraphic correlations. The example shown in figure 2.8.9 illustrates that, by doing so, one can progress from a simple static picture of the reservoir layering to a dynamic description of the reservoir architecture which takes fully into account the depositional history of the basin.

The following contains some case studies illustrating these applications.

3.4.2 Lithostratigraphic Correlation Using Spectroscopy Logs

Lithostratigraphic correlation has been defined above, and it was stated that most well correlations using logs are essentially lithostratigraphic correlations. No logs, however, reflect the lithological and the mineralogical composition better than nuclear spectroscopy logs, as they are practically insensitive to the fluids contained in the rocks.

The gamma ray is also sensitive to lithological changes, but among the naturally radioactive elements (thorium, potassium, uranium) only one occurs as a major element in the principal rock-forming minerals: potassium forms part, among others, of illites, micas and potassium feldspars. High amounts of thorium and uranium, on the other hand, can occur without any major change in the lithologies, for example through a slight increase in heavy minerals, or even through heavy oil or tar. The total gamma ray is therefore not always very useful in well correlation. A natural gamma ray measurement is already better suited as it breaks the natural radioactivity down into its components, and thus the potassium concentration can be separated from the rest. However, it is of course much better to have more major elemental concentrations available for a good lithological characterization. The nuclear spectroscopy logs fill this requirement very well.

The example presented here consists of two wells, both logged among others with the Elemental Capture Sonde (ECS, chapter 2.7). The gamma ray logs are displayed in figure 3.4.1 and show close similarities over a selected interval of

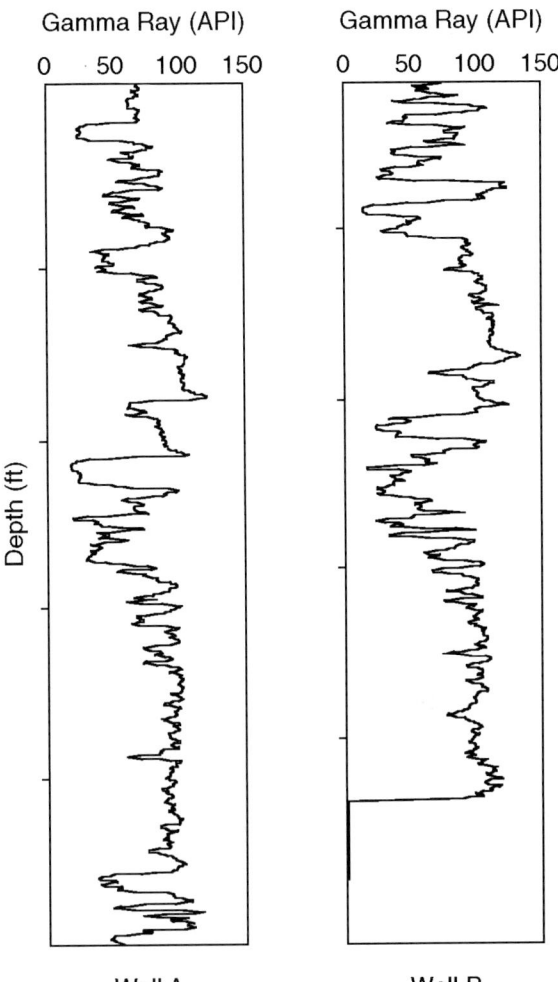

Figure 3.4.1. Gamma ray logs from two wells can easily be correlated except for some sandstone layers in the central part of the interval.

interest, with a correlation whose depth shift seems to be relatively constant. However, details of the correlations are more difficult to work out, particularly in the central section where some clean sands occur. The connecting lines are omitted on purpose because they often make correlations appear more obvious than they actually are.

The spectral data was processed to obtain a simple mineralogical breakdown into calcite, quartz, and clay volumes (anhydrite was included in the model but is not present in the interval shown). Figure 3.4.2 shows the results, plotted as percentages of the solid volume, again without correlation lines. Several carbonate streaks show up on both wells, and they correlate very well. Even the minor amounts of carbonates correlate both in shape and absolute volume such

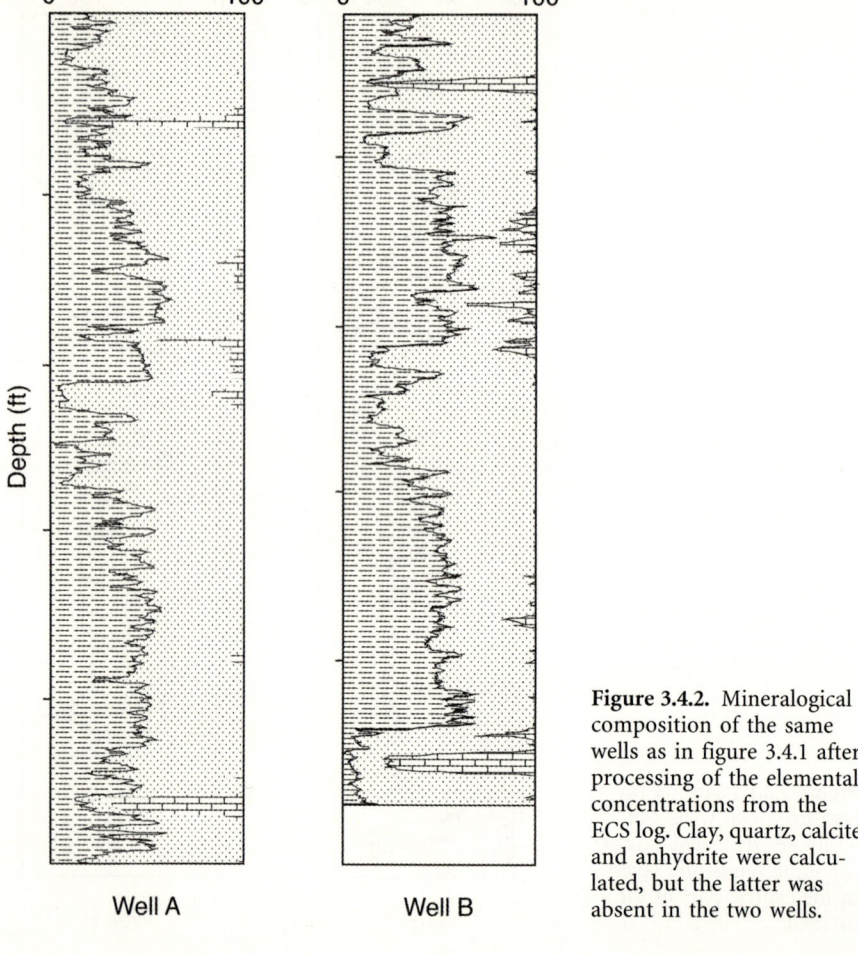

Figure 3.4.2. Mineralogical composition of the same wells as in figure 3.4.1 after processing of the elemental concentrations from the ECS log. Clay, quartz, calcite and anhydrite were calculated, but the latter was absent in the two wells.

that a fairly unambiguous and complete lithostratigraphic correlation is possible throughout the entire interval. When all mineral components are considered, the correlation covers also the carbonate-free intervals and can be done in great detail. Figure 3.4.3 shows these resulting correlations, with blue lines indicating correlations based on the carbonates, and black lines on the shales. Some sands and shales are seen to *pinch out*, and at other intervals the different mineral volumes suggest *facies changes*. These intervals are highlighted on figure 3.4.3 with P and F respectively.

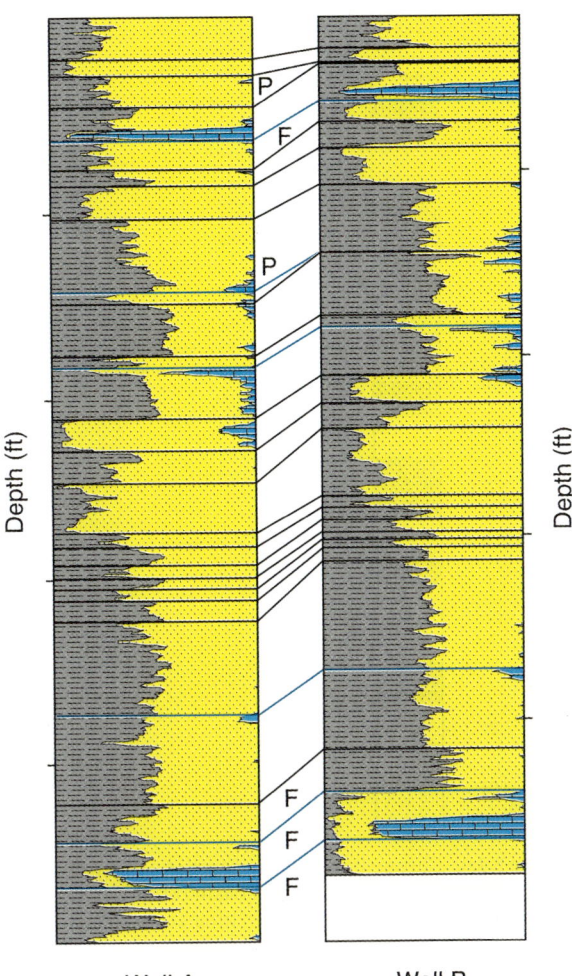

Figure 3.4.3. Lithological correlation of the two wells from figure 3.4.2 based on the mineralogical composition alone. Carbonate-based correlations are blue, clay-based correlations black. Detailed correlation is possible, showing intervals of facies changes (F) and layer pinch-outs (P).

3.4.3 Chronostratographic vs. Lithostratigraphic Correlations

This case study has already been touched on in the chapter on paleomagnetic logging (2.8). It consists of two nearby wells through an early Tertiary carbonate sequence in the south of France. Using the paleomagnetic borehole data from the two wells and the global polarity time scale (GPTS), Thibal et al. (1999) succeeded in making a chronostratigraphic correlation, shown in figure 2.8.8 and repeated in figure 3.4.4 in a slightly different form. There is good confidence in the correlation below the unconformity, where an almost unambiguous match with the GPTS was possible, albeit with the help of biostratigraphic information. In the upper section, covering the upper Eocene and overlying an unconformity of unknown duration, the chronostratigraphic zonation is less certain. Thibal et al.

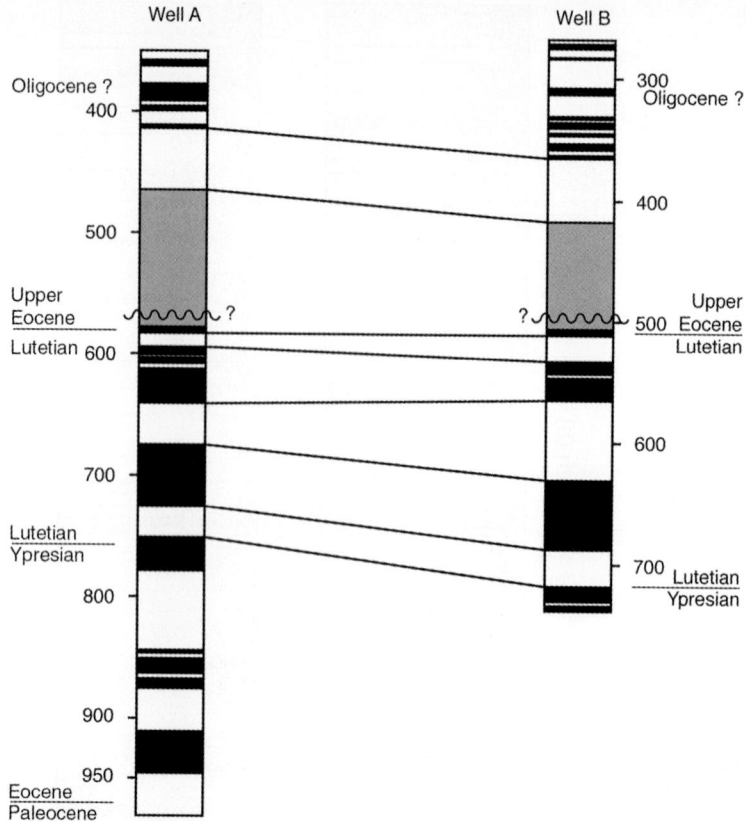

Figure 3.4.4. Chronostratigraphic correlation of paleomagnetic polarities from GHMT in two nearby wells in the Aquitaine basin (after Thibal et al., 1999). The correlation of these two wells with the global polarity time scale (GPTS) is shown in figure 2.8.8.

(1999) also made a lithostratigraphic correlation using the gamma ray and the susceptibility log. The latter responds to the amount of ferromagnetic and ferrimagnetic minerals, both usually occurring in accessory volumes but often with good lateral continuity because of the fine-grained nature of the particles. This is also observed in the present example, where the two susceptibility curves can be well correlated, as shown in figure 3.4.5. The correlation in fact can be worked out in great detail, similar to the example discussed in the previous paragraph, mostly because the carbonates present in these two wells have a fair amount of fines and therefore, the gamma ray and susceptiblity response is very good.

This example becomes particularly interesting when the chronostratigraphic and the lithostratigraphic correlations are compared. We focus on the interval below the unconformity where both correlations have a good density and a high confidence. As seen in figure 3.4.6, the chronostratigraphic correlation lines (the isochrones) have a strong angular disparity with the lithological correlation lines

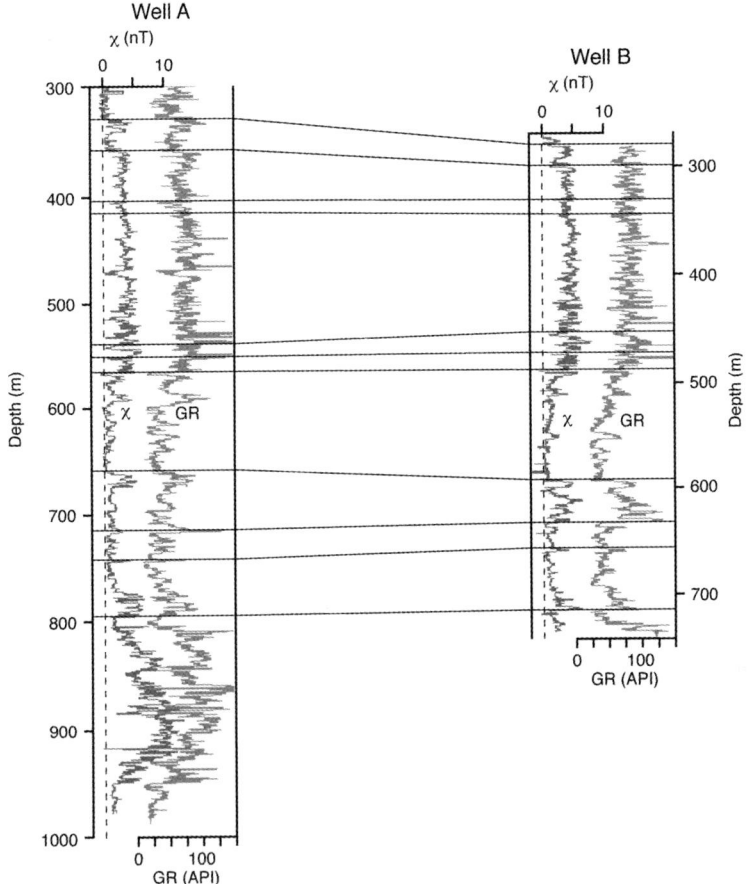

Figure 3.4.5. Lithostratigraphic correlation of the same wells as in figure 3.4.4 (after Thibal et al., 1999). The magnetic susceptibility (χ) and the gamma ray (GR) log have been used as lithological indicators for the correlation.

in the lower half of the interval, which covers about 200 meters. The relationship is such that lithologically equivalent layers become younger when going from well A towards well B. This can be caused by lateral accretion, or facies progradation, which in carbonates happens for example in the forereef area. By contrast, the upper half of the interval considered here shows good parallelism between the two sets of correlation lines. This, in turn, suggests vertical aggradation, i.e. a depositional situation where the facies in the two wells are the same and have equal deposition rates. The gamma ray log as well as the susceptibility measurement indicates that this upper part has more clastic fines. This is consistent with a depositional model where the lower, diachronous sediments are laid down in a relatively high-energy environment, while the upper part is a

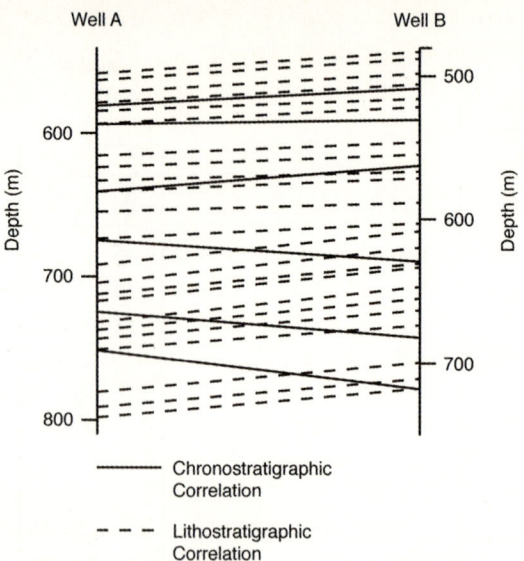

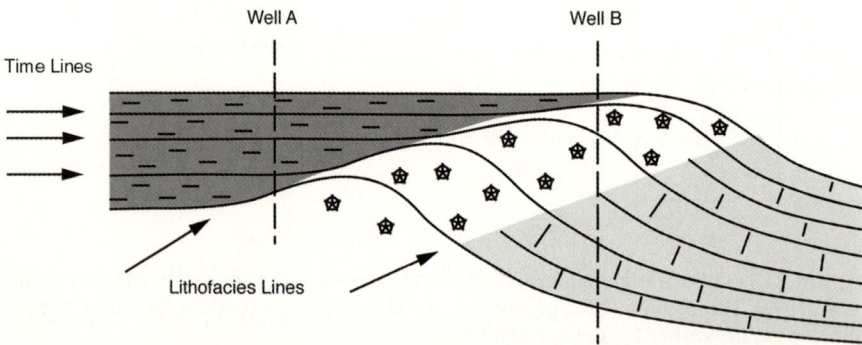

Figure 3.4.6. Comparison of the lithostratigraphic versus chronostratigraphic correlation of the two wells in the Aquitaine basin. Only lower interval is shown. Notice the diachronous nature of the correlation in the lower part, and the isochronous correlation in the upper part (from Thibal et al. 1999). The bottom sketch depicts a possible depositional scenario leading to this situation.

low-energy deposit such as found, for example, in a lagoonal setting. At the bottom of figure 3.4.6 a schematic sketch illustrates the scenario envisioned here.

When the data set to be correlated comprises a large suite of curves in each well, it is nearly impossible to take all this information into account while correlating visually, nor is it possible to follow each layer and to put each one into a coherent geological framework. As most of the task consists of sequence recognition, pattern matching and event detection, mathematical methods have been used from the early days of computing to assist in this task (Fang et al, 1992; Doveton, 1994). In the following, we discuss some newer computer-aided methods to correlate logs between wells.

3.4.4 Well Correlation Using Neural Networks

Neural networks (used here for *artificial* neural networks) have been used in a large variety of non-linear modeling and classification problems in engineering, medical and biological sciences for complex problems which could not be solved using first-principles. Hecht-Nielsen (1990) defines neural networks as having a *"parallel, distributed information processing structure consisting of processing elements (which can possess a local memory and can carry out localized information processing operations) interconnected via unidirectional signal channels called connections..."*

In multilayer feed-forward neural networks, such as the backpropagation networks, there is an input layer to which an input signal is given, and an output layer to which a corresponding output response is presented. The connection weights between the processing elements are adjusted in a manner such that after presentation of a large number of examples the mapping from input to output is satisfactory. When constructing a neural network, the following choices are of importance:

- The learning rules and algorithms to update the connection weights
- The number of layers in between the input and output layer (termed hidden layers)
- The number of processing elements in each layer, and
- The criteria to assess when the training of the network is sufficient.

These issues are discussed in the relevant literature (e.g. Haykin, 1994) where also guidelines are given for a proper network design.

When correlating geological layers using logs, the geologist first identifies markers of interest, identified on the exploration and early appraisal wells. These are the "datums" to which subsequent correlations, both vertically and laterally, are tied. Below and above the datums, the geologist identifies other boundaries of interest ("markers") which help build a reservoir model. This approach is generally tedious and iterative, and the geologist continuously re-examines his previous correlations and changes markers if necessary. In order to capture this process computationally, Luthi and Bryant (1994) adopted a method in which the neural network first gets trained on a datum in key wells, after which it identifies the datum in other wells. It then gets trained on other markers and on their vertical relationship to the datum in the key wells before, as a last step, it determines these markers in all other wells. The procedure thus involves two major steps, which are handled by two separate but similar networks.

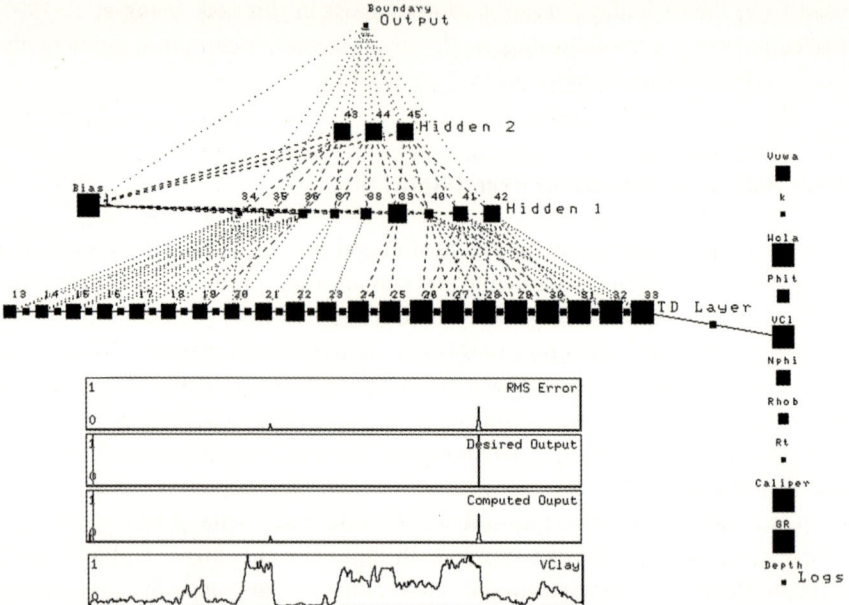

Figure 3.4.7. Neural network structure used for correlating seven wells in the Maracaibo area, Venezuela. This snapshot is from the training phase, whereby the clay volume (VCl, right) is presented in intervals of ten feet (21 samples) with the known output (1 for the marker to be present, 0 for it to be absent). The learning progress at this stage of the training is shown in the panels at the bottom (from Luthi and Bryant, 1994).

The logs correspond, in signal-processing parlance, to a stationary time-series and known procedures can be adapted to the problem. A section of the log is presented to the neural network together with the desired output response, which is 1 (one) when the boundary is present and 0 (zero) otherwise. Thus, there is only one example of the boundary present in the data set (if only one well is used), and the data have to be presented many times to achieve a statistically significant number of training samples. Figure 3.4.7 depicts the specifics of the neural network architecture used. The input data is first fed through a tapped delay layer of length p with fixed connections of value 1. With every iteration the data in the tapped delay is shifted forward by one sample and a new data point is entered at the beginning; the tapped delay therefore acts as a sliding window. The output response o corresponding to the sample at position $n = (p+1)/2$ is fed to the output node in the training phase. Two intermediate "hidden" layers are inserted with a number of processing elements intermediate to the previous and the following layer. The network is partially connected such as to give higher weights to the samples in the middle part of the tapped delay. During training, the difference $e(n) = \hat{o}(n) - o(n)$ is taken as the cost function which is minimized by adjusting the connection weights according to the delta learning rule (Haykin, 1994). A bias of value 1 is connected with variable weights to all processing elements in the hidden layers, a standard procedure for this type of neural network. The activity levels of the processing elements are scaled by a sigmoidal transfer function before being passed on to the next node.

Field data from the Lower Lagunillas Member (Miocene) of Bloque IV in Lake Maracaibo, Venezuela were used to test the procedure (Luthi and Bryant, 1994; Luthi et al., 1995). The sequence consists of delta plain and delta front deposits and contains three fining upward cycles in an overall transgressive setting. The top of a prominent lagoonal shale, known locally as layer VIII, was designated as datum A; in most wells, it is overlain by a high-permeability channel-fill sand (layer IX) and underlain by sandy lagoonal mouth bars and deltas (layer VII) which are of interest for future field development. Well 1112, centrally located in the field and featuring a complete logging suite, served as key well. The volume of clay (V_{cl}), computed from a suite of open-hole logs, was used as the principal log. The first network was trained on datum A in well 1112 for 200,000 iterations over a well log length of 654 samples, or 327 feet; the boundary itself was thus presented slightly over 300 times to the network for learning.

The network learned to recognize the boundary slowly but steadily. Further training did not improve the output, but impaired the global applicability of the network (it is said to be "overtrained"). The trained network was then applied to six wells from various parts of the field, and the levels with the highest output values were chosen as the most probable depths of occurrence of datum A. The results are shown in table 3.4.1 and indicate that the neural network answers match the expert answers in all cases to within one or two feet. Figure 3.4.8 compares the expert correlation and the neural network picks in fence panels. The confidence value, which reflects the probability the network assigns to a given depth to be the correct boundary, was found to be higher in some wells than in the training well 1112. This is explained by the competition of two similar-look-

Table 3.4.1. Comparison of expert versus neural network answers for datum A

Well	Expert Depth (ft)	NN Depth (ft)	Confidence
1112	10529.2	10528.1	0.598
991	10579.8	10578.5	0.781
922	10651.7	10650.3	0.720
893	10298.7	10298.8	0.581
866	10797.1	10797.2	0.864
706	10244.6	10242.5	0.014
670	10116.8	10015.5[a]	0.200
		10115.5[a]	0.193

[a] Two probable depths with similar confidence values

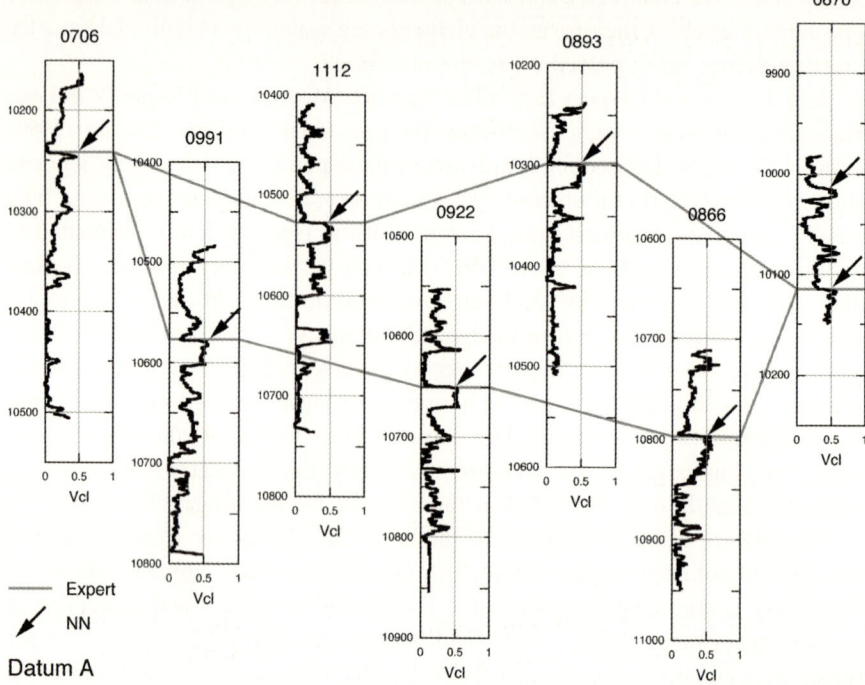

Figure 3.4.8. Results of expert correlation (solid lines) versus the most probable picks by the neural network (arrows), Datum A, Lagunillas Formation, Venezuela (from Luthi and Bryant, 1994). The wells are placed roughly according to their geographic position, with north pointing upwards. The neural network can handle the facies changes towards east and west as long as the layers do not disappear altogether.

ing boundaries present in the training well, and the difficulty of the network to separate them from each other. In well 706, layer VIII is very thin but the network still identifies it as the most probable location for the datum, albeit with greatly reduced confidence. In well 670, two boundaries were picked with almost the same confidence, with the boundary picked by the expert being considered slightly less probable by the network than another boundary higher up in the well.

In a second step, Luthi and Bryant (1994) developed a neural network that identifies additional markers. The boundary between layers VII and VIII was chosen for this and termed datum B. This boundary is much more variable than datum A due to lateral facies changes, and even an expert finds it difficult to correlate it confidently across the field. This network makes use of the depth relationship to datum A, just like a geologist uses relative depths of markers to assess their stratigraphic position. The rationale is that if a potential marker is close to where it is to be expected according to the training set, it should get higher confidence values than if it is far away. This network, therefore, features

Table 3.4.2. Comparison of expert versus neural network answers for datum B

Well	Expert Depth (ft)	NN Depth (ft)	Confidence
1112	10548.4	10548.1	0.281
991	10604.2	10603.0	0.297
922	10671.8	10671.3	0.708
893	10323.9	10322.8	0.479
866	10812.0	10791.2[a]	0.010
		10845.2[a]	0.009
706	10249.0	10249.5	0.073
670	10138.9	10089.5[a]	0.0051
		10081.5[a]	0.0046
		10025.0[a]	0.0046

[a] Two or more probable depths with similar confidence values

two tapped delays, one containing the log of the volume of clay, and the other the vertical distance to datum A. Table 3.4.2 contains the results for the most probable picks. For five wells (including the training well 1112) the expert answer and the neural network pick coincide, although in well 706 the confidence is low due to the thin nature of layer VIII. In well 866, neither of the two picks of the neural network matches the expert answer. Comparison of the log with the expert pick (figure 3.4.9) indicates that the expert had picked a subtle transition within a thick shale, applying geological reasoning pertaining to facies changes in the underlying layer VII. The lower pick by the network is lithologically reasonable, but far from the expected position, while the upper pick is both lithologically and stratigraphically difficult to accept. For both picks the confidence values are low. In well 670, which is also characterized by facies changes due to its lateral position with respect to the local depocenter, the boundary picked by the expert is similarly subtle as in well 866. All three answers by the neural network lie above the (younger) datum and are thus stratigraphically unreasonable (ruling out reverse faulting), and the confidence values are accordingly low.

Neural networks hold great promise as a help to the production geologist in interactive well-to-well correlation systems using well logs. The example discussed here shows that they can be readily trained and provide reasonable answers both for major datum identification as well as secondary marker correlation, even in the presence of considerable lateral changes.

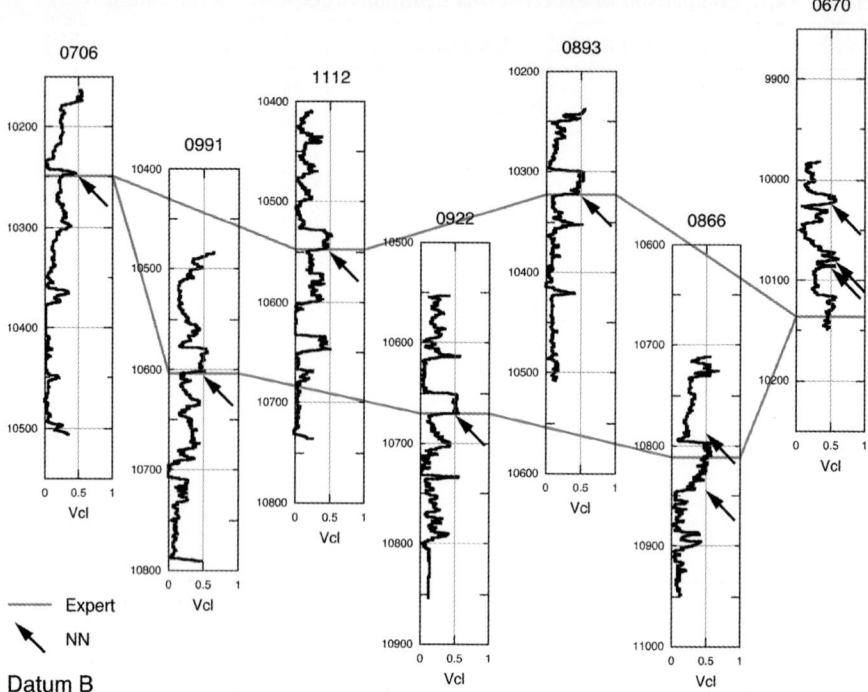

Figure 3.4.9. Results of expert correlation (solid lines) versus most probable picks by neural network (arrows), Datum B, Lagunillas Formation, Venezuela (from Luthi and Bryant, 1994). The correlations disagree in the eastern part of the field where facies changes are important.

3.4.5 Well Correlation Using Dynamic Programming

In the last decade there have been several publications on the adaptation of a general procedure known as Dynamic Programming (DP) to comparison or sequence alignment of strings, such as in biological DNA, RNA and protein sequence analysis, speech comparison and shape comparison. DP algorithms are used for finding shortest paths in graphs and in many other optimization problems. The method can handle multiple gaps, use any number of logs, accommodate stretching, and find a solution that is an optimum of all possible alternatives (Smith and Waterman, 1980; Fang et al., 1992). Information from a large number of logs may be used to correlate a number of wells.

Within a given sequence with common bottom and top markers in each wells, the objective is to construct first a graph (a cost matrix) and then to find a path through the graph. The latter is a well-to-well assignment of associated points such that the mismatch of the measurements at the associated points in the well is smallest (Le Nir et al., 1998). The graph represents for each depth point

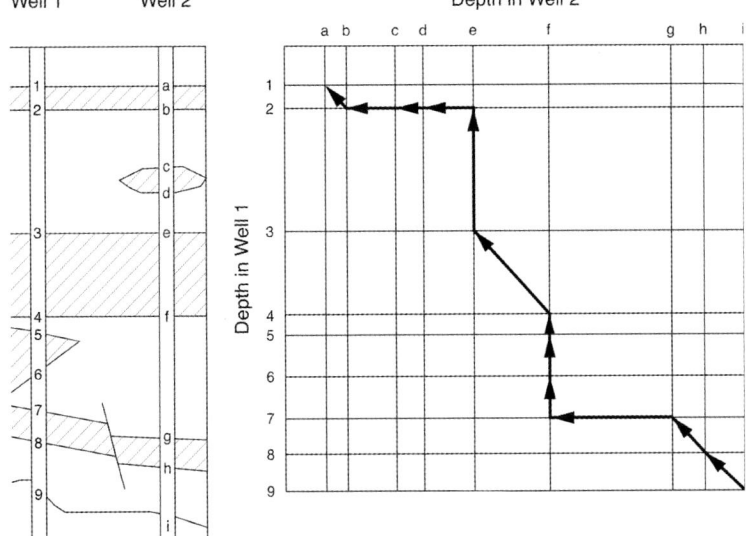

Figure 3.4.10. Matrix of correlation and path used in dynamic programming. Diagonal vectors indicate a match (correlated series). Horizontal vectors represent gaps in well 1, while vertical vectors indicate gaps in well 2 (from Le Nir et al. (1998).

of well 1 the cost of the match (or mismatch) of the log values of well 1 with the log value of well 2 at a depth point of well 2 (figure 3.4.10).

The method was applied to logs from the Stratton gas field in South Texas, a widely studied data set in the industry (Levey et al., 1994; Hardage et al., 1994). The map of the well locations (figure 3.4.11) shows well spacing to range from 800 feet (between well 18 and well 12) to 4600 feet (between well 9 and well 10). All wells penetrate the Oligocene Middle Frio, which consists of multiple stacked

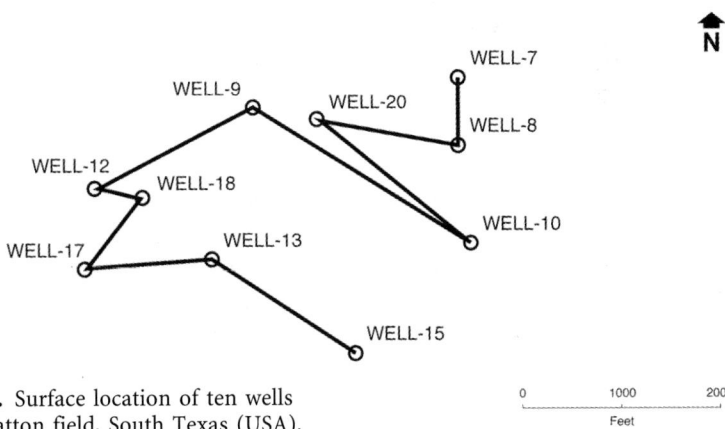

Figure 3.4.11. Surface location of ten wells from the Stratton field, South Texas (USA). The connecting line indicated the well arrangement in subsequent figures (from Le Nir et al., 1998).

331

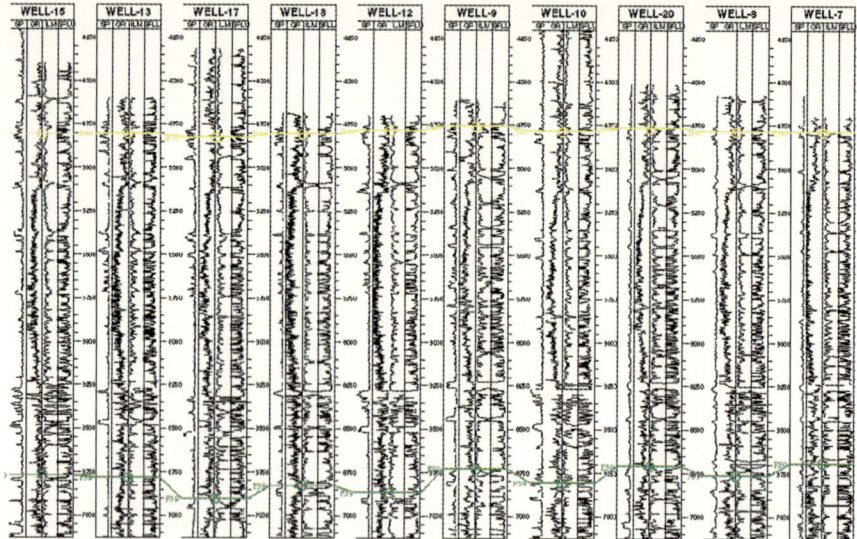

Figure 3.4.12. Wireline log data for ten wells in the Stratton field, illustrating the complexities of the changes from well to well (from Le Nir et al., 1998).

pay sandstones within a series of reservoir sequences referred to as B, C, D, E and F series reservoirs. In these amalgamated fluvial channel-fill and splay sandstones, channel fill deposits range from 10 to 30 feet in thickness and about 2500 feet in lateral extent, usually exhibiting a fining upward or blocky log signature. Crevasse splay deposits range from less than 5 feet to 20 feet and are coarsening upward.

Ten out of the 21 available wells have the following suite of logs in common: gamma ray, neutron porosity, density, spontaneous potential, spherically focused laterolog and induction medium log. Figure 3.4.12 shows a panel of the ten wells, aligned along the path indicated on the map in figure 3.4.11, but with only four logs displayed per well. Le Nir et al. (1998) selected as top and bottom limits of the correlation interval the regional markers B46 and F39, indicated with correlation lines in figure 3.4.11. They then used the six logs to compute the cost matrix and to backtrack the optimum correlation path.

Well 15 was selected as reference well from which all others were correlated, one after the other. Over long intervals such as the ones here, the DP method produces a large number of markers for each well, even if only markers common to all wells are retained. In this example, after correlating the ten wells 284 common correlations were obtained (figure 3.4.13). This large number of markers can be reduced by filtering them as a function of minimum and/or maximum value of a given log in one well, or by requiring a minimum distance between markers. Intervals with no visible correlation correspond to zones where the logs could not be correlated across all the selected wells.

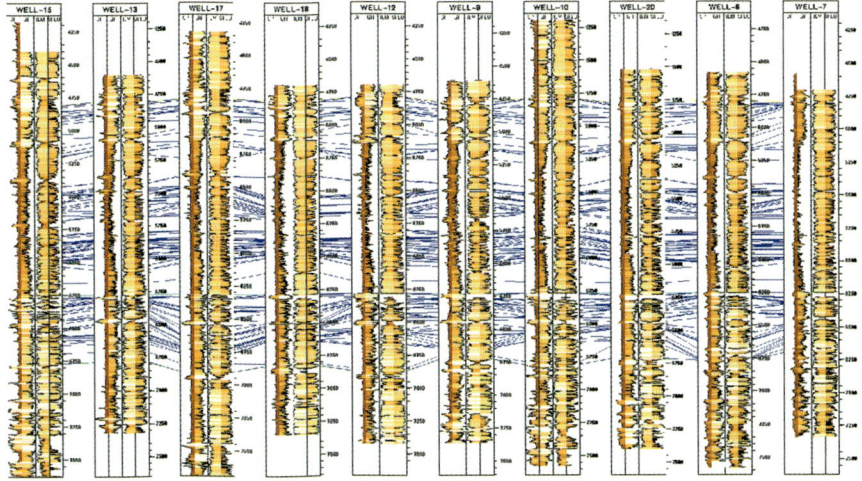

Figure 3.4.13. Display of correlations in the Stratton field resulting from automatic correlation with dynamic programming. This display is aligned on the correlation marked with D. Shading intensity reflects the relative log values (from Le Nir et al., 1998).

This method takes into account all available information (i.e. the logs common to all wells), and the interpreter can then check the quality of the correlations and select those that appear most relevant to his study. The method does not require a-priori knowledge, apart from the top and bottom markers of the series to be matched in all wells. Within the given interval, it produces the optimum correlation schema common to all the selected well. It is a fast procedure but the geologist remains judge of the final result. He will need to discriminate between non-geological noise, local geological events, and regional markers.

3.4.6 Well Correlation from Facies Zonation

We have discussed in chapter 3.3 how a facies (or "electrofacies") zonation can be obtained through classification of the log responses in a well. Depending on the types of logs used for this classification, the result will correspond more to a petrophysical, lithological or lithofacies zonation. With conventional, petrophysical open-hole logs as inputs, the result will be primarily a petrophysical segmentation. When more geological measurements are used, the zonation becomes more lithofacies-like.

Using again the Stratton data set in the Middle Frio formation, a segmentation into eight basic lithofacies types was carried out in all wells. Although simplistic, the purpose it to synthesize the information contained in the logs into a display which allows easier visual correlation. The identification of these lithologies is done first in the reference well and then propagated to neighboring wells using a neural network technique. Figure 3.4.14 shows the results in a similar panel dis-

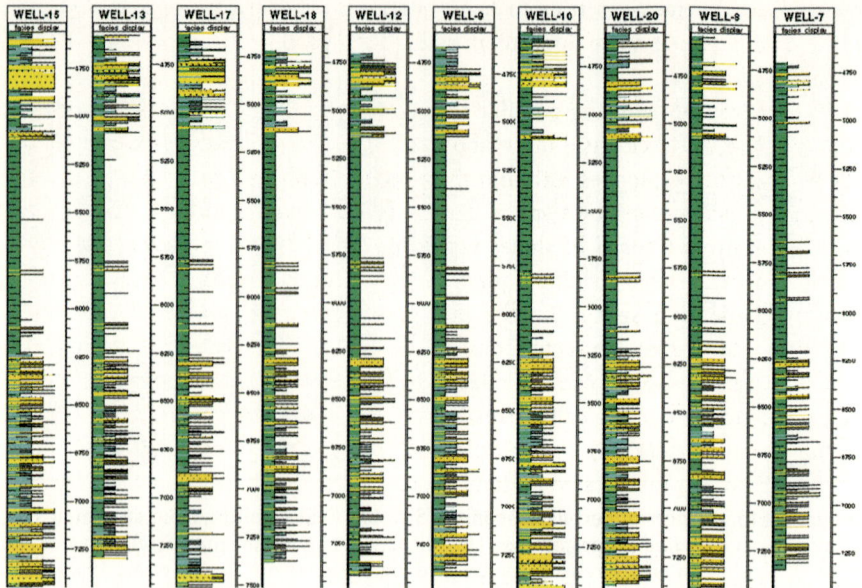

Figure 3.4.14. Cross-section of the lithofacies zonations in the ten wells of the Stratton field. Shales are towards left, sands towards right. Refer to discussion in text.

play of the Stratton data sets as in previous figures, with a coding corresponding to the rock types. Comparison with the original logs in figure 3.4.14 immediately shows how much easier the correlation task becomes, and how this can be related to a sequence-stratigraphic framework.

Current sequence-stratigraphic approaches to interpreting alluvial successions invoke base level control of fluvial erosional and depositional patterns: rivers adjust their slopes in response to changes in base level to maintain a graded longitudinal profile. These adjustments result in deep incision of valleys during relative sea level falls, followed by valley aggradation during transgression and highstands. The response to a base level fall is believed to be transmitted far upstream, and major unconformities within alluvial strata can be correlative with downward inflections on relative sea-level curves (Posamentier 1988, Posamentier et al 1988, Posamentier and Vail 1988). The definition of "base level" and its relation to the fluvial profile and sediment supply is somewhat controversial (Blum 1993; Schumm, 1993; Bloom, 1991; Wheeler 1964). Nevertheless, variations of the base level are widely used to interpret fluvial strata, representing a convenient framework to integrate sea-level fluctuations and subsidence variations into the concept of "changes in accomodation" (Jervey, 1988). As a general rule, stacked channels and coarse sequences indicate a decrease in accomodation space, while thick intervals of flood plain deposits are interpreted as increase in accomodation space (due to subsidence on the coastal plain or to relative sea-level rise). The river tries to keep pace with this increased accomodation space by depositing more flood plain sediments. Paleosoils may indicate breaks in these processes.

In the Stratton field it is possible to observe these types of stacking patterns (figure 3.4.14). A thick basal interval of over 1000 feet thickness presents mainly stacked channel sequences with thin shale intervals, interpreted to be caused by a decrease in accommodation space. The middle interval is about 1100 feet thick and is mainly composed of shales with only few thin silt and sandstone beds. These are interpreted as flood plain deposits, corresponding to a period of high accomodation space. Finally, the upper interval is about 600 feet thick and presents similar characteristics as the bottom interval. These variations in accomodation space can be related to variations of the relative sea level. Stacked sandstone levels can correspond to agradational/ progradational high-stand systems tract (HST). The middle part, dominated by flood plain deposits, is attributed to a transgressive systems tract (TST). The basal part of the upper interval can be interpreted as an early HST, while the upper part, showing thick stacked sandstones, is a late HST with accommodation being smallest. Thus, lower-order sequences can be subdivided through detailed examination of the major units.

The lateral sandstone/shale distribution shows that sandstone intervals are thickening eastwards, while floodplain shales are more developed westwards. Bebout et al. (1982) and Levey et al. (1994) have observed that the Frio sandstone

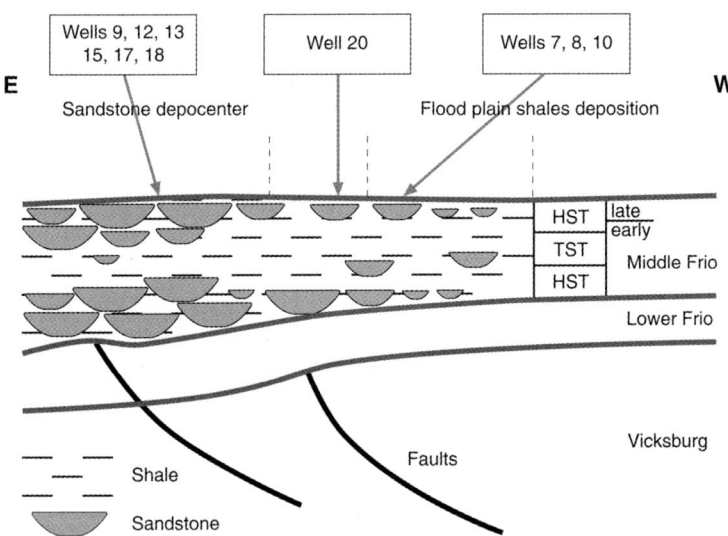

Figure 3.4.15. Schematic of the depositional situation in the Frio (Oligocene, South Texas). Fluvial channel position is controlled by underlying tectonics and base level changes. Approximate position of wells (see figure 3.4.14) is indicated. HST = Highstand System Tract; TST = Transgressive System Tract).

depocenters are generally located near faults, which in this part of Texas strike NNE-SSW. The Middle Frio formation is little affected by tectonics, but the underlying faults are likely to have a topographic expression at the surface, causing the fluvial channels to shift into depressions caused by fault rotation.

The sketch in figure 3.4.15 shows schematically how this sequence-stratigraphic model can look like, and where the wells of the Stratton field are approximately located. No bed-to-bed correlation is proposed here because many of the channels are amalgamated and/or smaller in width than the typical well spacing. The automated correlation discussed in the previous chapter, however, outlines the general correlation trend. It is compatible with the sequence stratigraphic model discussed here.

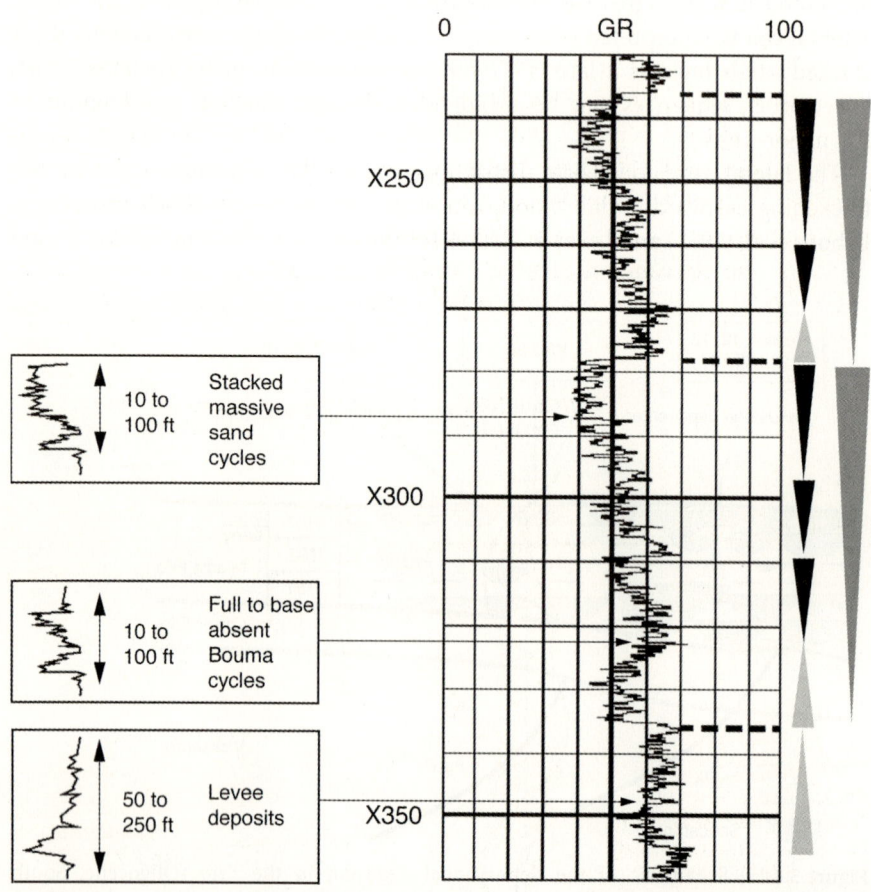

Figure 3.4.16. Log shapes in the turbidite sequence of well A, compared to standard patterns defined by Chapin et al. (1994).

3.4.7 Stratigraphic Units and Sequences

A stratigraphic unit is composed of relatively conformable succession of genetically related strata and is bounded at its top and bottom by unconformities or their correlative conformities. Sequence analysis aims at detecting patterns in the spatial arrangement of these units in order to unravel their depositional origin.

In general, three types of basic sequences can be recognized, each corresponding to a different sedimentary model: agradational, progradational and retrogradational. Retrogradational sequences are located near the maximum flooding surface and are in general slightly richer in clay and uranium than progradational sequences.

The example presented here is from a turbidite sequence in southern Europe and has been discussed in Ruiz and Le Nir (1999). The correlation procedure consists of a combination of conventional lithological zonation and sequence stratigraphy analysis in a key well, followed by extrapolation to other wells using neural networks.

Cyclicities. The key or reference well A exhibits patterns which match those discussed by Chapin et al. (1994). Notably, shale-rich sequences, most likely corresponding to levee deposits, and sand-rich deposits can be identified on the log patterns (figure 3.4.16). These sequences occur in irregular cycles, which may have been caused by a combination of several orbital cycles controlling the sedimentation rates. Figure 3.4.17 shows on the left three cycles of different frequencies, the convolution of which leads to the resultant cyclicity shown on the right. The sediment supply can be controlled by this overall cyclicity and

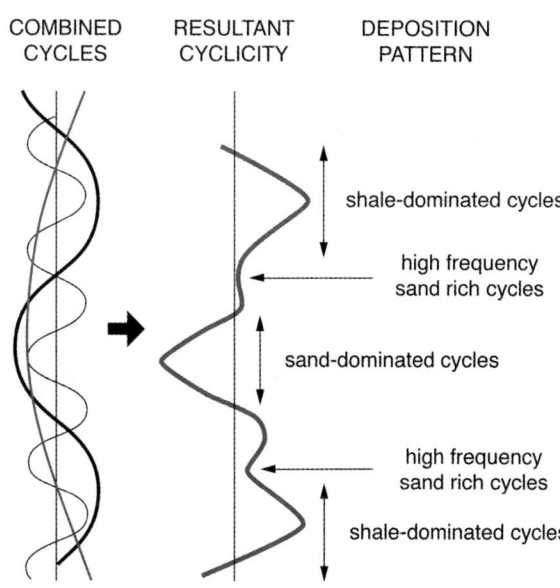

Figure 3.4.17. A combination of cycles can lead to a complex resultant cyclicity in the sedimentation control. Its expression varies across a turbidite system.

337

is reflected in the relative amounts of sand versus shale (Reading and Richards, 1994). If sedimentary sequences are therefore controlled by such cycles, we can assume that these cycles are also felt in other wells, albeit causing a different facies expression (Homewood et al., 1992). In order to investigate this, Ruiz and Le Nir (1999) undertook a zonation of the wells in question, which was then translated into an estimation of the paleoflow energy followed by a well correlation.

Stratigraphic Units. The gamma ray and sonic logs were used for a stratigraphic zonation (figure 3.4.18). Seven facies were defined based on the supposed clay volumes and the stratigraphic position in the key well. These facies were fed to a neural network for learning and were then extrapolated to the remainder of the key well, followed by a quality control and correction of the result where necessary. Seven facies were defined, from sand-rich (facies 1) to mud-dominated (shales, facies 7). Once the training of the neural network had been com-

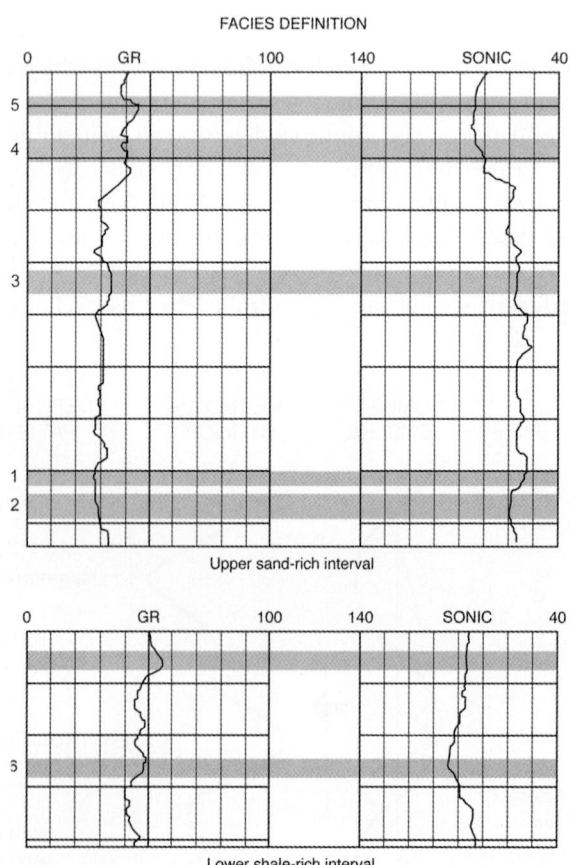

Figure 3.4.18. Intervals defining the seven lithologies in key well A for training of the neural network. Facies range from 1 (sand-rich) to 7 (clay-rich) (after Ruiz and Le Nir, 1999).

pleted to satisfaction, the two other wells were fed to the neural network and a zonation was obtained.

Correlation. The variation of the facies distributions in well A, B, and C is shown in figure 3.4.19. A paleoflow energy is assigned to each facies, from highest for facies 1 to lowest for facies 7. This admittedly very subjective curve is then smoothed in order to visualize better the trends (shown by the solid black lines in figure 3.4.19). It is interesting to notice that the lithofacies zonation is similar between wells B and C, but quite different in well A. However, their paleoenergy curves resemble each other very closely, although offset with respect to each other, making well correlation quite straightforward. The values of the curves are at different levels because of a different depositional setting, but their sedimentary trends and cycles are the same. Thus, a facies transition 1 – 2 – 1 in well A may correspond to a facies transition 2 – 3 – 2 in well B. Wells B and C seem to be more centrally located on the turbidite lobe than well A, but all three

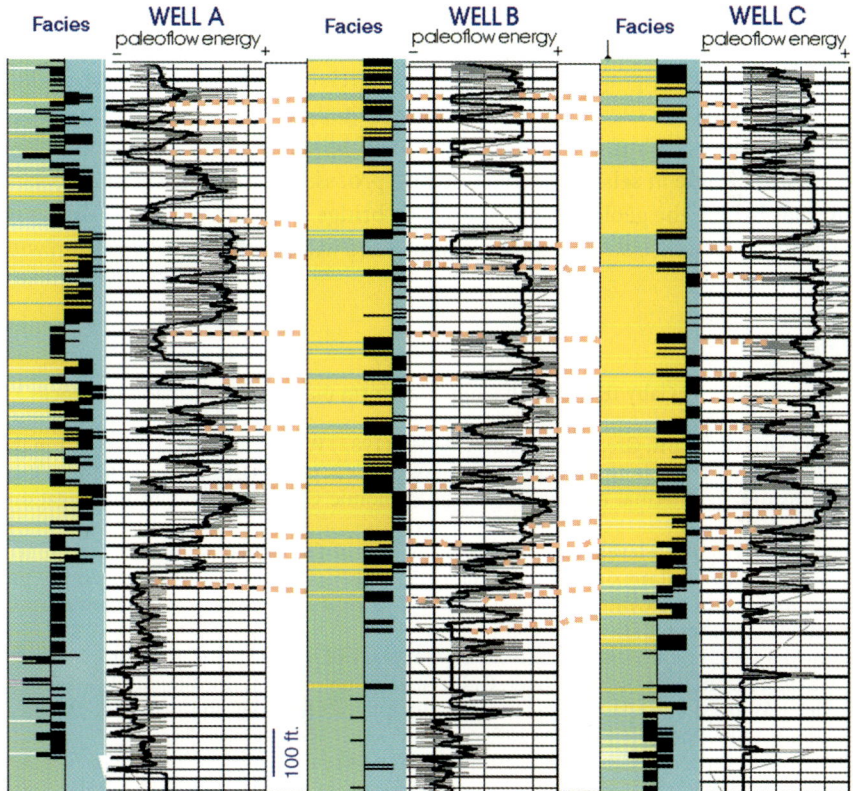

Figure 3.4.19. Resulting lithofacies zonation of wells A, B, and C (left columns) together with a measure of the paleoflow energy (right tracks). Correlation between wells is easy and quite unambiguous using the smoothed paleoflow energy curve. Although the absolute values are different, the trends are the same (after Ruiz and Le Nir, 1999).

have felt the same changes in sediment supply and other factors controlling the depositional pattern. The resulting correlation is thus entirely based on the sedimentary trends, not on the lithofacies themselves.

3.4.8 Summary

In this chapter, we have discussed several aspects of well correlation using logs. New geological logs can contribute significantly to a better lithostratigraphic zonation, with the paleomagnetic log even allowing for a chronostratigraphic zonation from well logs. Computer-aided correlation methods have been discussed with a focus on how the geologist's task can be made easier, faster, more reliable, and more consistent. All of these new methods, however, are only as good as the context within which they are used. The geologist still has to make the choice of which well logs and which correlation programs to rely on, and most importantly, which correlations to accept and which ones to reject.

Geological reasoning is difficult to automate, but there are good aids available, and they continue to be improved. One of the difficulties of well correlation within a sequence stratigraphic framework is the relative lack of formalism in this concept, which is crucial when working with only a few, isolated data sets as the subsurface geologist is often condemned to do. The great progress done in the last decade in seismic acquisition and processing, particularly in 3 – D seismic, has helped the geologist enormously to bridge some of these gaps. However, research in sequence stratigraphy continues and will eventually provide him with better tools.

References

Anderson KR, J. E. Gaby JE (1983) Dynamic waveform matching. Information Sciences 31: 221 – 242.

Bebout DG, Weise BR, Gregory AR, Edwards MB (1982) Wilcox sandstone reservoirs in the deep subsurface along the Texas Gulf coast; their potential for production of geopressured geothermal energy. Bureau of Economic Geology Report of Investigations 117.

Bloom AL (1983) Sea-level and coastal morphology of the United States through the late Wisconsin glacial maximum. In: Wright HC, and Porter SC (eds) Late Quaternary Environments of the United States. Volume 1, The Pleistocene. Minneapolis University of Minnesota Press, 215 – 230.

Blum MD (1993) Genesis and architecture of incised valley fill sequences: A late quaternary example from the Colorado River, Gulf coastal plain of Texas. In: Weimer P, Posamentier HW (eds) Siliciclastic Sequence Stratigraphy. Recent Developments and Applications. AAPG Memoir 58: 259 – 283.

Chapin MA, Davies P, Gibson JL, Pettingill HS (1994) Reservoir architecture of turbidite sheet sandstones in laterally extensive outcrops, Ross Formation, Western Ireland. Soc Econ Paleo and Mineral Gulf Coast Section, Proc 15th Ann Res Conf, 53 – 68.

Doveton JH (1994) Geologic Log Analysis Using Computer Methods, AAPG Computer Applications in Geology 2.

Eichenseer HL, Leduc JP (1996): Automated genetic sequence stratigraphy applied to wireline logs. Bull Centres de Recherches Exploration-Production Elf-Aquitaine. 20: 277 – 307.

Fang JH, Chen HC, Shultz AW, Mahmoud W (1992) Computer-aided well correlation. Bull Am Ass Petrol Geol 76: 307–317.

Hardage BA, Levey RA, Pendleton V, Simmons J, Edson R (1994) A 3–D seismic case history evaluating fluvially deposited thin-bed reservoirs in a gas-producing property. Geophysics 59: 1650–1665.

Haykin S. (1994) Neural Networks. MacMillan Press.

Hecht-Nielsen R (1990) Neurocomputing. Addison-Wesley.

Hedberg HD (1976) International stratigraphic guide. A guide to stratigraphic classification, terminology and procedure. Int Union Geol Sciences, Commission on Stratigraphy, Int Subcommission on Stratigraphic Classification. Wiley, New York.

Homewood PW, Guillocheau F, Eschard R, Cross TA (1992) Correlations haute résolution et stratigraphie génétique: une démarche integrée. Bull Centres de Recherches Exploration-Production Elf-Aquitaine. 16: 357–381.

Jervey MT (1988) Quantitative geological modelling of siliciclastic rock sequences and their seismic expression, In: Wilgus CK, Hastings BS, Posamentier HW, Van Wagoner J, Ross CA, Kendall C (eds) Sea-level changes: an integrated approach: Soc Econ Paleo Mineral Spec Pub 42: p 47–70.

Le Nir I, Van Gysel N, Rossi D (1998) Cross-section construction from automated well log correlation: a dynamic programming approach using multiple well logs. Trans 39th Symp Soc Prof Well Log Analysts: Paper DDD.

Levey RA, Hardage BA, Edson R, Pendleton V (1994) 3-D seismic and well log data set – Stratton Field, South Texas. Bureau of Economic Geology. 30 p.

Luthi, SM and Bryant, ID (1994) Well-Log Correlation Using a Backpropagation Neural Network. Mathematical Geology 29: 413–425.

Luthi, SM, Bryant, ID and Gamero de Villaroel H (1995) Well log correlation using backpropagation neural networks. Trans 4th Int Congr Braz Geophys Soc, Rio de Janeiro: 733–736.

Posamentier, HW (1988) Fluvial deposition in a sequence stratigraphy framework. In: James PD, and Leckie DA (eds) Sequences, stratigraphy and sedimentology: surface and subsurface. Can Soc Petrol Geol Memoir 15: 582–583.

Posamentier, HW, Jervey MT, Vail PR (1988) Eustatic controls on clastic deposition I: conceptual framework. In: Wilgus CK, Hastings BS, Posamentier HW, Van Wagoner J, Ross CA, Kendall C (eds) Sea-level changes: an integrated approach: Soc Econ Paleo Mineral Spec Pub 42: 110–124.

Posamentier HW, Vail PR (1988) Eustatic controls on clastic deposition II: sequence and systems tract models. In: Wilgus CK, Hastings BS, Posamentier HW, Van Wagoner J, Ross CA, Kendall C (eds) Sea-level changes: an integrated approach: Soc Econ Paleo Mineral Spec Pub 42: 125–154.

Posamentier HW, Allen GP, James DP, Tesson M (1992) Forced regressions in a sequence stratigraphic framework: concepts, examples and exploration significance. Bull Am Assoc Petrol Geol 76: 1687–1709.

Reading HG and Richards, MT (1994) Turbidite systems in deep water basin margins classified by grain size and feeder system. Bull Am Ass Petrol Geologists Bull 78: 792–822.

Ruiz G, Le Nir I (1999) Sequence stratigraphy and facies analysis – computer-aided interpretation. Trans 61st Conf Eur Ass Geosci Eng, P535.

Schumm SA (1993) River response to base level change: implications for sequence stratigraphy. J Geology 101: 279–293.

Smith TF, Waterman MS (1980) New stratigraphic correlation techniques. J Geology 88: 451–457.

Thibal J, Etchecopar A, Pozzi JP, Barthès V, Pocachard J (1999) Comparison of magnetic and gamma ray logging for correlations in chronology and lithology: example from the Aquitaine Basin (France). Geophys J Int 137: 839–847.

Wheeler HE (1964) Baselevel, lithostratigraphic surface and time stratigraphy: Bull Geol Soc Am Bulletin 75: 599–610.

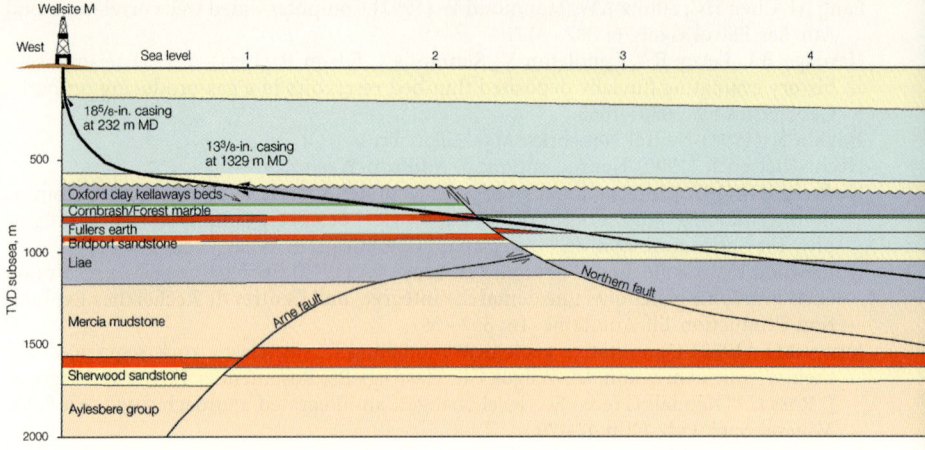

3.5 Geological Drilling

3.5.1 Directional Drilling

Perhaps no other field in the oil industry has seen as much progress in the last decade as directional drilling. It was initially developed as a technique to reach reservoir targets not accessible with vertical wells, for example fields – or parts of fields – situated offshore but close to the coastline where a drill rig could be placed. Several fields offshore California and in the Caspian Sea around Baku (Azerbaijan) were developed in the 1930s and 1940s with deviated drilling. With the rapid growth of offshore drilling from platforms, directional drilling became common as a large number of wells were drilled from the same platform into a field. Although geology dictated what objective the wells were targeting, the well trajectory was usually defined in strictly geometric terms. After an initial vertical or steep section of the well, deviation was built up at a prescribed depth and build-up angle, and once the desired deviation was reached the well was on its target, perhaps with a vertical drop-off in its last section where it penetrated the reservoir.

Two key technologies have moved directional drilling from a geometric drilling exercise towards a powerful field development technique: *Downhole steerable assemblies,* and *measurement- and logging-while-drilling (MWD/LWD).* Combined, these two technologies allow to design and drill a well that can target a particular geological objective in the reservoir, and adjust the well trajectory as soon as a deviation from the anticipated reservoir model is perceived[1]. Field de-

---

[1] The *Oilfield Review* contains several comprehensive updates on new developments in drilling technology, notably in the following issues: July 1990, Winter 1995, and Winter 1998. The material presented here has benefited from these articles.

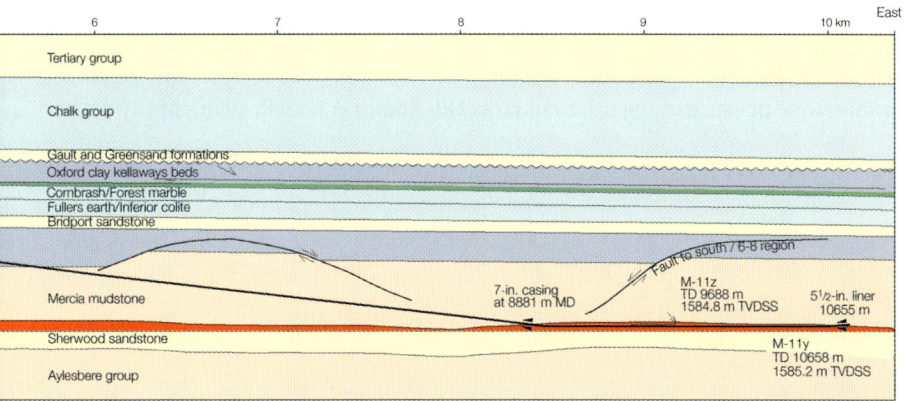

Figure 3.5.1. Sketch of the Wytch Farm extended reach well M-11, at true aspect ratio. The well is located beneath Poole Bay on the southern coast of England and drills, for mostly environmental reasons, from a surface wellsite into the Sherwood sandstone. Total measured depth of the main well is 10,658 meters.

velopments such as the one sketched in figure 1.3, featuring vertical, deviated, and horizontal well, are only possible with these advanced directional drilling methods.

Highly deviated and horizontal wells can follow the target layer in the reservoir for a longer interval than vertical wells and therefore have usually higher production rates than vertical wells. In the offshore Wytch Farm field in England, for example, drilling has been done from the mainland for environmental reasons, and the operator, British-Petroleum, chose to develop the more remote part of the field with extended-reach drills, i.e. very long horizontal or subhorizontal wells. Some of their wells have an offset, or horizontal displacement, of up to 10,000 meters, with a true vertical depth at the bottom of the well of slightly over 1600 meters below sea level. Figure 3.5.1 shows the trajectory and reservoir structure of one of these wells, at a scale that demonstrates the precision required in drilling this well and "landing" it accurately in the target. Similar wells are drilled elsewhere, for example by Total in the Ara field inTierra del Fuego, Argentina. In the Austin chalk, horizontal wells are drilled at a right angle to the dominant fracture system, thereby crossing a much larger number of fractures than vertical wells and making previously marginal fields economically viable through larger production rates per well. In other parts of the world, horizontal wells are drilled into thin pay zones, or into oil rim reservoirs where nearby gas or water might make a vertical well prone to coning or cusping.

Multilateral drilling, originally developed in the Soviet Union, is a method whereby several branches are drilled from a main well, often with short build-up radii (figure 3.5.2). Two general types of multilaterals can be distin-

343

guished: vertically staggered laterals, and horizontally spread laterals. The first type targets reservoirs with multiple reservoir layers, while the second type targets different areas of the same layer. Drilling and completing multilaterals requires complex technology, and although its use is on the rise, it is still not commonly done because of the inherent risks (El-Khatib & Ismail, 1996). Many of the

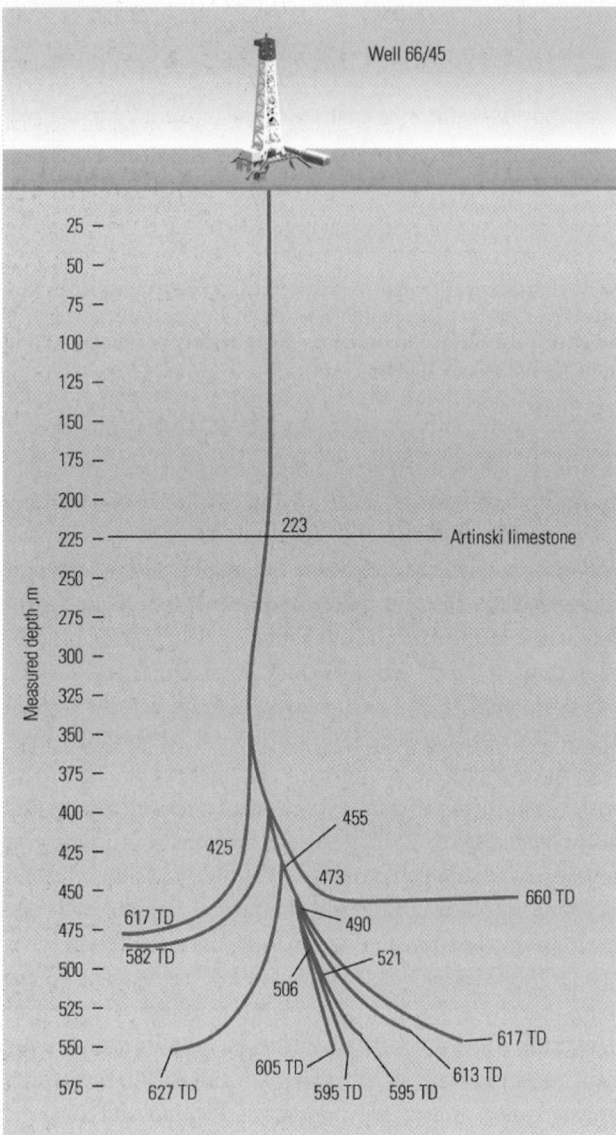

Figure 3.5.2. Well sketch of an early multilateral development in Bashkortostan, South Ural, Russia. Drilled in 1953 with a downhole turbodrill motor, the branches are designed to spread out like roots into the Carboniferous reefal build-up in order to increase oil production rates.

multilaterals currently drilled have a relatively simple well path design with open-hole completions. However, their potential benefits are large, and it can be anticipated that there will be a substantial increase in multilaterals in the years to come (Ehlig-Economides et al., 1996; Longbottom & Herrera, 1997).

Today, it is estimated that around 40 % of all drilled wells are highly deviated or horizontal. The benefits of these wells can be summarized as follows:

- Geological layers can be accurately followed even if dip changes and faults occur;
- The productive interval is significantly longer than in vertical wells, leading to higher flow rates;
- Production can be at lower drawdown pressure than in vertical wells, therefore decreasing the danger of water coning or gas cusping;
- In fractured reservoirs, a larger number of fractures can be penetrated if the well is oriented correctly, thus increasing well productivity.

These benefits have to be carefully weighted against the increased cost of horizontal wells (see e.g. Mukherjee & Economides, 1991), and oftentimes considerably more time is spent in planning a horizontal well than in drilling it.

Directional Drilling Techniques

There exists a large and sometimes bewildering number and types of directional drilling techniques. The vocabulary given in table 3.5.1 should help to understand the terminology.

Common rotary drilling is done either from the rotary table or with a top-drive system, i.e. a motor at the top of the drill pipe. Changing the deviation of the well is made by placing a whipstock at a certain depth, which deflects the drill bit into the desired direction. Further adjustments in the well deviation can be made with downhole-adjustable stabilizers, whereby the gauge of the stabilizer is changed by varying the weight on the bit and locked in place by a given flow rate of the mud[2]. This in essence changes the angle of the lowermost part of the drill string, from the stabilizer down to the drill bit.

Downhole motors, or positive displacement motors (PDM), are placed immediately above the drill bit in the bottom hole assembly. They are powered by mud displacement, whereby the borehole mud circulating inside the drill string causes a helical steel shaft inside a rubber housing to rotate (figure 3.5.3). This shaft is connected to the drill bit, both of which rotate at several hundred revolutions per minute, while the remainder of the drill string assembly remains stationary. This mode of drilling is called *sliding*, because the entire drill string moves into the hole without rotating.

---

[2] For example in the Andergauge system (see www.andergauge.com)

Table 3.5.1. Some common terms in directional drilling.

Term	Definition
Horizontal well	A well with a horizontal or subhorizontal section, usually in the last part
Extended reach well	A well with a long horizontal or subhorizontal section, typically over 5 kilometers in length (figure 3.5.1)
Multilateral wells	A well consisting of several branches in the lowermost part (figure 3.5.2)
Short radius	A horizontal well drilled with a build rate of 1.5° to 3° per foot (typical horizontal section less than 300 meters)
Medium radius	A horizontal well drilled with a build rate of 5° to 8° per 100 foot (typical horizontal section about 800 meters)
Long radius	A horizontal well drilled with a build rate of 2° to 6° per 100 foot (typical horizontal section 1000 meters and more)
Build rate	The rate at which curvature is built in a directional well
Tangent	A section of constant inclination and azimuth of the wellbore
Offset	The horizontal distance between the rig location and the bottom of the well
True vertical depth (TVD)	The vertical distance between the rig floor and the bottom of the well
Total depth	The drilled distance measured along the well trajectory. Also: Measured Depth, Along Hole Depth (AHD).
Toolface	The angle between a tool reference axis and gravity ("gravity toolface") or magnetic north ("magnetic toolface") in a vertical or horizontal projection plane respectively.
Reference point	A measure point in a MWD tool to which all measurements are related
Measurement While Drilling (MWD)	A set of downhole measurements such as azimuth, inclination, tool orientation, weight, torque, etc. Can be recorded downhole or transmitted to surface.
Logging While Drilling (LWD)	A set of downhole measurements related to formation properties. Can be recorded downhole or transmitted to surface.
Rate of penetration (ROP)	The interval drilled per unit time, either averaged or instantaneous.
Whipstock	A packer with an angled upper face that deflects the BHA into a desired direction
Bottom hole assembly (BHA)	All equipment mounted in the lowermost part of the drill string, typically consisting of drill bit, drill collars, and stabilizers. It can include MWD/LWD tools, PDM and adjustable stabilizers.
Bent housing	A bend in the BHA. It can be fixed or adjustable and is typically in the order of 1.5° to 3°

Table 3.5.1. Some common terms in directional drilling. (Continued)

Positive displacement motor (PDM)	A motor placed in the BHA that turns the drill bit while the remainder of the drill string remains stationary (figure 3.5.3).
Stabilizer	A bladed device used in BHAs to control the hole trajectory, eliminate vibration, and prevent the parts of the BHA above the drill bit to touch the borehole wall (figures 3.5.4, 3.5.5). Adjustable stabilizers contain buttons that can be extended or retracted to increase or decrease the gauge and influence the well deviation

Downhole motors can be used to build up well deviation in a different manner from rotary drilling. This is done by inserting a *bent housing* into the bottom-hole assembly, providing a permanent bend of typically 0.5° to 1.5°. In rotary mode, the bent housing causes the drill bit to cut a hole slightly larger than the nominal bit size because of the outward tilt of the drill bit. When drilling is switched from rotary to sliding mode, only the drill bit rotates. Since it is pointed in a direction slightly away from the well axis, the well starts deviating from its previous axis and follows the drill bit (figure 3.5.4). Its build-up is determined by the angle of the bent housing and its distance from the drill bit, and it continues as long the drilling is in sliding mode. The hole during this time is in-gauge, i.e. its diameter is equal to the bit size. As soon as the desired well direction is achieved, drilling returns to rotary because of the superior penetration rates achieved in this mode, which can be up to 100 meters per hour in a horizontal well. Rotary drilling provides also fewer dog-legs (bends in the wellbore) and thus lower sliding friction exerted by the borehole walls.

Schlumberger's Geosteering Tool features an additional bent in the housing that can be adjusted when the tool is at the surface (figure 3.5.5). This allows building up angle at a higher rate than with the fixed bent housing alone. Additional controls of the well direction can be achieved with downhole adjustable stabilizers, mounted between the drill bit and a fixed stabilizer higher up in the bottom hole assembly. By changing the gauge of the adjustable stabilizer, the bend defined by the lower contact points of bit, adjustable, and fixed stabilizer

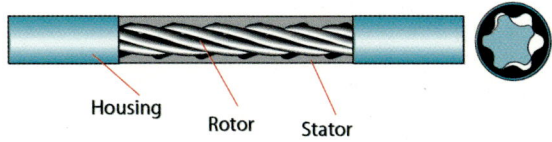

Figure 3.5.3. Sketch of a positive displacement motor (PDM) used for drilling in sliding mode. The eccentered rotor turns in the rubber stator as the drilling mud is pumped through a channel in between the two.

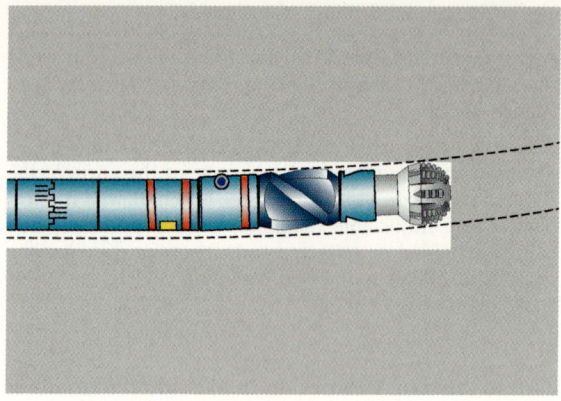

Figure 3.5.4. In rotary mode, the drill bit cuts a straight but enlarged path because of its tilt at the bent housing. In sliding mode, only the drill bit rotates. The hole follows the direction of its tilt and is in gauge.

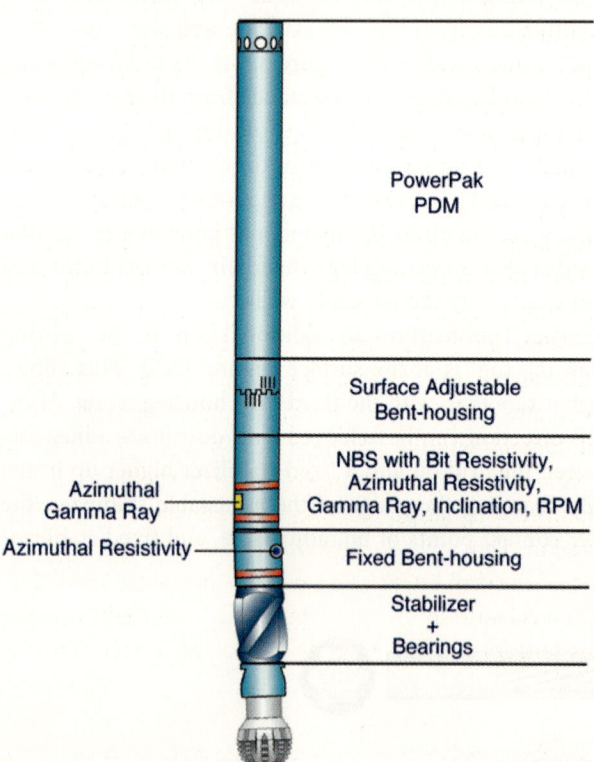

Figure 3.5.5. The Geosteering tool consists of gamma ray and resistivity measurements made below the PDM. A fixed and a surface-adjustable bent housing are used to steer the bit in sliding mode. MWD measurements critical to the steering operations are inclination, rotations per minute, and weight on the bit.

Figure 3.5.6. The bias unit of Schlumberger's PowerDrive steerable rotary system. Three pads are actuated in a sequentially coordinated ways so that the drill bit get pushed into the desired direction.

determines the new drill direction (Poli et al., 1996). Careful monitoring of the bit inclination is mandatory in this mode to maintain the desired drill direction.

Measurement of the drill bit direction includes the inclination from the horizontal, and the azimuth with respect to north. This is done by MWD equipment mounted immediately behind the drill bit. Among others, it monitors the *toolface*, or the orientation of the drill bit to the left or right of the high side (top) of the hole. If the drill bit is not pointing in the desired direction, the directional driller can make minor adjustments by changing the weight on the bit and thus changing the reactive torque on the bit in sliding mode. Larger adjustments are made by lifting the drill string off the drill bit and by reorienting the entire assembly.

In extended-reach drilling such as Wytch Farm (figure 3.5.1), the friction on the drill string becomes too large to be able to drill in sliding mode after about 8 kilometers total depth. Therefore, a *steerable rotary assembly* has to be used, whereby the drilling direction is controlled by a near-bit system that pushes the drill bit in the desired direction[3]. There are two units involved in these systems, the bias unit, and the control unit. The bias unit applies a lateral torque to the bit while it constantly rotates. This is done through a number of exterior pads that are kept in constant contact with the borehole wall (figure 3.5.6). The extension of these pads is controlled by mud-powered actuators. When the hole direction needs to be changed, the actuators extend the pads in the direction opposite the desired deviation. Thus, each pad extends and retracts in sequence, once with each revolution of the drill bit, pushing the drill bit in the desired direction. The control unit is mechanically linked to the bias unit and sets direction and deviation. It contains sensors and control eletronics inside a case mounted

---

[3] The following refers to Schlumberger's PowerDrive system. Other systems on the market use non-rotating sleeves that can be eccentered to steer the bit (AutoTrak system by Baker-Hugher INTEQ; SmartSleeve by Rotary Steerable Tools) or driveshafts that can be deflected to point the bit in the desired direction (GeoPilot system by Speery Sun, a division of Halliburton Energy Services). All are combined with MWD sensors and are controlled from the surface.

on bearings within a non-magnetic drill collar. The control unit can be programmed from the surface with a sequence of mud pulses, and a surface system monitors the new settings and the new direction.

Steerable rotary drilling is still new, but its potential benefits are considerable. Weight on the bit and rotary speed can be adjusted within wider limits, and as a consequence the rate of penetration in this mode of directional drilling is often much higher than with a donwhole motor. The toolface fluctuations caused by reative torque and drag when drilling in sliding mode are considerable reduced in rotary drilling. Consequently, directional control is more accurate, and fewer unwanted bends in the wellbore occur.

3.5.2 Geosteering

Combining directional drilling techniques with a geological model of the subsurface is called geosteering. The goal is to hit a geological target within prescribed deviation errors. It is sometimes contrasted to "geometric drilling", where the goal is to drill a well within prescribed geometric constraints.

Technique

In geosteering, MWD and LWD are crucial components. Their data are transmitted to the surface by mud pulses, and are the key information used by the directional driller in his decision-making. A common geosteering method consists of comparing the log measurements acquired downhole to a model computed beforehand (McCann et al., 1994; White, 1995; Roberts et al., 1998). The model is obtained from the geological structure of the field, and the properties of its layers, known either from a previous well or a pilot hole[4]. These petrophysical properties of the model are generally in the form of squared or averaged curves for each layer (see chapter 3.3). An example of such a layered model and its properties is shown in figure 3.5.7. When the planned well trajectory is overlain onto the layered reservoir model, the log responses can be calculated through the tool response functions and knowledge of the layer dips and well deviation. Typically the gamma ray, resistivities, and more recently the density and neutron LWD measurements are used (Prilliman et al., 1995). The apparent dip, i.e. the angle between layer dip and the plane perpendicular to the borehole axis, is particularly important in modeling the resistivity tools, because at the generally high apparent dips strong polarization effects occur (Anderson et al., 1990; Lüling et al., 1994). Modeled log response curves for the layered model of figure 3.5.7 are shown in figure 3.5.8. They include the gamma ray, and the phase shift and attenuation of 2–MHz LWD resistivity tool.

---

[4] A pilot hole appraises a planned horizontal well. It can be vertical or deviated.

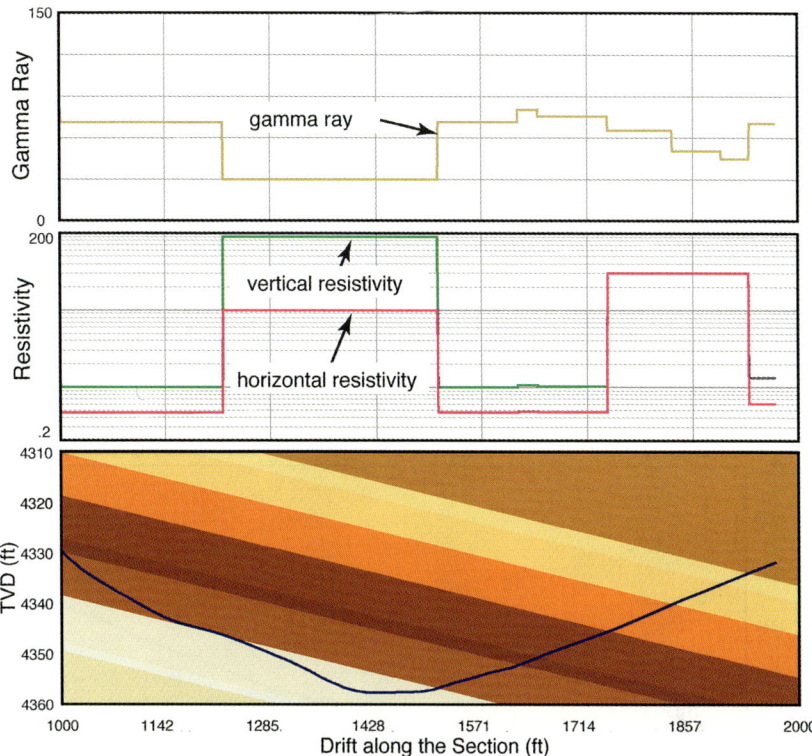

Figure 3.5.7. Geosteering model consisting of a layered, two-dimensional model through which the directional well is drilled (INFORM* system). The gamma ray and resistivities for each layer are shown as a function of the horizontal displacement. The well targets two oil-bearing layers, identified by their higher resistivities. The upward swing of the well towards the upper target at its end suggests this to be an appraisal target.

These LWD measurements are acquired with a tool such as the Geosteering tool shown in figure 3.5.5. Tool features sensors close to the drill bit, such as a gamma ray and resistivity measurement and an adjustable bent housing to steer the bit in sliding mode. The acquired logs are continually compared to the computed or expected curves (White, 1995). When a discrepancy occurs, two possibilities exist: Either the well deviation or layer dip are different from the model, or the properties have changed such as in the case of a facies change, or when crossing a fault. The wellsite team has to figure out which case applies. A fault generally causes the formation properties to change abruptly, while crossing a layer contact shows smoother transitions. Azimuthal LWD measurements such as resistivity imaging (chapter 2.2) or the azimuthal density/porosity log (chapter 2.4) will not only tell if a layer contact has been crossed, but also in which sense, i.e. from below or from above. Thus, if the wellbore follows a sandstone layer but

---

* Mark of Schlumberger

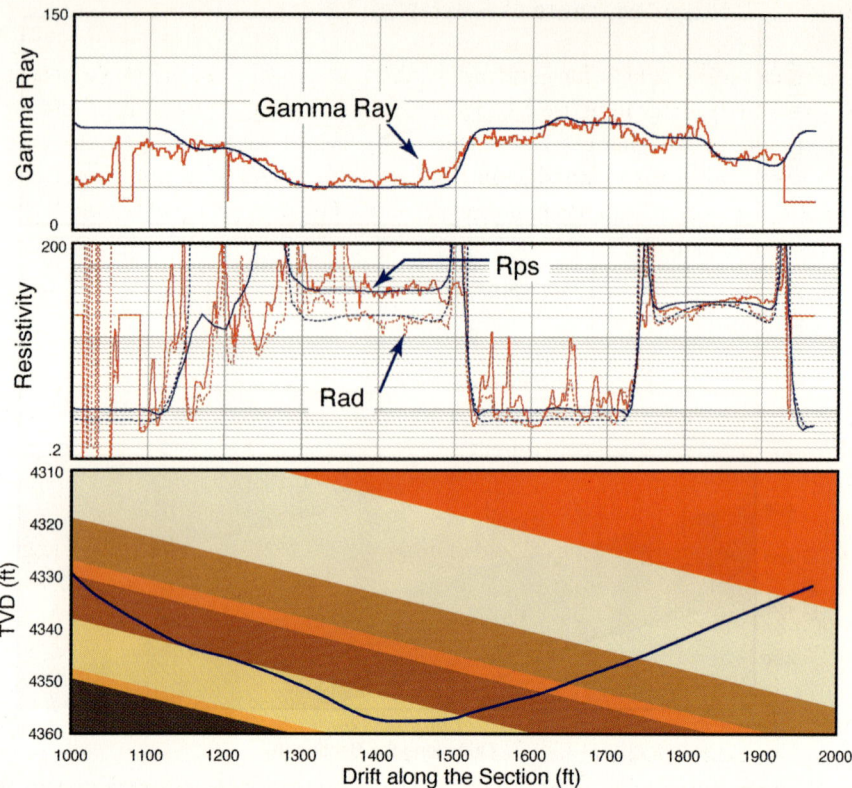

Figure 3.5.8. Geosteering model and actual measurements. In blue, the modeled gamma ray and resistivities obtained from the tool response functions and the apparent dips. In red, the actual measurements from the LWD gamma ray and resistivity measurements (Rps: Resistivity from the phase shift; Rad: Resistivity from the attenuation). Notice the resistivity polarization horns where the well enters and exits the two oil-bearing layers. The second target was found fully oil-bearing.

gradually approaches a denser shale layer located above, the density measurement in the upper quadrant is seen to increase first, followed by the right and left quadrant. The lower sandstone density may still affect the lower quadrant when the sensor is already completely in the shale.

This analysis can also take place after the well is drilled and corrective action – if needed – can be made in the form of sidetracks or redrills. Figure 3.5.8 shows the modeled tool response together with the actual measurement, from which it is seen that there is a very good match, although reality is certainly more complicated than the model. The well in figure 3.5.8 had two targets, one in the lower sand, and, after an upward leg of the well, another one in a second sand. This second sand was an appraisal target. The acquired logs show a higher gamma ray and less separation of the two resistivities than the first sand, suggesting that it is oil-bearing but probably of lesser quality.

Borehole Images: The Future of Geosteering?

The most powerful help for geosteering may come from LWD borehole images. The resistivity image in figure 2.2.25 and the density image in figure 2.4.5 attest to this. Apparent dips that can be extracted from them may be used to verify if the well stays within the planned angle relative to the layers. If not, the amount at which the well deviation has to be corrected can be immediately obtained. It is of course important that

- the sensor is as close as possible to the drill bit, so that the delay time is minimal; and that
- the data is quickly transmitted to the surface, preferably in real-time mode.

The high data rates associated with borehole images make the latter requirement difficult to achieve, but experiments using compression algorithms are in progress and the first results are promising. LWD dips processed downhole and transmitted in realtime to the surface are already available from the density and resistivity imaging tools (Rosthal et al., 1995; 1997). LWD borehole images, so far, are usually retrieved from downhole memory at regular intervals when drilling is stopped.

The potential contribution of LWD borehole images to geosteering can be appreciated from figures 3.5.9 to 3.5.11. These are all electrical images recorded with the RAB tool (chapter 2.2). Figure 3.5.9, shown in a cylindrical projection, is from the Austin chalk and shows layering dipping towards the right, and a distinct fracture crossed at a right angle. The information gleaned from this example is that the well is heading updip but in the right azimuth as far as fracture geometry is concerned, i.e. it crosses it orthogonally. However, the well might exit the reservoir soon and therefore a slight downward correction might be at order. Figure 3.5.10 shows a 12.5 meter long section of a horizontal well

Figure 3.5.9. LWD resistivity from the RAB tool in the Austin chalk. The images, projected onto a cylinder representing the borehole, show layers dipping in the direction of the well azimuth, and a single, vertical fracture (black). Length of interval circa 6 meters.

Figure 3.5.10. LWD images from the RAB tools through an evaporite layer showing flow folds. Length of interval circa 12.5 meters.

through an evaporite sequence, also in cylindrical projection. The objective of this well was to follow this evaporite layer which shows characteristic plastic flow folds. Realtime images would allow immediate correction as soon as the wellbore direction differs significantly from the layer dip.

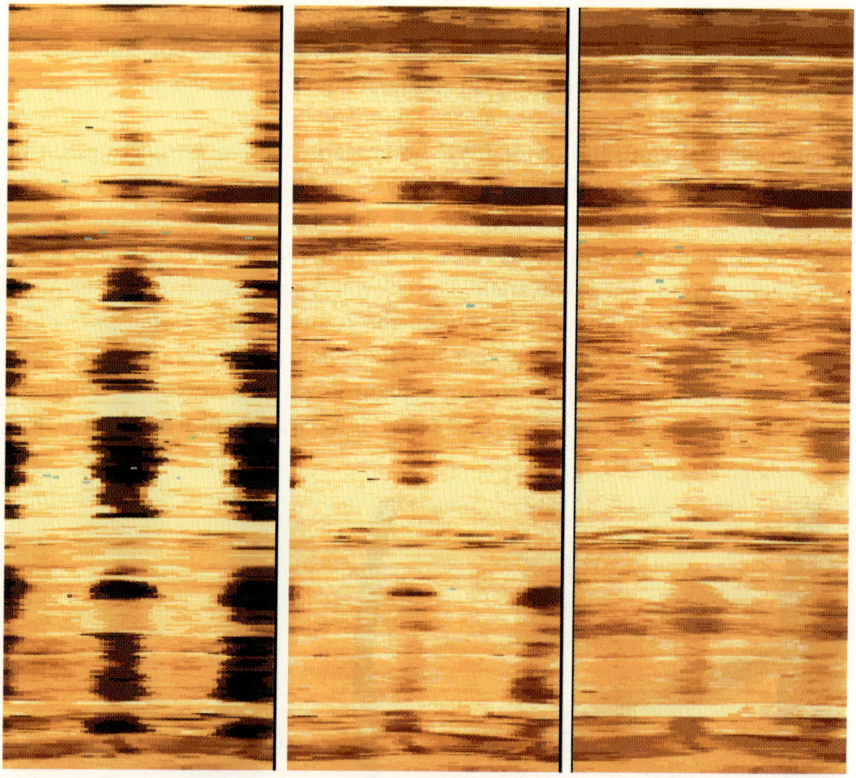

Figure 3.5.11. Three LWD images recorded with the RAB tool, with the shallow, medium and deep button of the RAB tool (from left to right). The hole is highly deviated, and the images are all displayed with the top of the hole on the left, going through right, bottom, left and top again on the right margin. Total interval shown is 26 meters. The dark, conductive streaks on the shallow image are caused by breakouts and are not as clearly seen on the other two images.

Figure 3.5.11 is from a highly deviated well through a clastic sequence in the Gulf of Mexico. It shows, over a 26-meter long section, three RAB images recorded with the shallow, medium, and deep button respectively. They are displayed as a function of depth versus orientation within the hole, whereby the top of the hole is on the left of the images, followed by right, bottom and left of the hole. Darker colors represent lower conductivities. The layering can be seen as subhorizontal wavy streaks, suggesting high dips (because the hole is horizontal), but because of the strongly compressed aspect ratio the layers dip is actually fairly low. The dark, conductive streaks down the shallow image on the left are not seen on the other image, or at least not as prominently. They are caused by borehole breakouts, i.e. shear failure and collapse of the borehole wall in the direction of minimum stress (see chapter 2.2.5 for a discussion, and figures 2.2.30, 2.3.9 and 2.3.11 for other examples). Because the breakouts are shallow, they are not seen as clearly on the images recorded with the medium and deep button. Since the breakouts are found at the top and bottom of the hole over most of the interval, we interpret the hole to be under lateral compressional stress. Such information is very important to the driller as the hole is unstable, and cuttings may accumulate in the hole, causing the drill pipe to get stuck. Having these images available while the well is drilled would allow the driller to take immediate remedial action, such as increasing the mud weight so that borehole damage can be avoided.

Geostopping

Geostopping is a wellsite decision-making method that involves the resistivity-at-the-bit tool (RAB*), in which one of the resistivity measurements is made very close to the drill bit (figure 2.2.5). Toroids located inside the tool body force a current out of the drill bit, and its intensity is converted into a resistivity of the formation volume around the drill bit. The measurement, therefore, can be used to sense changes at the bit. If the drill bit is in a shale and approaches the reservoir sand, the resistivity will increase as soon as the sand is part of the sensitive volume of the measurement. This increase may signal the wellsite geologist that drilling should stop (whence the name), and casing can be set, or coring can start. Like in geosteering, an *a priori* knowledge of the anticipated response should be available, for example on the level of resistivity expected in the layer of interest.

The example in figure 3.5.12 illustrates a case where the RAB resistivity was compared to the resistivity at a step-out well. As soon as the resistivity was seen to increase towards levels within the range of expectation, and at the expected depth, the decision to stop drilling was made.

---

* Mark of Schlumberger

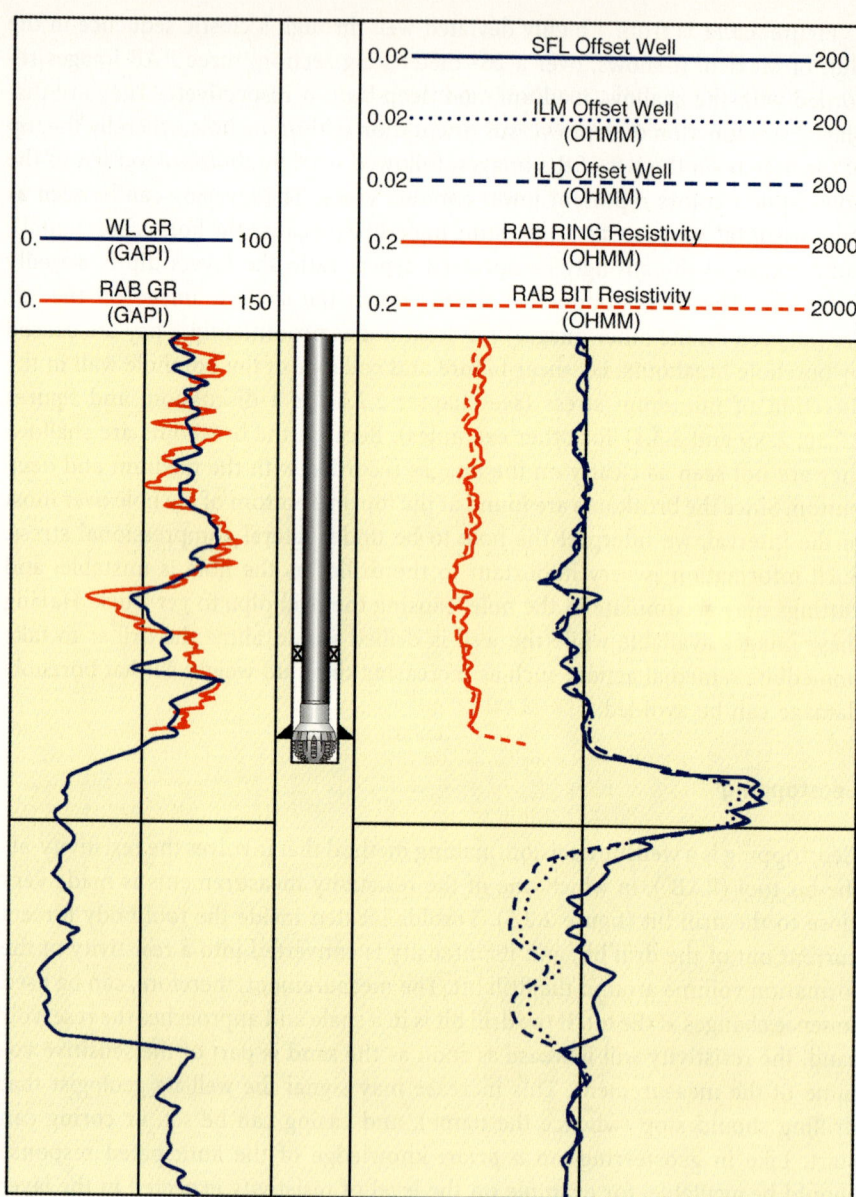

Figure 3.5.12. Geostopping, a particular application of the resistivity-at-the-bit (RAB) measurement. Wireline logs from an offset well (blue, showing only partial oil in the pay zone) are used to anticipate the reservoir and set casing at the appropriate depth. As soon as the RAB bit resistivity increases significantly, drilling is stopped. Notice that at this time the ring resisitivity and the gamma ray resisitivity of the RAB tool are still in the shale because of their higher location on the tool string.

3.5.3 Summary

Directional drilling, and particularly geosteering, is a powerful tool for the geologist to develop a field in an accurate and optimal way. Together with 4 – D seismic and smart completions, geosteering belongs to a new generation of reservoir development techniques that allow the reservoir specialists to perform their work in an accurate and more efficient way.

References

Anderson B, Bonner S, Lüling M, Rosthal R (1990) Response of 2-MHz LWD resistivity and wireline tools in dipping beds and laminated formations. Trans 31st Symp Soc Prof Well Log Analysts: Paper A.

Ehlig-Economides CA, Mowat G, Corbett C (1996) Techniques for multibranch well trajectory design in the context of a three-dimensional reservoir model. Soc Petr Eng Petroleum: Paper 35505.

El-Khatib H, Ismail G (1996) Multilateral horizontal drilling problems and solution experienced offshore Abu Dhabi. Soc Petr Eng Petroleum: Paper 36252.

Longbottom J, Herrera I (1997) Multilateral wells can multiply reserve potential. American Oil & Gas Reporter 40: 53 – 58.

Lüling MG, Rosthal RA, Shray F (1994) Processing and modeling 2-Mhz resistivity tools in dipping, laminated anisotropic formations. Trans 35th Symp Soc Prof Well Log Analysts: Paper QQ.

McCann D, Kashkar S, Austin J, Woodhams R, Siddiqui S (1994) Geological steering keeps horizontal well on target. World Oil 215: 37 – 43.

Mukherjee H, Economides ME (1991) A parametric comparison of horizontal and vertical well performance. Soc Petr Eng Petroleum Formation Evaluation: 209 – 216.

Poli S, Donati F, Oppelt J, Ragnitz D (1996) Advanced tools for advanced wells: rotary closed loop drilling system – results of prototype field testing. Soc Petr Eng Petroleum: Paper 36884.

Prilliman JD, Allen DF, Lehtonen LR (1995) Horizontal well placement and petrophysical evaluation using LWD. Soc Petr Eng Petroleum: Paper 30549.

Roberts MJ, Kirkwood, A, Bedford J (1998) Real-time geosteering in the Tern field for optimum multilateral well placement. Soc Petr Eng Petroleum: Paper 50663.

Rosthal RA, Young RA, Lovell JR, Buffington L, Arceneaux CL (1995) Formation evaluation and geological interpretation from the resistivity-at-the-bit tool. Soc Petr Eng Petroleum: Paper 30550.

Rosthal RA, Borneman T, Ezell JR, Schwalbach JR (1997) Bear-bit resistivity tool calculates dip real time. Soc Petr Eng Petroleum: Paper 38647.

White J (1995) Geological steering assists cost effective exploitation of marginal reservoirs. Soc Petr Eng Petroleum: Paper 30365.

Conclusions

The scope of this book is to demonstrate how modern geological well logs can be used in building a reservoir model. The examples presented in parts 2 and 3 illustrate how relevant reservoir properties can be extracted from this large, diverse, and perhaps underutilized family of logs. Modern reservoir modeling is strongly driven by seismic, which today is generally three-dimensional and, therefore, produces a unique data set that can fill the entire reservoir volume. The data collected from wells can provide calibration points for seismic attributes so that the reservoir volume can be populated with meaningful reservoir parameters. An example is shown in figure 4.1: It shows a base map with the location of the wells on it, all of which are strongly deviated. The isopach contours delineate the net pay thickness, established from well-to-well correlation and supported by seismic. They show an elongated sand body running north-south whose thickness rapidly decreases away from it, particularly toward east

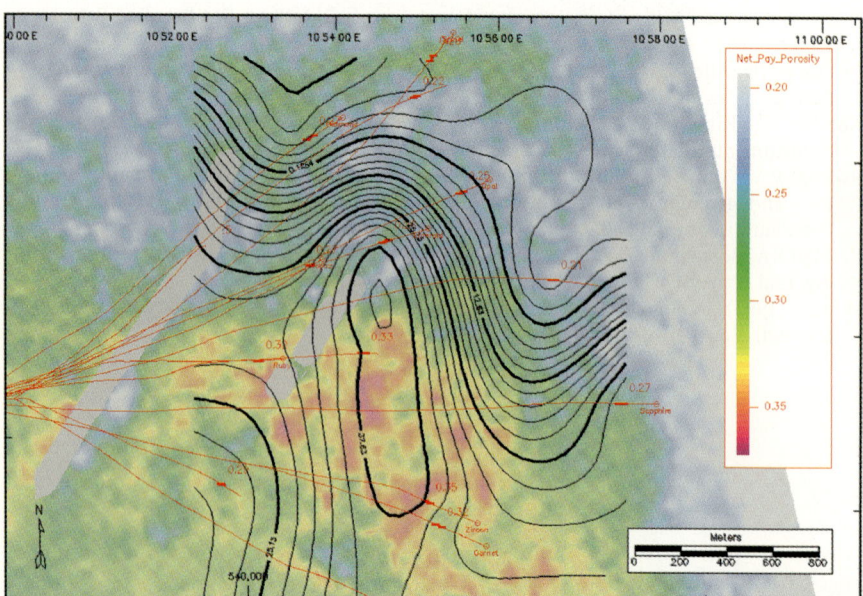

Figure 4.1. Map showing net pay thickness (contours) and porosity (color shading) of a reservoir developed with deviated wells. For discussion see text.

and north. The color shading indicates the net pay porosity. It has been obtained by comparing the acoustic impedance from the 3-D seismic data at the well locations with the net pay porosity obtained from petrophysical analysis of the well logs. The linear regression of the two showed a good correlation (figure 4.2) and was then used to map the porosity for the entire layer from the seismic data. Thus, the well data provided local control, and the seismic data provided two-dimensional coverage.

Processing and understanding. Figure 4.1 is a common product of the reservoir modeling process. In fact, it is a vital ingredient for reservoir simulation since porosity and net sand thickness are among the fundamental input parameters. However, both the isopach as well as the porosity map are lifeless, they do not tell a story. Or rather, one guesses that there must a reason why there is such a sand body, oriented the way it is, and that there must be a reason why the highest porosities are generally – but not always – found where the reservoir is thickest. One wonders perhaps why the porosity is distributed in such a patchy way: Is this a processing artefact, or is it reality? Thus, the map provides no more clues to the reservoir than a raw CAT scan of the brain tells the patient about a potential problem. It simply contains data from a chain of acquisition and processing steps. Without interpreting and evaluating it, there is no understanding, and no diagnosis is possible. In the case of the reservoir, the abrupt thickness changes towards north and east might correspond to small-scale faulting, or to erosion. Or perhaps the shape of the reservoir merely reflects an ancient barrier bar. The porosity changes might be primary, but the drop in porosity in the northern part of the reservoir could be diagenetic, implying perhaps that unfavorable clay types occur that reduce the permeability more than would be expected. If that is so, what caused the increased diagenetic alteration? And how does it affect a prospect located immediately to the north? What other heterogeneities can we expect in this reservoir?

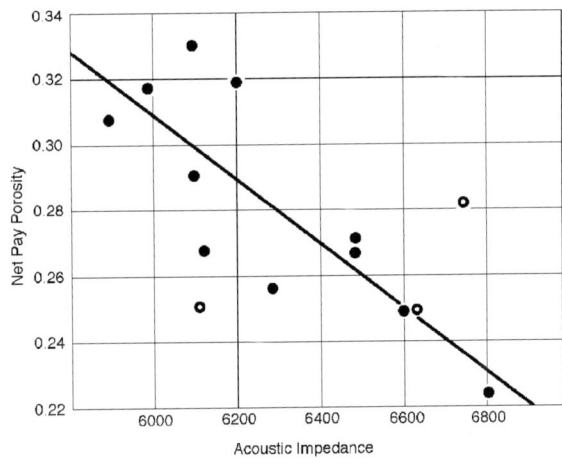

Figure 4.2. The correlation established between acoustic impedance (from seismic) and net pay porosity (from well logs) used to obtain the porosity map in figure 4.1.

The topics discussed in the main parts of this book tried to demonstrate how geological well logs, if properly used and analyzed, can answer many of these questions. The best example may be the carbonate fractured reservoir case study (chapter 2.3.2) where careful analysis of the fracture types and fracture orientation provides an explanation for each of the producing intervals. Interestingly, two of the producing intervals are *not* fractured, but produce from vugs. There are fundamental differences between the behavior of a vuggy and a fractured reservoir: Vuggy reservoirs can have low permeabilities but high porosities and often very low water cuts. Fractured reservoirs, on the other hand, are often very permeable but provide low storage and may water out very soon. This has to enter into a reservoir model and crucially affects the flow simulation. From the observation that most producing fractures have a preferential orientation, a better *field development plan* can be made, since well direction and spacing are strongly influenced by the fracture network.

Data reduction. The art is, therefore, to scrutinize the wealth of data in order to extract the significant pieces of information. Just how much this represents in data reduction can be seen again in the case of the fractured reservoir: The well data acquired may amount to many megabytes in volume, much of which is accounted for by the borehole imaging tools and perhaps a sonic waveform tool. The relevant data, after interpretation, may amount to merely a few hundred bytes: Depths, width and orientation of fractures and faults, depths of the vuggy intervals and unconformities. Thus, the data is reduced by many orders of magnitude. For a reservoir whose properties are facies-controlled, the data reduction may not be as large, but still significant. Each facies can be represented in a symbolic form ("Facies A"), but the symbol has to be related to reservoir properties, or preferably, to an expected range of such properties. Additionally, their spatial distribution has to be described in either a deterministic or a probabilistic way.

Table 4.1 summarizes the principal information that can be gathered from geological well logs and other methods common in the oil industry.

Interactive tools. Modern computing aids such as interactive workstations are an important aid in extracting relevant information. The interpreter can switch from one data set to another, zoom in or out, call up programs to perform a certain task and analyze the results on-screen. Figure 4.3 shows the screen of such an interactive system. Raw and processed well logs as well as borehole images are displayed on the backdrop of a processed and interpreted seismic line. This hypothetical example serves to illustrate how the presence of a fault can be evaluated, by comparing the well trajectory on the seismic line with the borehole images, where the fault may be identified, and to the logs, which may tell the interpreter on which side of the fault the reservoir is located. Based on this analysis, the geologist may recommend remedial action if needed, for example to drill a sidetrack well. He can design the proposed well directly onto the screen and deliver the necessary information such as kick-off point, build-up rate, and length of tangent to the drilling department (figure 4.4).

Table 4.1 A list of principal reservoir characteristics, and the logs and other methods from which they can be determined.

Reservoir characteristics	Well logs	Other methods
Reservoir structure	Dipmeter, borehole images	Seismic, well correlation, some potential methods
Fault identification	Borehole images, dipmeter	Seismic, well correlation, well tests
Fracture analysis	Borehole images, sometimes dipmeter	Cores, well tests
Sedimentary structures	Borehole images (electrical)	Cores
Grain size, pore types	NMR logs, optical imaging	Cores
Porosity	Combination of petrophysical logs	Cores, seismic attributes
Permeability	NMR logs, wireline tester	Cores, well tests, production data
Saturations	Combination of petrophysical logs, wireline tester	Cores (special analysis), well tests, sometimes seismic attributes
Mineralogical composition	Nuclear spectroscopy logs, combination of other logs	Cores
Rock mechanical properties	Array sonic logs, combination of other logs	Cores, drilling data, sometimes seismic
Facies analysis	Combination of logs	Cores, sometimes seismic
Age dating	Paleomagnetic log	Cores (biostratigraphy, paleomagnetism)
Sequence stratigraphy	Combination of logs	Cores, sometimes seismic

Integrated models, integrated teams. The most important aspect in reservoir modeling is the proper and efficient integration of all available data. Reservoir characterization is perhaps simplistically sometimes considered as having flow simulation as its only goal. In reality, it involves close collaboration of geologists, petrophysicists, geophysicists and reservoir engineers in order to make sure a coherent reservoir model is established that considers all available data (figure 4.5). In this exercise, expertise has to transcend boundaries, for if the specialists do not understand each other, they cannot share their efforts and results. Very often, this leads to a blurring of previously existing boundaries, best illustrated perhaps by the profession of production seismologists, i.e. specialists who work on relating production patterns to seismic properties. Petrophysicists are often asked to perform tasks previously done by geologists, and vice versa. Reservoir engineers learn about faults and facies, and while not too long ago they lamented that the geologists would not give them enough quantitative information about

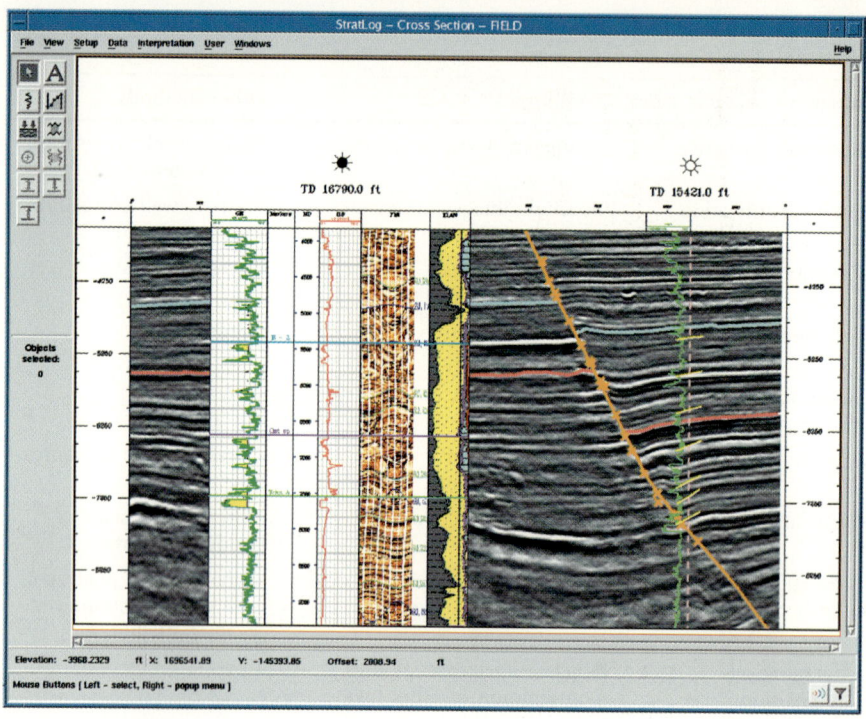

Figure 4.3. Screen view of an interactive interpretation system (Schlumberger's StratLog). Well logs, borehole images, and seismic data are simultaneously displayed and interpreted.

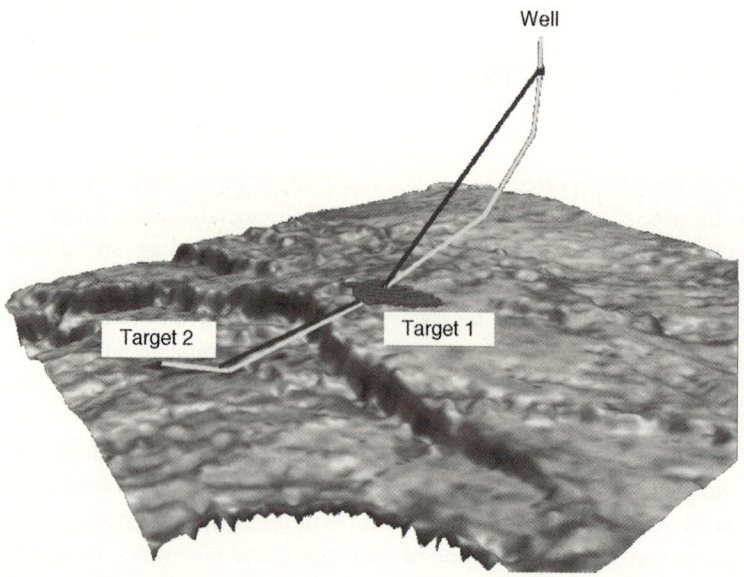

Figure 4.4. Planning trajectories for a deviated well on a workstation. The well has two targets in the reservoir, one on each side of a normal fault.

Figure 4.5. Reservoir characterization requires multidisciplinary teamwork.

fractures, they are now asking to reduce them to an amount their simulators are able to handle. These asset teams are a guarantee for the operating company to manage data in a responsible fashion. Their results are the basis for deciding whether an asset is economically attractive, and, if yes, how to develop it. As field development proceeds, more data are gathered and incorporated into the reservoir model (figure 4.6). When production declines, the economics are reassessed and, if the conclusion is positive, the loop continues until the field is abandoned or farmed out. Thus, reservoir modeling forms a crucial link in the reservoir management process that, thanks to modern technology, has become ever more reliable.

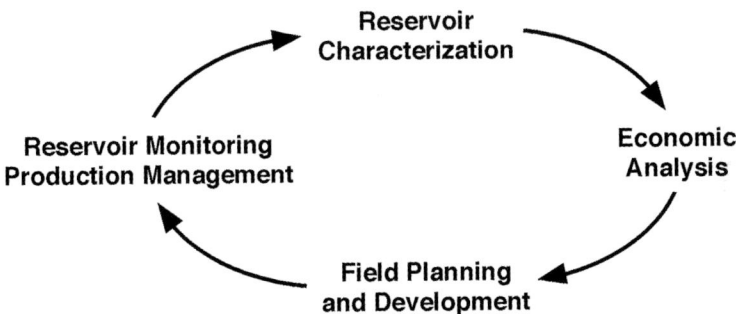

Figure 4.6. The reservoir management cycle.

Figure Credits

All figures are by the author or by permission of Schlumberger except for the following:

1.1.3   By permission of Schlumberger and Statoil

1.2.5   Courtesy N. Russell (Schlumberger)
1.2.6   Courtesy N. Russell (Schlumberger)

1.3.3   By permission of the Society of Exploration Geophysicists
1.3.6   By permission of Prof. G.V. Chilingarian

2.1.1   By permission of the Society of Petroleum Engineers
2.1.7   By permission of the Society of Petroleum Engineers
2.1.15 By permission of Schlumberger and AGIP
2.1.20 Reprinted with permission of the American Association of Petroleum Geologists whose permission is required for further use.
2.1.22 By permission of Romgaz
2.1.27 Reprinted with permission of the American Association of Petroleum Geologists whose permission is required for further use.

2.2.14 By permission of the Society of Professional Well Log Analysts
2.2.17 By permission of the International Association of Mathematical Geology
2.2.19 By permission of the International Association of Mathematical Geology
2.2.21 By permission of the Society of Petroleum Engineers
2.2.22 By permission of Halliburton Energy Services
2.2.23 By permission of Springer Verlag
2.2.28 Courtesy J. Lovell, Schlumberger (top); by permission of the Society of Exploration Geophysicists (bottom)
2.2.33 By permission of the Geological Society of London
2.2.34 By permission of the Society of Professional Well Log Analysts

2.3.2   By permission of the Society of Petroleum Engineers
2.3.8   By permission of the Society of Petroleum Engineers

2.4.1   By permission of the Society of Petroleum Engineers
2.4.2   By permission of the Society of Petroleum Engineers
2.4.3   By permission of the Society of Professional Well Log Analysts
2.4.4   By permission of the Society of Petroleum Engineers

2.5.1   By permission of the Canadian Geotechnical Society
2.5.2   By permission of the Canadian Geotechnical Society
2.5.3   By permission of DHV International
2.5.4   By permission of DHV International

2.6.3   right: By permission of Halliburton Energy Services Inc.
2.6.9   By permission of the Society of Professional Well Log Analysts
2.6.11 Courtesy S.Farooqui and R. Kleinberg (Schlumberger)

2.7.1   By permission of Elsevier
2.7.2   By permission of Academic Press
2.7.5   By permission of M. Rider
2.7.6   By permission of the Society of Petroleum Engineers
2.7.7   By permission of the Society of Petroleum Engineers
2.7.9   By permission of the Society of Professional Well Log Analysts

2.7.10 By permission of the Geological Society of London
2.7.12 By permission of the Society of Petroleum Engineers

2.8.1   By permission of Cambridge University Press
2.8.4   By permission of the American Geophysical Union
2.8.7   By permission of Elsevier
2.8.8   By permission of the Royal Astronomical Society
2.8.9   Anonymous, permission on file

2.9.2   Baker-Hughes document
2.9.3   Baker-Hughes document
2.9.4   Baker-Hughes document
2.9.6   Baker-Hughes document

3.1.3   Courtesy H. Anxionnaz (Schlumberger)
3.1.4   By permission of the University of Chicago Press
3.1.6   Reprinted with permission of the American Association of Petroleum Geologists whose permission is required for further use
3.1.8   By permission of the University of Chicago Press

3.2.8   By permission of the Society of Petroleum Engineers
3.2.10 Anonymous (permission on file)
3.2.13 By permission of Blackwell Science
3.2.14 By permission of Blackwell Science
3.2.18 By permission of Elsevier

3.3.2   Anonymous, permission on file
3.3.3   Anonymous, permission on file
3.3.4   Courtesy B. Newberry and M. Grace (Schlumberger)
3.3.5   Anonymous, permission on file
3.3.6   Anonymous, permission on file
3.3.7   Anonymous, permission on file
3.3.8   Anonymous, permission on file
3.3.10 Anonymous, permission on file
3.3.11 Anonymous, permission on file /courtesy J. van Doorn (Schlumberger)

3.4.1   Courtesy M. Herron (Schlumberger)
3.4.2   Courtesy M. Herron (Schlumberger)
3.4.3   Courtesy M. Herron (Schlumberger)
3.4.4   By permission of the Royal Astronomical Society
3.4.5   By permission of the Royal Astronomical Society
3.4.6   Top: By permission of the Royal Astronomical Society
3.4.7   By permission of the International Association of Mathematical Geology
3.4.8   By permission of the International Association of Mathematical Geology
3.4.9   By permission of the International Association of Mathematical Geology
3.4.10 By permission of the Society of Professional Well Log Analysts
3.4.11 By permission of the Society of Professional Well Log Analysts
3.4.12 By permission of the Society of Professional Well Log Analysts
3.4.13 By permission of the Society of Professional Well Log Analysts
3.4.18 By permission of the European Association of Geoscientists and Engineers
3.4.19 By permission of the European Association of Geoscientists and Engineers

Title Page, background photograph: Courtesy Rick Donselaar

Abbreviation Index

(Logging tool mnemonics, trademark names etc.)

ADN   148, 151
AHD   346
AIT   285
API   190
ARI   78, 83, 299, 304, 311–313
BGO   194, 204, 205
BHA   346
BHTV   124
BVI   178
CALI   149
CAST   128
CEA   222
CBIL   128
CDM   38
CGR   194, 196, 197
CLUSTER   47, 59
CMR   160, 163, 164
CNRS   222
CORIBAND   184
CPI   184
CPMG   163
CSNG   194
DSI   299, 311–313
DHV   154
DP   330
ECS   188, 204–206, 318–321
ELAN   30, 187
ELIAS   80
EMI   78, 105
EPT   70, 287
FFI   178
FMI   18, 76–78, 83, 87. 99, 101, 111–112, 243, 285–289, 291–294, 298–316
FMS   77, 271–274, 278–279
GEODIP   45, 48, 58
GHMT   222–223
GLOBAL   30, 186
GLT   188
GPTS   216–217, 228–233, 321–322
GR   194, 197
GST   187, 202, 207
GTF   140

HALS   299
HDIP   39
HDT   37–39
HNGS   194
HRI   284
HST   335
INFORM   351
LETI   222
LWD   4, 17, 342, 346, 350–357
MAXIS   16
MDT   299
MRIL   160, 163, 164
MSCT   240–244
MWD   17, 342, 345–350
NGS   194, 195, 197
NMR   159–180, 223–224
NPHI   149
OBDT   40–41
ODP   229
PDC   241
PDM   345, 346
PEF   149
PLT   311–313
PMT   205
RAB   18, 30, 80–82, 83, 353–356
RCOR   240–244
RHOB   149
RMT   188
ROP   346
RST   188, 205
RSCT   240–244
SCAT   61–62
SED   39
SEM   178
SGR   194
SHDT   38–39, 74, 75
SP   13, 37
STAR   78
TLC   132
TST   335
TVD   346
UBI   127–128, 299, 311–313

Subject Index

A

Accomodation space 334–336
Acoustic impedance 129
Acoustic properties 129
Acoustic velocity 129
Activities
– relative 190
Albite 184–185, 200
Alpha processing 285–287
Aluminum 188
– computation of 208–209
Analog-digital conversion 16
Angle of migration 64, 254–255
Anhydrite 184–185, 200
Anisotropy tool 15, 37
Anorthite 184–185, 200
Aquitaine basin 230, 322
Archie's law 21–24, 29, 211
Archie, Gus 15
Arrow plot 46
Azimuth frequency plot 46

B

Barite, in mud 147
Base level 334
Basement 311–316
Bayesian approach 30
Bed boundaries 96, 101–102, 270–271, 287–288, 291–294
Bedding
– analysis 94, 96–103, 143–144
– from image texture 290–294
– graded 98
– massive 57, 98
– surfaces 251–252
– types 96–97
– zonation 96–101, 259–295
Bedform ellipticity or curvature 254–258
Bedforms 64, 253–258
Bent housing 346
Biostratigraphic analysis 240
Biotite 191
Bioturbation 98
Bitumen 173
Blind spot 220–221
Borehole cross-section 133
Borehole displacement 141
Borehole elongation 139–140
Borehole imaging 17

– acoustic, or ultrasonic 124–144
– artefacts 83, 130
– azimuthal resistivity 78–80
– bedding analysis 94, 96–103, 290–294
– calibration 102
– color schemes 86
– density 147–152
– electrical 74–118
– electrodes 77, 79
– for geosteering 353–356
– fracture analysis 96, 108–116, 137–138, 152, 156–157, 297–316, 353
– graphic representation 89–91, 132–136
– heterogeneities 96, 116–118, 291
– interpretation 94–118, 137–144
– interpretation scheme 96
– lateral coverage 77
– LWD resistivity 80–82
– micro-electrical 76–78
– optical 154–158
– processing 83–89, 132–136
– reference frame 89
– scales 89–90
– scanning arrays 77
– sedimentological analysis 152
– signal 77
– structural analysis 95–96, 103–107
Borehole radius 131
Borehole shape 139–141
Borehole Televiewer 124
Bottom hole assembly (BHA) 346–347
Bouma sequence 259
Bound fluid logging 162, 171
Bounding surfaces 102, 142, 250
Breakouts 113–116, 136, 156, 313, 354–355
Breccias 98, 301–311
Brine 21
Build rate 346
Burrowing 99

C

Cable 3
– design 154
– fibreoptic 154
Calcite 29, 184–185, 200
Cameras 154
– side-viewing 155

367

Capture
- cross-section 185
- of gamma rays 187–188
Carbon
- inorganic, from logs 212
- organic, from logs 212
Carbon/Oxygen tool 187
Carbonates 208, 298–311, 319–325
Cement evaluation 128
Cementation exponent 22, 118
Channel
- accretion 57
- fill 56–58, 99
- fluvial 251, 334–336
Chlorite 27
Clay volume 197, 327
Clays 25–28, 208
Closure model 202
Clustering 262–265, 292–293
Compositional collinearity 207
Compton scattering 147, 192
Computer-Processed Interpretation CPI 184–185
Concretions 98
Conductance
- counterion equivalent 26
- surface 26
Conductivities 26
- double layer 26
Conglomerates 57, 98–99
Connectivity 22, 31
Contour maps 61–62, 358–360
Contour spacing 61
Convolution 281
Core
- applications 240, 244
- sampling 236–244
- slicer 236, 243
Coring
- conventional 236, 240
- sidewall percussion 238–240
- sidewall rotary 240–244
- wireline 236–244
Corkscrew 139
Correlation
- biostratigraphic 318
- chronostratigraphic 318
- from facies zonation 333–336
- interval 43–44
- lithostratigraphic 231–233, 317–325
- using dynamic programming 330–333
- using neural networks 325–330
- well 198, 213, 248–250, 317–340
- window 226
Correlation coefficient 43

Corrosion monitoring 128
CPMG (Carr, Purcell, Meiboom, Gill)
- measurement cycle 163–166
Cross-bedding 57, 142
- deterministic modeling 252–253
- eolian 57, 59, 254–258
- geometric constructions 64
- modeling 250–258
- planar 253, 257
- shapes 255
- statistical modeling 253–258
- tidal 99
- trough 253, 257
Cross-correlation 41–45, 75
Cross-section
- absorption (table) 204
- geological 61, 250
- volumetric 186
Cyclicity (sedimentary) 337–340

D

Datum (in well correlation) 325–330
Dead time 126
Deconvolution 285
Deep sea fan 199
Deep water reservoirs 5
Deformation
- postdepositional 98
- syndepositional 98
Dendrogram 263
Density 129, 313
- bulk 29, 183
- electron 185
Dephasing 165
- time 159
Differential spectrum method 170
Diffusion 169
Digital petrographic image analysis (PIA) 176–177
Dip 41–43
- apparent 41, 61, 248, 350
- azimuth 41
- calculation 41–48, 91, 136
- definition 41, 43
- from borehole images 91–94
- longitudinal component 62
- planarity 46
- quality 44
- removal of structural 56
- structural 56, 151
- transverse component 62
- true 41
Dip calculation 41–48, 91
- automatic 92–94

- interactive 91
- using Hough transform 92, 136
- using line-tracing 92
Diplog 39
Dipmeter
- curve character 67
- expert system 60-61
- fast channels 44
- for petrophysical evaluation 70
- graphical representation 46-49
- interactive interpretation 61
- interpretation with human expertise 51-61
- near salt dome 49-51
- patterns 55, 247-248
- resolution 67
- sedimentary interpretation 55-60
- six-arm tool 39, 78
- slow channels 44
- statistical methods to interpret 65-70
- stratigraphic interpretation 51, 55
- structural interpretation 51-55, 247-250
- synthetic curves 67
- tool 15, 37-71
Doll, Henri 14
Dolomite 29, 184-185, 200
Downhole processor 127
Drilling
- directional 342-357
- geometric 7, 342
- steerable rotary 349-350
Dual porosity systems 297
Dune
- ellipticity 254
- migration 252-258
- tranverse 255
Dynamic programming 330-333

E

Eccentering
- correction for 133-134
Echo spacing 160, 165
Electrical coring 12, 94
Electrofacies 260, 264, 291, 333
Electronics, role of 4, 16
Elemental
- concentrations 205-206
- yields 202
Elemental capture spectroscopy sonde 204-206, 318-321
Equalization 84, 132
Evaporites 191, 208, 354
Extended reach well 346

F

Facies change 320-321, 329-330
Failure 113-116
- shear 113
- tensile 113-114
Fast Fourier transform 281
Faults 96, 104-105, 156, 302, 315
- normal 52-53
- reverse 52
- rollover 52-53, 62
- thrust 52, 249-250
Field development 363
- offshore 7
Flap 77
Fluid
- bound 171-173
- capillary-bound 168
- clay-bound 168
- free 168, 171-172
Focussing
- active 79
- azimuthal 79
- passive 38, 76
Folds 96, 104-107
- apparent, in horizontal wells 106-107
- concentric 52-53, 62-63
- flow 354
- similar 52-55, 62-63
- slump 143
Foreset azimuths 253-258
Formation evaluation 16, 21-33
Formation factor 21
Fractured basement reservoir 311-316
Fractured carbonate reservoir 298-311
Fractured reservoir analysis 297-316
Fractures 96, 108-116, 297-316
- aperture calculation 96, 110-113, 301, 304-310
- cemented 108
- detection threshold 112, 137
- from Stoneley waves 113, 298, 312-315
- from ultrasonic imaging 131, 137-138, 297
- hydraulic 113-114
- induced 113-116, 138, 156, 302-304, 313
- morphology 138, 300-303
- open 108-116
- permeability 110
- production from 304, 315
- resolution threshold 112
- solution-enhanced 304-310
Free induction decay time 159
Frio sandstone 331-336

G

Gadolinium 188
Gamma Ray 189
– compensated 194, 197
– corrections 193
– detectors 191–192
– emission spectra 195
– logging 191–197
– LWD tool 193
– on cores 193
– on outcrops 193
– tools 192–193
Gamma Ray spectroscopy
– induced 187, 201–213
– natural 184, 194–201
– tool 187
Geochemical logging 187, 201
Geological services 9
Geomagnetic polarity time scale
   (GPTS) 216–217, 228–233, 321–322
Geometric reconstructions 61–65,
   248–258
Geosteering 5, 81, 347–355
Geostopping 5, 81, 355–356
Gilreath, Al 55
Grain size 31, 174
Gravity tool face 149, 346
Gullfaks field 7
Gypsum 184–185, 200

H

Halite 184–185, 200
Heavy hydrocarbons 173–174
Heterogeneities 96, 116–118, 291
– resistive, from Hough transform 118
Horizontal wells 343–346
Hot sands 200
Hummocky cross-stratification 57
Hydrogen 159
– index 173

I

Illite 27, 184–185, 191, 200, 207
Image compression 82
Image texture 290–294
Inclinometry 38, 74
Incoherence function 31, 186
Inelastic scattering 187
Invasion 24
– diameter of 29

K

Kaolinite 26–27, 184–185, 207
Karst 302, 306–311
Key seat 139–140, 313
Kink fold 52
Koenigsberger's ratio 226

L

Lake Maracaibo (Venezuela) 327
Lambda parameter 210–211
Laminated sand analysis 287
Laminations 98, 102
Larmor
– precession constant 224
Larmor frequency 163
Lateral accretion 323
Lithofacies
– data base 266–267
– from logs 260–267
– from neural networks 338–340
– zonation 333–340
Lithology 29–31
– compositions 183
– computations 183–188
Log analyst 187
Log shapes 198–200, 336
Logging speed 126, 162, 194, 204
Logging-While-Drilling (LWD) 4, 17, 342,
   346, 350–357

M

M-N 183
– plot 183
– values (table) 184
M/N plot 29
Magentotactic bacteria 219
Magnet
– field strength 163
– permanent 160–161
Magnetic
– declination 216
– dipole 217
– field or Earth 216, 220–223
– inclination 216
– induction 218, 220, 223
– polarities 216–217, 225–233
– quantities (table) 218
– stratigraphy 220
– susceptibility 218, 224–226, 322–325
– values of relevance (table) 220
Magnetic moment 159
Magnetization

– induced 218
– remanent 218–222
Magnetometer 74
Marker (in well correlation) 325–333
Mass flow deposits 57
Measurement-While-Drilling (MWD) 17, 342, 345–350
Microlaterolog 37
Microlog 37
Microporosity 176
MID plot 29, 184
Mineral
– concentrations 206–210, 319–321
– groups 208–210
Mobility 24–25
Model
– depositional 200
– structural 247–258
Monazite 191
Monocline 52
Monte Carlo modeling 253
Montmorillonite 27
Morphofacies 291–293
Mud
– attenuation 127
– clarity 155
– conductive 74
– filtrate 24
– oil-based 40
Mudcracks 98
Multiexponential fit 168
Multilateral drilling 343–346
Muscovite 191

N

Net thickness pay 358
Neural networks (artificial) 325–330, 338–340
Normalization 85, 134
Nuclear induction 159
Nuclear magnetic resonance (NMR) 159–180
– coupling effects 176
– depth of investigation 160
– frequency of measurement 160
– in carbonates 175
– in sandstones 174
– sensitive volume 160, 162
– wait time 160, 162
Nuclear spectroscopy
– for correlation 318–321
– logging 183–213

O

Oil-based mud 40
– dipmeter 40–41
Orthoclase 184–185, 191, 200, 207
Ovalization 139

P

Paleoflow energy 339
Paleomagnetic logging 216–233
– corrections 226
– interpretation 228–233
– measurement principle 222–224
– processing 224–228
Pattern
– analysis of logs 198–200, 336–338
– recognition 45, 75
Periodicals 9
Permeability 31–33
– calculation from logs 32–33
– from NMR 178–180, 211
– from nuclear spectroscopy 210–212
– intrinsic 210
Petrophysical evaluation 151, 312
Petrophysics 16, 21–33
Photoclinometer 37
Photoelectric absorption 147, 192
– factor 185–186, 313
Pinch-out 320–321
Point bar 199
Pore
– shape 174
– size 162, 167, 174–178
– structure 21
Porosity 21, 23
– measurements 183
– neutron 29, 183, 313
Potassium 189–190
– occurrence 191, 200
Pressure
– build-up 32
– draw-down 32
– test 299
Principal component analysis 260–264
Processing chain 84, 132
Production logging 298, 312–316
Professional societies 9
Projection
– three-dimensional 133, 136
Proton 159, 161
– precession 163

Q

Quantization 85
Quartz 184–185, 200

R

Radiation 188
Radioactive decay (law of) 189
Radioactivity
– units of 190
Ramped anticline 52–53, 249
Rate of penetration (ROP) 83, 346
Real-time transmission 7, 82, 151, 353–354
Reef 56
– paleozoic 60
– prograding 57, 323–324
Reflected amplitude 124, 129
Relaxation
– bulk 162, 180
– diffusion 180
– mechanisms 180
– surface 162, 180
– transverse bulk 165
– transverse surface 166
Relaxation time 159
– longitudinal 159, 162, 180
– spin-lattice 159
– spin-spin 159
– thermal 159
– transverse 32, 159, 165–166, 180
Reservoir
– characterization 5–8, 358–363
– economic analysis 363
– elements 6
– management 363
– monitoring 363
– quantities 6, 358, 361
– saturation tool 205
– simulation 360–363
– zonation 259–295
Resistivity
– invaded zone 24
– lateral 37
– radial 81
– true 23
– uninvaded zone 23, 29
Resolution
– azimuthal 79, 82
– enhancement 285–290
Response equations 29
Ripple laminae 57
Rock resistivity 21
Rootlets 98

S

Sample
– azimuthal position 85
– interpolation 89, 94
– optimum estimation 84, 89
Sampling
– azimuthal or lateral 77, 79, 82, 124, 148, 150
– vertical 77, 79, 82, 124
Sand bar 56, 199
– prograding 57
Sand count 271–273, 290
Saturation
– exponent 22
– fluid 21, 29–31
– irreducible water 32
– partial 168–170
– water 22–23
Scale-space filtering 100
Scanning
– arrays 77
– ultrasonic 124
Schlumberger
– Conrad and Marcel 12
– logging crews 15
– Well Services company 13–15
Schmidt equal area plot 46–47, 49, 308–309
SDR equation 178–179
Secular equilibrium 189
Sedimentation rate 232–233
Segmentation
– marker-controlled 118
– of image 97
Seismic
– attributes 358–359
– integration with dip data 248–250
Seismic stratigraphic units 60
Sequence 198
– analysis 337–340
– stratigraphy 198–200, 233, 334–340
Service companies, integrated 19
Shale
– draping 56, 59
– indicators 197
Shaly formations 25–28
Shifted-spectrum method 169
Sinusoid 89–90
Slide (drilling mode) 82, 345
Slump 99
Smectites 26, 184–185
Sonic slowness 312–313
Sonic travel time 29, 183
Source rocks 212–213, 240

Spin echos 165
Spiral plot 133
Spot size 124
Squared logs 264–265, 288, 292
Stabilizer 345, 347
Statistical curvature technique (SCAT) 61–62
Stereographic plot 46–47
Stick plot 46, 61, 248–250
Stochastic modeling 253
STOIIP (stock tank oil initially in place) 282
Stoneley wave
– attenuation 32, 312
– reflections 298, 312–315
Stresses
– around borehole 114
Structure
– large-scale 247–250
– small-scale 250–258
Stylolites 98–99
Supervised methods 266–267, 290–291
Surface relaxivity 162
Susceptibility measurement 222–225, 228

T
Tadpole plot 46
Tar 173
Teleclinometer 14
Texture energy 98–100
Thin beds 102–103, 268–290
Third interface 128
Thorium 189–190
– occurrence 191, 200
Time constant 193
Timur-Coates equation 178–179
Tool
– resolution 268–271
– sharpening of response 284–290
Tool acceleration 88
Tool velocity 85, 87–88
Tool zero 84
Toolface 149, 346, 349–350
Tornado chart 29
Tortuosity 22, 31
Transducer
– focussed 127–128
– ultrasonic 124

Transit time 131
– interval 124, 131
Translation plane 62–63
Trough fill 56–57
Turbidites 271–284, 295
– correlation of 336–340
– depocenters 275–276, 284
– experimental 280–283
– power-law distributions 274–275
– volumetric distributions 274–284

U
Unconformities 52–53, 96, 102, 104, 201, 233
Unsupervised methods 259–266, 291
Uranium 189–190
– in carbonates 196
– occurrence 191, 200

V
Vertical aggradation 323
Video
– downhole still picture 157
– logging with cameras 154
Vugs 96, 98, 178
– connectedness 118

W
Washout 130, 136
Waxman-Smits model 26–28
Well deviation 342–357
– build-up rates 151
Well logging 3
– acoustic 4
– companies 8
– electrical 4
– first operation 12
– history 12–20
– nuclear 4
West Africa 233
Whipstock 346
Wireline logging 3
Wireline testing 298–299
Workstations 49, 90, 360–362

Printing (Computer to Film): Saladruck, Berlin
Binding: Stürtz AG, Würzburg